Atmospheric Dispersion of Heavy Gases and Small Particles

International Union of Theoretical
and Applied Mechanics

Atmospheric Dispersion of Heavy Gases and Small Particles

Symposium, Delft, The Netherlands
August 29 – September 2, 1983

Editors
G. Ooms and **H. Tennekes**

With 210 Figures

Springer-Verlag
Berlin Heidelberg New York Tokyo
1984

Prof. Dr. Ir. Gijsbert Ooms

Delft University of Technology
2628 AL Delft – The Netherlands

Prof. Dr. Ir. Hendrik Tennekes

Royal Netherlands Meteorological Institute
3730 AE de Bilt – The Netherlands

ISBN-13:978-3-642-82291-9 e-ISBN-13:978-3-642-82289-6
DOI: 10.1007/978-3-642-82289-6

Library of Congress Cataloging in Publication Data

Main entry under title:
Atmospheric dispersion of heavy gases and small particles.
At head of title: International Union of Theoretical and Applied Mechanics.
"Contains papers that have been presented at the IUTAM-Symposium on Atmospheric Dispersion of Heavy Gases and Small Particles which was held from August 29th to September 2nd 1983 in Scheveningen in The Netherlands" -- Pref.
1. Gases -- Environmental aspects -- Congresses.
2. Air -- Pollution -- Congresses.
3. Atmospheric diffusion -- Congresses.
4. Smoke plumes -- Environmental aspects -- Congresses.
I. Ooms, G. (Gijsbert).
II. Tennekes, H. (Hendrik).
III. International Union of Theoretical and Applied Mechanics.
VI. IUTAM-Symposium on Atmospheric Dispersion of Heavy Gases and Small Particles
(1983 : Schevenigen, Netherlands)
TD 885.A85 1984 628.5'32 84-5543

Softcover reprint of the hardcover 1st edition 1984

2061/3020 5 4 3 2 1 0

Preface

The present book contains the papers that have been presented at the IUTAM-Symposium on Atmospheric Dispersion of Heavy Gases and Small Particles, which was held from August 29th to September 2nd 1983 in Scheveningen in The Netherlands. Attendees from many countries were present; 4 review lectures and about 25 research papers were presented. The realization of the symposium was made possible by the support of

Delft University of Technology
Koninklijke/Shell-Laboratory Amsterdam
Prins Maurits Laboratory/Institute for Chemical and Technological Research
Royal Netherlands Meteorological Institute.

The symposium organization was carried out by the local organizing committee consisting of

F.G.J. Absil - Delft University of Technology
G.W. Colenbrander - Koninklijke/Shell-Laboratory Amsterdam
G. Ooms - Delft University of Technology
G. Opschoor - Prins Maurits Laboratory/Institute for Chemical and Technological Research
H. Tennekes - Royal Netherlands Meteorological Institute
A.P. van Ulden - Royal Netherlands Meteorological Institute.

The work of the organizing committee was supported in many respects by the sientific committee, which consisted of

H. Fiedler - Technische Hochschule Karlsruhe, Fed. Rep. of Germany
F.N. Frenkiel - Naval Ship Research and Development Center, Bethesda, Maryland, U.S.A.
J.C.R. Hunt - University of Cambridge, England
J. Kondo - National Institute for Environmetal Studies, Japan
J.L. Lumley - Cornell University, Ithaca, New York, U.S.A.
G. Ooms - Delft University of Technology, The Netherlands
H. Tennekes - Royal Netherlands Meteorological Institute, The Netherlands
J.S. Turner - The Australian National University, Canberra, Australia

The editors would like to express their thanks to the numerous people who helped to make the symposium so successful. Particular thanks are due to the secretary Mrs. H.J. van der Brugge-Peeters. The cooperation with Springer-Verlag was very much appreciated.

Delft, February 1984 G. Ooms and H. Tennekes

Contents

Dispersion of a Stack Plume Heavier than Air

G. OOMS and N.J. DUIJM

Delft University of Technology
Laboratory for Aerodynamics and Hydrodynamics
Delft, The Netherlands

Summary

Some theoretical models have been developed during the last twenty five years for the dispersion of a stack plume with a density larger than air. Also experiments have been carried out to investigate the properties of such heavy plumes. A critical review about these models and experiments has been written.

Introduction.

In the past much attention has been devoted to the study of a stack plume with a density equal to or smaller than the density of the surrounding air. Although most industrial stack plumes are indeed lighter than air, it is also possible that the density of a plume is considerably larger than the air density. In such a case, instead of rising, the plume comes down and can cause a high ground level concentration of a possible pollutant present inside the plume. So it is of practical importance also to be able to predict the plume path and dispersion of a plume heavier than air. We have studied the literature on heavy stack plumes and have written the following review about it. Also recommendations for further work will be given.

Some of the theoretical models developed for light plumes can without much work be extended in such a way, that they are also valid for heavy plumes; see for instance the model of Schatzmann [1]. However, in this paper we have left such models out of consideration; only models already used for the study of a heavy plume are discussed.

Literature review on theoretical models.

1) Bosanquet ([2]).

In a discussion belonging to Bosanquet's paper on the path of a plume lighter than air it is mentioned that an advantage of his theoretical model is, that by a slight modification the path of a plume heavier than air can also be determined. To understand the modification Bosanquet's model for a light plume will first be summarized. In the first part of his paper he calculates the path of a plume with the same density as the surrounding air. He treats such a plume as a vertical jet, which is gradually accelerated in horizontal direction. It is assumed that the velocity distribution in a plume cross section is linear; the maximum velocity at the plume axis is three times the mean velocity over a cross section. Reynolds' analogy is assumed to hold. If U_o is the plume exit velocity and if c^* is the effluent volume fraction at an arbitrary point on the plume axis, then according to Bosanquet the axial velocity at that point is c^*U_o and with the assumed distribution the mean velocity u in the cross section through that point is given by

$$\bar{u} = c^*U_o/3 \ . \tag{1}$$

So the total flow rate Q through that cross section is equal to

$$Q = \pi b^2 c^* U_o/3 \ , \tag{2}$$

in which b is the plume radius. The "partial" flow rate of effluent at an arbitrary point inside the plume is according to Bosanquet equal to c^2U_o, where c is the local effluent volume fraction. Integration over the cross section through that point gives the following expression for Q_o, the volume of effluent emitted in unit time,

$$Q_o = \pi b^2 c^{*2} U_o/6 \ . \tag{3}$$

(2) and (3) yield

$$Q_o/Q = c^*/2 \ ; \tag{4}$$

and with (1) the following expression is found

$$\frac{\bar{u}}{U_o} = \frac{2}{3}\frac{Q_o}{Q} \ . \tag{5}$$

The factor 2/3 is a consequence of the change from a uniform velocity at the stack exit to a linear distribution.

The plume has initial velocity components relative to a reference system moving with the wind velocity equal to U_o upwards and U_a upwind. U_a is the wind velocity. In the absence of external forces the direction of relative motion with respect to this system is unchanged by entrainment. Since the upward velocity at any time is equal to $(2Q_o/3Q)U_o$, the rate of drift of the plume in downwind direction relative to this system is

$$-\frac{2}{3}\frac{Q_o}{Q}U_a . \tag{6}$$

So relative to a system moving with the wind the total velocity v' is

$$v' = \{(\frac{2}{3}\frac{Q_o}{Q})^2 U_o^2 + (\frac{2}{3}\frac{Q_o}{Q})^2 U_a^2\}^{\frac{1}{2}} , \tag{7}$$

and with respect to a system at rest with respect to the earth surface the total velocity v" is

$$v'' = \{(\frac{2}{3}\frac{Q_o}{Q})^2 U_o^2 + (1 - \frac{2}{3}\frac{Q_o}{Q})^2 U_a^2\}^{\frac{1}{2}} . \tag{8}$$

Bosanquet assumes that the rate of entrainment per unit area is equal to

$$\alpha\{(\frac{2}{3}\frac{Q_o}{Q})^2(U_o^2 + U_a^2) + U_a^2\}^{\frac{1}{2}} , \tag{9}$$

in which α is the entrainment coefficient. The value of α is

$$\alpha = 0.13 . \tag{10}$$

The first part of (9) is due to turbulence generated by the relative velocity of the plume with respect to the atmosphere; the second part is due to the atmospheric turbulence. The entrainment coefficients for both parts are assumed to be the same. The surface area of the plume containing one second's flow is equal to

$$2(\pi Q v'')^{\frac{1}{2}} . \tag{11}$$

Combining (8), (9) and (11) the following expression for the rate of increase of Q is found

$$\frac{dQ}{dt} = \alpha\{(\frac{2}{3}\frac{Q_o}{Q})^2(U_o^2 + U_a^2)\}^{\frac{1}{2}} 2(\pi Q)^{\frac{1}{2}} \{(\frac{2}{3}\frac{Q_o}{Q})^2 U_o^2 + (1 - \frac{2}{3}\frac{Q_o}{Q})^2 U_a^2\}^{\frac{1}{4}} . \tag{12}$$

This equation can be integrated numerically and yields the total flow rate

Q as function of time. From (4) and (5) c* and $\bar{u}$ can then be found, and hence the concentration - and velocity distribution as function of time can be calculated. If ΔH is the plume height with respect to the stack exit and L the distance travelled downwind at time t, then from (5) and (6) the following equations can be found

$$\frac{d\Delta H}{dt} = \frac{2}{3}\frac{Q_o}{Q}U_o \quad \text{and} \quad \frac{dL}{dt} = U_a - \frac{U_o}{U_a}\frac{d\Delta H}{dt} . \tag{13}$$

These equations can also be solved numerically and all plume properties are known.

In the second part of his paper Bosanquet investigates the influence of a (positive) density difference between the surrounding air and the plume. To that purpose (5) is extended in the following way

$$Q\bar{u} = \frac{2}{3}Q_oU_o + \frac{gQ_o(T_o - T_a)}{2T_a}t , \tag{14}$$

in which g is the acceleration due to gravity, T_o the plume temperature at the exit and T_a the atmospheric temperature. The first part represents the constant plume momentum due to the emission from the stack; the second part is the increase of plume momentum with time due to buoyancy.

In Bosanquet's model the buoyancy is only due to a temperature difference between plume and atmosphere; if this temperature difference vanishes the buoyancy vanishes also. From (14) the average plume velocity $\bar{u}$ can be calculated.

In the case of a density difference there is an external force working on the plume. So it may no longer be assumed that the direction of relative motion of the plume with respect to axis moving with the wind remains unchanged. Using a point source instead of a stack exit of finite dimension Bosanquet shows that it seems reasonable to assume that the plume immediately drifts downwind with the wind velocity U_a after emission from the stack. The entrainment equation (12) can then be written as

$$\frac{dQ}{dt} = \alpha[\{\frac{2}{3}\frac{Q_o}{Q}U_o + \frac{g\,Q_o(T_o - T_a)t}{2Q\,T_a}\}^2 + U_a^2]^{3/4}\,2(\pi Q)^{\frac{1}{2}} . \tag{15}$$

The equations for ΔH and L now become

$$\frac{d\Delta H}{dt} = \frac{2}{3}\frac{Q_o}{Q}U_o + \frac{g\,Q_o(T_o - T_a)}{2Q\,T_a}t \quad \text{and} \quad \frac{dL}{dt} = U_a . \tag{16}$$

Equations (15) and (16) can be solved numerically and yield again the total the total flow rate, plume height and distance travelled downwind as function of time t.

For the case of a heavy plume the plume first rises in the atmosphere due to its exit momentum and then falls down due to its negative buoyancy. To calculate the falling of the plume it is assumed that the plume is emitted from an imaginary reverse stack located above the actual stack at a height twice the plume rise calculated from the model for the neutral plume.

In Bosanquet's model many simplifying assumptions are made. For instance, the entrainment coefficient α is assumed to be the same for both entrainment processes. Moreover it is supposed that the plume immediately drifts downwind with the wind velocity after emission from the stack; and the emission is supposed to be taking place from a reverse stack. The entrainment due to atmospheric turbulence is only dependent on the average wind velocity; temperature- and velocity gradients are not taken into account. So it is clear that using this model for the dispersion of a heavy plume only rough estimates can be made.

2) Hoot, Meroney and Peterka ([3]).

Assuming a cross wind of constant velocity, a "top hat" velocity profile inside the plume, the model of Hoult, Fay and Forney [4] for the entrainment (where entrainment due to parallel and perpendicular velocity differences are supposed to be additive), the following basic equations are derived

$$\frac{d}{ds}(b^2\bar{u}) = 2\alpha_1 b(\bar{u} - U_a \cos\theta) + 2\alpha_2 bU_a|\sin\theta| \tag{17}$$

$$\frac{d}{ds}(\bar{\rho}b^2\bar{u}^2 \cos\theta) = \rho_a U_a \frac{d}{ds}(b^2\bar{u}) \tag{18}$$

$$\frac{d}{ds}(\bar{\rho}b^2\bar{u}^2 \sin\theta) = (\rho_a - \bar{\rho})b^2 g \tag{19}$$

$$\frac{d}{ds}(\bar{\rho}b^2\bar{u}) = \rho_a \frac{d}{ds}(b^2\bar{u}) , \tag{20}$$

in which s is the distance measured along the plume axis, b the plume radius, $\bar{u}$ the mean velocity in the direction of the plume axis, θ the angle between the plume axis and the horizontal, α_1 and α_2 entrainment coefficients, U_a again the wind velocity, $\bar{\rho}$ the average plume density and

ρ_a the atmospheric density. (17) represents the conservation of volume; the first term on the right-hand side represents entrainment due to turbulence generated by a longitudinal velocity difference between plume and atmosphere,and the second due to turbulence generated by the normal velocity difference. By adapting model predictions to experimental results the values for α_1 and α_2 have been determined. For instance, the following values can be found

$$\alpha_1 \simeq 0.045 \tag{21}$$

$$\alpha_2 \simeq 0.45 \ . \tag{22}$$

However, by adapting model predictions to other experimental results also quite different values for α_2 have been found. (18) and (19) are the equations of conservation of momentum in horizontal and vertical direction respectively. The horizontal momentum changes due to entrainment; the vertical due to the negative buoyancy force acting on the plume. (20) represents the conservation of mass. In this model the influence of temperature differences is not taken into account. The negative buoyancy force is only due to the fact that the density of the stack gases at atmospheric temperature is different from the atmospheric density.
In (17), (18), (19) and (20) four unknown quantities occur, viz. b, $\bar{u}$, $\bar{\rho}$ and θ. So the plume width, velocity, density and angle of inclination can be calculated as function of s (and therefore also as function of horizontal distance x and vertical distance z) by numerical integration of these equations starting from the initial conditions at the stack exit. Hoot, Meroney and Peterka do not solve the equations numerically. They divide the plume path in three regions and solve the equations analytically in each of them. To calculate the concentration the following relation between density and concentration is used

$$\bar{\rho} = \rho_a\{1 - \frac{c}{c_o}(\frac{\rho_o}{\rho_a} - 1)\} \ , \tag{23}$$

in which c_o is the exit concentration and ρ_o the exit density of the plume. Hoot, Meroney and Peterka's model is an improvement with respect to Bosanquet's model. The entrainment coefficients of the entrainment processes due to horizontal and vertical velocity differences are different, whereas in Bosanquet's model they are equal. It is known from experiments that these coefficients are, indeed, very different; the coefficient in

the vertical direction being much larger than in the horizontal direction. The plume velocity in horizontal direction is gradually increasing from zero at the stack exit to the wind velocity at some distance downwind of the stack. In Bosanquet's model it is assumed that the plume immediately drifts away with the wind velocity. Also the plume is no longer assumed to be emitted from an imaginary reverse stack, as is the case in Bosanquet's model.
An important disadvantage of Hoot, Meroney and Peterka's model is, however, that the entrainment due to atmospheric turbulence is not taken into account. It is well known that this turbulence is an important factor in the dispersion of stack plumes. Moreover, the use of a similarity profile is only valid at a certain distance from the stack. Immediately after the stack exit the velocity- and concentration profiles are developing towards similarity.

3) Ooms, Mahieu and Zelis ([5]).

In the model of Ooms, Mahieu and Zelis the plume is divided into two parts, viz. the zone of flow establishment and the zone of established flow. In the zone of flow establishment the similarity assumption for the profiles does not hold yet, the profiles in the region developing from their shape at the exit to similarity profiles. In the zone of established flow the similarity assumption does hold; the model is only valid in this region. The characteristics of the zone of flow establishment are taken into account by using Keffer and Baines' [6] results. The origin of the coordinate system is chosen on the plume axis and on the line between the zones of flow establishment and established flow. s, r and θ are the plume coordinates. s is again the distance along the plume axis, r the radial distance to this axis in a normal section of the plume and θ the angle between the axis and the horizontal.
The similarity profiles of plume velocity, plume density and pollutant concentration are assumed to be cylindrically symmetric and to be of Gaussian shape

$$u(s,r,\theta) = U_a \cos\theta + u^*(s)\, e^{-r^2/b^2(s)} \tag{24}$$

$$\rho(s,r,\theta) = \rho_a + \rho^*(s)\, e^{-r^2/\lambda^2 b^2(s)} \tag{25}$$

$$c(s,r,\theta) = c^*(s)\, e^{-r^2/\lambda^2 b^2(s)} \; . \tag{26}$$

$u(s,r,\theta)$ represents the plume velocity relative to a coordinate system at rest with respect to the earth surface at an arbitrary point of the plume in the direction of the tangent to the plume axis, and $u^*(s)$ is the plume velocity relative to the surrounding atmosphere on the plume axis in the direction of the tangent to this axis. $b(s)$ represents the local characteristic width of the plume. The radius of the plume is defined as $b\sqrt{2}$. $\rho(s,r,\theta)$ is the density at an arbitrary point of the plume, ρ_a the atmospheric density and $\rho^*(s)$ the density difference between the plume on the axis and the atmosphere. $c(s,r,\theta)$ represents the concentration (= mass per unit volume) of a certain pollutant at a point inside the plume and $c^*(s)$ the value of this concentration on the plume axis. $\lambda^2(\approx 1.35)$ is the so-called turbulent Schmidt number; this quantity represents the small difference in plume radius which exists between the velocity profile on the one hand and pollutant concentration and density on the other.

The equation of conservation of mass is given by

$$\frac{d}{ds}\left(\int_0^{b\sqrt{2}} \rho u\, 2\pi r dr\right) = 2\pi b \rho_a \{\alpha_1 |u^*(s)| + \alpha_2 U_a |\sin\theta|\cos\theta + \alpha_3 (\varepsilon b)^{1/3}\}, \quad (27)$$

in which ε represents the eddy energy dissipation due to atmospheric turbulence. The first part of the right-hand-side represents the entrainment due to plume turbulence close to the exit, the second part the entrainment due to plume turbulence at a large distance from the stack and the third part the entrainment due to atmospheric turbulence. The values of the entrainment coefficients are

$$\alpha_1 = 0.057\ ,\ \alpha_2 = 0.5 \ \text{ and }\ \alpha_3 = 1.0\ . \quad (28)$$

The equation of conservation of a certain pollutant in the plume reads

$$\frac{d}{ds}\left(\int_0^{b\sqrt{2}} c u\, 2\pi r dr\right) = 0\ . \quad (29)$$

The equation of conservation of momentum in the x-direction is

$$\frac{d}{ds}\left(\int_0^{b\sqrt{2}} \rho u^2 \cos\theta\, 2\pi r dr\right) = 2\pi b \rho_a U_a \{\alpha_1 |u^*(s)| + \alpha_2 U_a |\sin\theta|\cos\theta + \alpha_3 u'\} +$$

$$+ c_d\, \pi b \rho_a U_a^2 |\sin^3\theta|\ , \quad (30)$$

in which the first term on the right hand side is due to entrainment and the second due to a drag force exerted on the plume by the normal component of the wind velocity. From experiments is found that the drag coefficient is equal to

$$c_d = 0.3 \ . \tag{31}$$

The equation of conservation of momentum in the z-direction is

$$\frac{d}{ds}\left(\int_0^{b\sqrt{2}} \rho u^2 \sin\theta \, 2\pi r dr\right) = \int_0^{b\sqrt{2}} g(\rho_a - \rho) 2\pi r dr +$$

$$\pm c_d \, \pi b \rho_a U_a^2 \sin^2\theta \cos\theta \ , \tag{32}$$

in which the + sign is valid for $-\frac{\pi}{2} \leq \theta < 0$ and the - sign for $0 \leq \theta \leq \frac{\pi}{2}$. The first term on the right hand side is due to negative buoyancy; the second term due to the drag force.

The equation of conservation of heat is given by

$$\frac{d}{ds}\left(\int_0^{b\sqrt{2}} \rho u \, c_p (T - T_{a,0}) 2\pi r dr\right) = 2\pi b \rho_a \, c_{p,a} (T_a - T_{a,0}) \{\alpha_1 |u^*(s)| +$$

$$+ \alpha_2 U_a |\sin\theta| \cos\theta + \alpha_3 u'\} \ , \tag{33}$$

in which T is the temperature at an arbitrary point inside the plume, T_a the atmospheric temperature, $T_{a,0}$ the atmospheric temperature at the plume exit, c_p the specific heat at an arbitrary point in the plume and $c_{p,a}$ the specific heat of air. The stack gases and air are assumed to satisfy the ideal gas law

$$T = \frac{\mu p}{R\rho} \quad \text{and} \quad T_a = \frac{\mu_a p}{R\rho_a} \ , \tag{34}$$

in which μ represents the molecular weight of the plume at an arbitrary point, μ_a the molecular weight of air, p the atmospheric pressure and R the universal gas constant. If the molecular weight and specific heat of air are different from those of the plume, two extra equations are required. These are

$$\mu = \mu_o \frac{cT}{c_o T_o} + \mu_a \left(1 - \frac{cT}{c_o T_o}\right) \tag{35}$$

and

$$\mu c_p = \mu_o c_{p,o} \frac{cT}{c_o T_o} + \mu_a c_{p,a} \left(1 - \frac{cT}{c_o T_o}\right) , \tag{36}$$

in which μ_o represents the plume molecular weight at the exit, T_o the plume exit temperature and $c_{p,o}$ the plume specific heat at the exit.
In the original paper the atmosphere was assumed to be linearly stratified in density. However, in a later development of the model also arbitrary density- and wind velocity distributions were taken into account.
The equations of conservation are solved by substituting the similarity profiles (24), (25) and (26) into these equations and calculating the integrals.
In principle the model of Ooms, Mahieu and Zelis seems to give a rather realistic description for the dispersion of a heavy stack plume. An important factor is that with this model a plume with a temperature and molecular weight different from the atmospheric temperature and molecular weight can be investigated. A disadvantage of their model is, that the plume cross section is assumed to be circular. It is known from atmospheric measurements that the shape of the cross section is dependent on the atmospheric stability and in general is elliptical with the principal axis in the vertical and lateral direction. Therefore, in a later paper Ooms and Mahieu [7] changed the circular plume cross section into an ellipse. This was done in such a way that the area of the elliptical cross section is equal to that of the circular cross section, and that the ratio of the principal axis of the ellipse is the same as the ratio of the dispersion coefficients as measured by Singer and Smith [8] on real plumes.

4) Chu ([9]).

As Chu's model is rather similar to the two foregoing models, it will not be discussed in detail. In this model only the entrainment due to the vortex circulation at large distances behind the stack is considered. The effects of jet spreading and pressure drag are not taken into account.

5) Bloom ([10]).

The model of Bloom includes simple mechanisms for the removal of plume material by condensation and by reaction at the ground surface. Also chemical reactions and their heat release are taken into account. As in Bosanquet's model the plume properties are averaged over a cross section normal to a given point at the plume axis. In this way Bloom derives a set of ordinary differential equations. The first equation is the mass balance

$$\frac{d}{ds}(b^2 \bar{u} \bar{\rho}) = 2b \rho_a u_e - b^2 \bar{\rho} \sum_{n=1}^{N} K_n (c_n - c_{gn}) + \sum_{n=1}^{N} S_n , \qquad (37)$$

in which b is again the plume radius, $\bar{u}$ the average velocity parallel to the centreline, $\bar{\rho}$ the average plume density, and ρ_a the atmospheric density. c_{gn} is the mass fraction of component n in the gaseous phase in the plume, c_n the total fraction of that component (both gaseous and condensed) and K_n is the rate constant for the removal of the condensed material of component n. S_n is the rate at which component n reacts with the ground surface. So the first term on the right hand side of (37) is due to entrainment, the second due to condensation and the last is caused by reaction at the ground surface.

The vertical momentum balance reads

$$\frac{d}{ds}(b^2 \bar{u} \bar{\rho} \bar{w}) = gb^2(\rho_a - \bar{\rho}) , \qquad (38)$$

in which $\bar{w}$ is the average plume velocity in vertical direction.

The horizontal momentum balance is

$$\frac{d}{ds}(b^2 \bar{u} \bar{\rho} \bar{v}) = U_a \frac{d}{ds}(b^2 \bar{u} \bar{\rho}) , \qquad (39)$$

in which $\bar{v}$ is the average plume velocity in the horizontal direction. So the influence of a possible drag force as in the model of Ooms, Mahieu and Zelis is not taken into account. The energy balance is given by

$$\frac{d}{ds}(b^2 \bar{u} \bar{\rho} \bar{e}) = 2be_a \rho_a u_e + b^2 \sum_{j=1}^{J} R_j H_{Rj} -$$

$$- b^2 \bar{\rho} \sum_{n=1}^{N} H_{cn} K_n (c_n - c_{gn}) + \sum_{n=1}^{N} H_{gn} S_n , \qquad (40)$$

in which $\bar{e}$ is the average total energy per unit mass of the plume, e_a the energy per unit mass of the atmosphere, H_{Rj} is the energy released by reaction j, R_j the rate of reaction j, H_{cn} the specific enthalpy of the condensed phase of component n and H_{gn} the specific enthalpy of the gas phase of component n. The first term on the right hand side of (40) represents change of energy due to entrainment, the second is due to chemical reaction, the third is due to condensation and the last is caused by reaction with the ground surface. The total energy $\bar{e}$ is related to average plume enthalpy $\bar{h}$ by means of the following relation

$$\bar{e} = \bar{h} + gz \; , \tag{41}$$

and $\bar{h}$ can be written as

$$\bar{h} = \sum_{n=1}^{N} \{c_n H_{cn} + c_{gn}(H_{gn} - H_{cn})\} \; . \tag{42}$$

H_{cn} and H_{gn} are calculated as function of the temperature and hence as function of the density by means of auxiliary relations. The mass balance of component n reads

$$\frac{d}{ds}(b^2 \bar{u} \bar{\rho} \bar{c}_n) = 2b\rho_a c_{an} u_e + b^2 \sum_{j=1}^{J} F_{nj} R_j -$$

$$- b^2 \bar{\rho} K_n(c_n - c_{gn}) + S_n \quad , \tag{43}$$

in which F_{nj} is the mass of component n produced per unit mass by reaction j. The meaning of the terms is obvious from the earlier equations. From (37), (38), (39), (40), (41), (42) and (43) and the auxiliary relations the plume properties b, $\bar{u}$, $\bar{\rho}$, $\bar{e}$ and $\bar{c}_n$ can be calculated.

Of course the real concentration distribution inside the plume is not constant, but changes in radial direction. To take this effect into account Bloom assumes that the distribution is Gaussian with standard deviation $b_y/3$ and $b_z/3$. The plume cross section of the Gaussian plume is approximated by an ellipse with principal axis given by b_y and b_z. Like Ooms and Mahieu [7] the circular plume cross section (radius b) is changed into the ellipse, keeping the area constant and calculating the ratio of the principal axis of the ellipse from the ratio of the dispersion coefficients σ_y and σ_z of Gifford [11]. In this way the following relations between b_y and b_z on the one hand and b, σ_y and σ_z on the other are found

$$b_y = b\sigma_y^{\frac{1}{2}}/(\sigma_y|\frac{\bar{w}}{\bar{u}}| + \sigma_z|\frac{\bar{v}}{\bar{u}}|)^{\frac{1}{2}} \tag{44}$$

and

$$b_z = b(\sigma_y|\frac{\bar{w}}{\bar{u}}| + \sigma_z|\frac{\bar{v}}{\bar{u}}|)^{\frac{1}{2}}/\sigma_y^{\frac{1}{2}} \; . \tag{45}$$

In order to be able to solve (37) - (43) the entrainment velocity u_e has to be known. Bloom uses for the first part of the plume an expression derived by Hirst [12]

$$u_e = \{0.08061 + |0.5\ gb\bar{w}(\frac{\rho_a}{\bar{\rho}} - 1)/\bar{u}^3|\}\{|(\bar{u}^2 - \bar{v}U_a)/\bar{u}| + $$

$$+ 4.5|\bar{w}U_a/\bar{u}|\} . \tag{46}$$

This equation is applied up to a certain distance downwind. Several criteria are considered for this distance. All of the criteria assume first that before another expression for u_e is used the plume has entrained enough air to cause at least a 150-fold dilution. The transition is then made if either the derivative of the vertical velocity ($d\bar{w}/ds$) or the derivative of the energy ($d\bar{e}/ds$) changes sign. If neither of these terms change sign by a 500-fold dilution, the transition is made at that point.

After the transition is made, the entrainment velocity is based on a dilution rate deduced from the Gifford dispersion coefficients. For the entrainment velocity the follwoing expression is then used

$$u_e = \bar{u}\, b_T \frac{d}{ds} \left[\frac{\{\sigma_y(\sigma_y|\frac{\bar{w}}{\bar{u}}| + \sigma_z|\frac{\bar{v}}{\bar{u}}|)\}}{\{\sigma_y(\sigma_y|\frac{\bar{w}}{\bar{u}}| + \sigma_z|\frac{\bar{v}}{\bar{u}}|)\}_T} \right]^{\frac{1}{2}} , \tag{47}$$

in which the index T refers to conditions at the transition point.

Bloom's model is very comprehensive. A remarkable fact, however, is that a drag force on the plume due to the wind is not taken into account. Ooms, Mahieu and Zelis found it necessary to incorporate such a force in their model in order to find a good agreement between theoretical predictions and experimental results. It would be very interesting to compare predictions made with Bloom's model with these same experimental results. It is perhaps due to his entrainment equation (46) which is quite different from the one of Ooms, Mahieu and Zelis, that such a drag force is not necessary in his model. Further work on this matter is required. In Bloom's model also no attention is given to the influence of the zone of flow establishment; the model starts at the stack exit. This fact also needs further study.

Literature review on experiments.

We will restrict ourselves to experiments of heavy plumes in a cross wind. Experiments with plumes in a quiescent environment, see for instance Turner [13], are left out of consideration.

1) Bodurtha ([14]).

Bodurtha performed wind tunnel experiments on heavy stack plumes. Stack plumes of a certain density were achieved by means of a proper mixture of Freon-114 and air. Velocity- or concentration measurements were not made. The plumes were made visible by an oil-fog smoke, and for the evaluation of the experiments the visible plumes were photographed. From the photographs the maximum initial rise, the distance to the touch-down point and the velocity of plume descent were determined. Also a rough estimate of the plume concentration was made by measuring the plume radius. For the maximum initial plume rise ΔH_{max} the following relation is given

$$\Delta H_{max} = 3.00(2b_o)^{0.5}(\frac{U_o}{U_a})^{0.75} , \qquad (48)$$

in which b_o represents the stack radius.
As the dimensions of the left hand side and right hand side of (48) are not the same, this equation may only be used for the case that SI-units are used. The results for the velocity of plume descent and the ground level concentration are given in the form of graphs. The distance to the touch-down point can be read from a table.
A difficulty with Bodurtha's results is that they are given after scaling to full-scale stacks. For the scaling the dimensionless Froude number $U_a^2/2b_og$ is used.
It is known that the plume properties become independent of the exit Reynolds number, if the value of this number is sufficiently large. For real stack this is always the case; for the Bodurtha experiments the values of the Reynolds number are probably too low. All theoretical models are only valid for sufficiently large values of the exit Reynolds number; Reynolds number independency is assumed.
Another difficulty with these experiments is that, as mentioned earlier, the plumes were made visible with an oil-fog smoke. It can be doubted, especially when the plume velocity is low, whether the oil droplets really follow the plume. Also the influence of atmospheric turbulence was not investigated.

2) Hoehne and Luce ([15]).

Hoehne and Luce investigated the dispersion of various diameter plumes of methane, ethane, butane and heptane gas in a wind tunnel. Through the selection of exit temperature and molecular weight the ratio of the exit plume density to the air density ranged from 0.28 to 2.44; so both light and heavy plumes were studied.
A disadvantage of these experiments is, that the Froude number was always too large (>> 1) to simulate in a correct way the dispersion of a full-scale heavy plume. If the Froude number is much larger than unity the density difference between the plume and the air has almost no effect on the plume path. Indeed, during the experiments the heavy plumes did not fall down after an initial rise, contrary to observations on actual plumes. The exit Reynolds number was always larger than 10000, and Reynolds number independency was observed.
Velocity-, temperature- and concentration measurements were carried out by Hoehne and Luce. Particular attention was paid to the concentration decay along the plume axis. Some relations for this concentration decay are given in the paper. The coordinates of the plume axis are made dimensionless by means of the factor $b_o(U_o/U_a)(\rho_o/\rho_a)^{\frac{1}{2}}$; this emphasizes the fact that only exit momentum is important and that the influence of buoyancy is negligable for these experiments.
Another disadvantage of these experiments is that the plume was emitted via a circular opening in a flat plate. It is well-known that such a plate causes a growing boundary layer, which can effect the plume path considerably. Moreover, the influence of the atmospheric turbulence was not studied, and this influence is important.

3) Holly and Grace ([16]).

The experiments of Holly and Grace were carried out in a water channel; the heavy plume consisted of water mixed with salt. The values of the Froude number were small enough for buoyancy effects to be important. The values of the exit Reynolds number were large enough for Reynolds number independency.
Velocity- and concentration measurements were carried out; particular attention being paid to the maximum initial plume rise, the distance to the touch-down point, the width of the plume and the plume concentration. For the maximum initial plume rise the following relation is given in the paper

$$\frac{\Delta H_{max}}{2b_o} = [3.4 \cdot 10^{-0.148U_a/\{2b_o g(\rho_o - \rho_a)/\rho_a\}^{\frac{1}{2}}}] \cdot$$

$$\cdot [U_o/\{2b_o g(\rho_o - \rho_a)/\rho_a\}^{\frac{1}{2}}] \ . \qquad (49)$$

For the case of $U_a \to 0$ this relation compares well with the relation of Turner [13] for a heavy plume in a quiescent environment. For the distance x_d to the touch-down point the following relation is given

$$x_d = 9.62\ \Delta H_{max} \log[2U_a/\{2b_o g(\rho_o - \rho_a)/\rho_a\}^{\frac{1}{2}}] \ . \qquad (50)$$

The concentration decay along the plume centreline can be determined from

$$\frac{c_o}{c^*(s)} = [31 \cdot 10^{0.4U_a/\{2b_o g(\rho_o - \rho_a)/\rho_a\}^{\frac{1}{2}}}] \ (\frac{x}{x_d})^{0.68} \ . \qquad (51)$$

As mentioned earlier the Froude number is small enough and the Reynolds number large enough for a realistic simulation of a full-scale heavy stack plume. So at first sight the Holly and Grace experiments seem very suited for the testing of theoretical models.
A disadvantage of the experiments is, however, that the plumes were not emitted from a stack but from a diffuser located on the water channel floor, from which the plume is discharged vertically through a circular port. The diffuser consisted of copper tubing extending across the full width of the floor and plugged at one end. The reason for this experimental set-up is, that Holly and Grace were not interested in the dispersion of a heavy stack plume, but in the dispersion of waste brine of a desalination plant. We have no idea in what way the diffuser influenced the dispersion process.
Another disadvantage is, that again the environmental turbulence is not taken into account.

4) Hoot, Meroney and Peterka ([3], [17]).

Heavy stack plume experiments with and without a laminar cross wind were carried out by Hoot, Meroney and Peterka; also the dispersion of a ground source in a turbulent boundary layer was investigated. Only the stack plume experiments with a cross wind will be discussed.
The tunnel had a 24 × 24 inch cross section. The exit Reynolds number was not always sufficiently large to justify the assumption of turbulent entrainment. Therefore the turbulence was artificially generated. A short-

edged orifice was placed in the stack 8 diameters upstream from the exit. According to the investigators this measure warrants Reynolds number similarity.

The experiments were conducted as follows. Tunnel and stack rates were set. The plume density was obtained by mixing Freon 12 with air in appropriate proportions. The effluent was bubbled through $TiCl_4$ to produce smoke. Extended time expose photographs were taken against a blackboard divided into marked horizontal and vertical increments. In this way the plume path and width were determined.

Concentration measurements at the maximum rise height were made of 18 plumes injected into the laminar cross wind. The concentrations at the points where these plumes touched the floor were also measured. Also some detailed cross sectional concentration measurements were taken.

From the experiments it was found that the maximum initial rise can be described by

$$\frac{\Delta H_{max}}{2b_o} = 1.32\left(\frac{U_o}{U_a}\right)^{1/3}\left(\frac{\rho_o}{\rho_a}\right)^{1/3}\left(\frac{U_o^2}{2b_o g}\cdot\frac{\rho_o}{(\rho_o-\rho_a)}\right)^{1/3} . \tag{52}$$

The downwind distance to the touch-down point is given by

$$\frac{x_d}{2b_o} = \left[\left(\frac{U_o^2}{2b_o g}\cdot\frac{\rho_o}{(\rho_o-\rho_a)}\right)\left(\frac{U_o}{U_a}\right)^{-1}\right] + 0.56\left[\left(\frac{\Delta H_{max}}{2b_o}\right)^3\left\{\left(2+\frac{H}{\Delta H_{max}}\right)^3-1\right\}\right]^{\frac{1}{2}} .$$

$$\cdot\left[\left(\frac{U_a^2}{2b_o g}\cdot\frac{\rho_o}{(\rho_o-\rho_a)}\right)\left(\frac{U_o}{U_a}\right)^{-1}\right]^{\frac{1}{2}} , \tag{53}$$

in which H is the stack height.

The plumes exhibited a somewhat skewed Gaussian distribution of concentration during the rising portion of the trajectory and a more symmetric distribution during the falling portion. For the maximum concentration in a cross section the following relations were derived

$$\frac{c^*}{c_o} = 1.688\left(\frac{U_o}{U_a}\right)\left(\frac{\Delta H_{max}}{2b_o}\right)^{-1.85} \tag{54}$$

for the point of maximum rise, and

$$\frac{c^*}{c_o} = 9.434\left(\frac{U_o}{U_a}\right)\left(\frac{\Delta H_{max}+H}{2b_o}\right)^{-1.95} \tag{55}$$

at the touch-down point.

The relations (52), (53), (54) and (55) can qualitatively also be derived with the model of Hoot, Meroney and Peterka discussed earlier; however the constants in these relations must then be determined by adapting model predictions to experiments.
The experiments of Hoot, Meroney and Peterka are reliable and suited to test theoretical models. A disadvantage however is that no detailed velocity and turbulence measurements in plume cross sections have been carried out. Also the influence of the environmental turbulence on the velocity- and concentration distributions has not been investigated.

5) Anderson, Parker and Benedict ([18]).

Experiments somewhat similar to those of Holly and Grace were carried out by Anderson, Parker and Benedict. Their water channel was 60 feet long, 1.0 feet deep and 2.0 feet wide. During most experiments the height of the water in the flume was about 10 to 12 inch. The cross flow was not stratified and consisted of fresh water. The plume consisted of water mixed with salt at the same temperature as the cross flow. The plumes were injected at several initial angles. In the case of injection at 60° or 45° the exit was at 0.08 to 0.10 feet above the bottom. At 90° the plume was injected through a tap in the bottom of the flume; so in that case there can be an influence of the boundary layer along the bottom. The exit diameters varied between 0.72 and 0.95 cm.
During the experiments concentration measurements were carried out with the aid of a conductivity probe. The plume centreline, the concentration at the plume centreline and the width of the concentration profile were measured at several distances downstream. The values of the Froude number were low enough to show buoyancy effects. The ratio of the jet exit velocity to the cross flow velocity varied between 5 and 20. The Reynolds number reached values of only 1000 to 5000. So Reynolds number similarity is questionable.
The experimental results were compared with the model of Abraham [19], which was adjusted to the case of negatively buoyant plumes in a way similar to that of Ooms, Mahieu and Zelis. For not too small values of the Froude number reasonable agreement between model predictions and experiments was found as long as the drag coefficient c_d was set to zero.

6) Chu ([9]).

The experiments of Chu were performed in a 30 cm wide, 45 cm deep and 9 m long flume. The plume was simulated by injecting a dyed saline solution vertically upward into a uniform open channel flow through an injection pipe of 1 cm inside diameter at 5 cm above the channel floor. The values of the exit Reynolds number seem high enough to assure Reynolds number independency, and the values of the Froude number are small enough for negative buoyancy effects to be important. The primary objective of the experiments was to determine the path of the plume. For each experiment a photograph of 8 exposures of the plume at approximately 10 s intervals were taken. A multiple exposure of the photographs gives the plume boundaries. The path of the plume was defined to be the line of maximum concentration, i.e. the darkest region in the multiple exposure of the photographs. These experiments can be used for the testing of theoretical models.

6) Badr ([20]).

Badr also performed experiments in a flume. Fresh water was discharged through a chimney into flowing warm water. Temperature measurements were carried out and the following quantities were determined from the measurements: the thermal axial trajectory, the jet width, the vertical mean temperature profiles and the vertical temperature mean standard deviation profiles. As Badr's paper is published in this book we will not discuss it here in more detail. We believe that these experiments are very valuable for the testing of models.

Recommendations for further work.

As was shown some rather detailed theoretical models for the dispersion of a heavy stack plume have been developed. However, their reliability has to be checked by comparing predictions made with these models with accurate and detailed experimental results. A first step in this direction is the work of Hoot, Meroney and Peterka and of Badr. Although some good experiments have already been carried out by them, there is still a need for more data. For instance, data about the velocity distributions in plume cross sections for different values of the relevant parameters are of importance. The Froude number must be low enough for negative buoyancy forces to be important; the Reynolds number must be high enough for Reynolds number independency to hold.
Also the influence of atmospheric turbulence on the plume dispersion needs

further study. Although this influence is taken into account in some models, the correctness of the modelling of this influence has to be checked. Recently an interesting study of the influence of a stable stratification on turbulent diffusion was reported by Britter, Hunt, Marsh and Snyder [21]. They carried out experiments in which a grid is towed horizontally along a large tank filled first with water and then with a stably stratified saline solution. The turbulent diffusion from a point source located 4.7 mesh lengths downstream of the grid was studied. σ_y, σ_z, the horizontal and vertical plume widths, were measured by a rake of probes. σ_y was found to be largely uneffected by the stratification and grew like $t^{\frac{1}{2}}$, while σ_z was found in all cases to reach an asymptotic limit $\sigma_{z\infty}$ where $0.5 \leq \sigma_{z\infty} N/w'_s \leq 2$, w'_s being the r.m.s. vertical velocity fluctuations at the source and

$$N = \left(-g \frac{\partial\bar{\rho}/\partial z}{\bar{\rho}}\right)^{\frac{1}{2}}$$

the buoyancy frequency. These results are largely in agreement with the theoretical model of Csanady [22].

If a heavy plume is injected in a stable atmosphere it is possible that at a certain distance from the stack the lower part of the plume becomes unstably stratified. This can, of course, considerably influence the dispersion process. As far as we know this effect has not been studied so far.

Immediately after the stack exit secondary flows occur inside the plume due to the influence of the wind. At a certain distance from the stack a light plume behaves as a line thermal rising in the atmosphere, in which secondary flows are also present. The direction of rotation of the secondary flows close to the stack and at large distances are the same. However, a heavy plume behaves at large distances as a "thermal" falling in the atmosphere and the direction of rotation of the secondary flow is then opposite to the direction of rotation at the stack exit. So in that case somewhere along the plume path the secondary flow changes its rotation direction. This difference between a light and a heavy plume can be important for the plume properties and needs further study.

In the model of Ooms, Mahieu and Zelis a drag force on the plume due to the wind occurs; however in the other models such a force is not present. Therefore the necessity of this drag force in the modelling of a heavy plume has to be investigated.

The influence of the zone of flow establishment has to be studied in more detail. A model for this zone will give the initial conditions for the model in the zone of established flow.
Finally, all models for a heavy plume are integral models. However, it is possible nowadays to solve the relevant partial differential equations for a heavy plume using a turbulence model. For a light or neutral plume such a calculation has already been carried out, see for instance Jones and McGuirk [23]. It would be interesting to perform such calculations also for a heavy plume and compare the results with those of the integral models.

References

1. Schatzmann, M.: Auftriebstrahlen in natürlichen Strömungen, Entwicklung eines mathematischen Modells. Dissertation, Universität Karlsruhe 1976.

2. Bosanquet, C.H.: The rise of a hot waste gas plume. J. of the Institute of Fuel 30 (1957) 322.

3. Hoot, T.G.; Meroney, R.N.; Peterka, J.A.: Wind tunnel tests of negatively buoyant plumes. Fluid Dynamics/Diffusion Laboratory, Colorado State University, Colorado 80521, 1973.

4. Hoult, D.P.; Fay, J.A.; Forney, L.J.: A theory of plume rise compared with field observations. J. Air. Poll. Control Ass. 19 (1969) 585.

5. Ooms, G.; Mahieu, A.P.; Zelis, F.: The plume path of vent gases heavier than air. First International Symposium on Loss Prevention and Safety Promotion in the Process Industries. The Hague 1974.

6. Keffer, J.F.; Baines, W.D.: The round turbulent jet in a cross wind. J. Fluid Mech 15 (1963) 481.

7. Ooms, G.; Mahieu, A.P.: A comparison between a plume path model and a virtual point source model for a stack plume. Appl. Sci. Res. 36 (1981) 339.

8. Singer, I.A.; Smith, N.E.: Atmospheric dispersion at Brookhaven National Laboratory. Int. J. Air and Water Pollution 10 (1966) 125.

9. Chu, V.H.: Turbulent dense plumes in laminar cross flow. J. Hydraulic Research 13 (1975) 263.

10. Bloom, S.G. A mathematical model for reactive negatively buoyant atmospheric plumes. Symposium on Heavy Gas. Frankfurt 1980.

11. Gifford, M.F.: An outline of theories of diffusion in the lower layer of the atmosphere. U.S. AEC Report No. TID-24190, 1968.

12. Hirst, E.A.: Analysis of round, turbulent, buoyant jets discharged to flowing stratified ambients. U.S. AEC Report No. ORNL-4685, 1971.

13. Turner, J.S.: Jets and Plumes with negative or reversing buoyancy. J. Fluid Mech 26 (1966) 779.

14. Bodurtha, F.T.: The behaviour of dense stack gases. J. Air Poll. Control Assoc. 11 (1961) 431.

15. Hoehne, V.O.; Luce, R.G.: The effect of velocity, temperature and molecular weight on inflammibility limits in wind blown jets of hydrocarbon gases. Proc. Div. of Refining, American Petroleum Institute, 35th Mid Year Meeting 1970.

16. Holly, F.M.; Grace, J.L.: Model study of dense jet in flowing fluid. J. of Hydr. Div., Proc. of the ASCE 98/9365 (1972) 1921.

17. Meroney, R.N.: Wind-tunnel experiments on dense gas dispersion. J. Hazardous Materials 6 (1982) 85.

18. Anderson, J.L.; Parker, F.L.; Benedict, B.A.: Negatively buoyant jets in a cross flow, EPA-Report No. 660/2-73-012, 1973.

19. Abraham, G.: Round buoyant jet in cross flow. 5th International Conference on Water Pollution Research. San Francisco 1970.

20. Badr, A.: Temperature measurements in a negatively buoyant round vertical jet issued in a horizontal cross flow. IUTAM Symposium "Atmospheric Dispersion of heavy gases and small particles", Scheveningen, The Netherlands 1983.

21. Britter, R.E.; Hunt, J.C.R.; Marsh, G.L.; Snyder, W.H.: The effects of stable stratification on turbulent diffusion and the decay of grid turbulence. J. Fluid Mech. 127 (1983) 27.

22. Csanady, G.T.: Turbulent diffusion in a stratified fluid. J. of Atmospheric Science 21 (1964) 439.

23. Jones, W.P.; McGuirk, J.J.: Computation of a round turbulent jet discharging into a confined cross flow. Second Symposium on Turbulent Shear Flow, London 1980.

Inclusion of Particles in Second-Order Modeling of the Atmospheric Surface Mixed Layer

J. M. LOTTEY, J. L. LUMLEY, T.-H. SHIH

Sibley School of Mechanical and Aerospace Engineering
Cornell University
Ithaca, NY 14853 USA

Abstract

We present a second-order model for the motion of atmospheric aerosol particles. Third moment terms are constructed from first principles and contain no new adjustable constants; models for drift terms and particle flux relaxation terms are constructed from realizability considerations, and have general validity. The results compare favorably with Csanady's simple model for the crossing-trajectories effect of Yudine when the particles are passive. It is suggested how the dynamical effect of the particles on the turbulence may be included.

Introduction

Second order modeling of the atmospheric surface mixed layer with stable or unstable stratification, wind shear and yaw and passive pollutants has been remarkable successful [1,2,3,4]. In many respects such modeling is no longer simply empiricism: for example, the role of buoyancy in modifying the vertical turbulent transport is now well understood [5];the return to isotropy has been examined in detail [6]; the relaxation of the heat flux has been carefully examined [7]; a rationally based form for the third moments (without adjustable constants) has been obtained [8,9,10].

The presence of particles in the atmospheric surface mixed layer is often of importance: these may be dust or sand, aerosols or fog droplets, salt particles or ocean spray. Similar situations occur in the ocean, although the density difference is usually quite small (except in the case of air

*Supported in part by the U.S. Office of Naval Research under the following programs: Physical Oceanography (Code 422PO), Fluid Dynamics (Code 438), Power (Code 473); in part by the U.S. National Science Foundation under grant no. ATM 79-22006; in part by the U. S. Air Force Office of Scientific Research; and in part by the U.S. Air Force Geophysics Laboratory. Prepared for presentation at IUTAM Symposium "Atmospheric Dispersion of Heavy Gases and Small Particles", August 29 - September 2, 1983, Delft University of Technology.

bubbles). We will confine ourselves to the atmosphere.

The presence of particles has been considered in [1], but these were completely passive; that is, they neither fell out of the eddy with which they started (the crossing-trajectories effect of Yudine, [11]), which reduces the diffusivity, nor did they have an influence on the turbulence. Both these effects can be included in a second order model with relatively little difficulty.

It is well-known that the inertia terms can generally be neglected for atmospheric aerosols and dust, as can particle-particle interaction [12]; this leaves only the crossing-trajectories effect to distinguish particulate motion from that of a heavy gas. In Lottey [13] the influence of this effect on particle dispersion was investigated, in the absence of buoyancy effects (either due to stability, or to the particles themselves), using extensions of second order models developed in [14]. In particular, it was necessary to develop models for the terms which are proportional to the terminal velocity in the second and third order governing equations. The predictions from this model were compared with the predictions of Csanady [15] for two simple cases, both with an isotropic homogeneous turbulence: one in which the particles initially uniformly filled the left half-space, and the other in which they initially uniformly fill the upper half-space.

We now present new results, showing by an order-of-magnitude analysis that the inclusion of the terms in the terminal velocity in the equations for the third moments is unnecessary, so long as the terminal velocity is not greater than the r.m.s. turbulent velocity, the principal effect on the diffusivity arising from the term in the (second order) particle flux equation. We present, in addition, a more general, and more fundamentally based, form for the model of this term, including an exact calculation of the initial return of disorder following artificially organized initial conditions. The general method of derivation is similar to that presented in [7].

Finally, we show how the effect of particle loading on the turbulence can be simply included in a calculation by definition of a pseudo-virtual potential temperature; the downward drag on the fluid due to the particles, which appears as a pseudo-buoyancy [14], can be included with temperature (and with water vapor) just as water vapor is. Although it can be shown by an order-of-magnitude analysis that it is unnecessary to include particle pseudo-buoyancy in the third moment terms, since this effect is not primarily responsible for the turbulence production, it is more convenient to do so, and use as primary variable net density anomaly.

The equations of motion

We consider the case of atmospheric aerosol particles, for which the effect of particle inertia is negligible [12]. We consider first the case of dynamically passive particles; we will consider later the influence of the particles on the turbulence. By using the Reynolds decomposition, we can form the equations for the first and second moments. We do not reproduce here the equations for the velocity field, which are unchanged by the presence of the particles.

$$\partial_t\langle C\rangle + \partial_j\langle C\rangle(U_j + V_j) + \partial_j\langle cu_j\rangle = 0 \tag{1}$$

$$\partial_t\langle c^2\rangle + 2\,\partial_i\langle C\rangle\langle cu_i\rangle + \partial_i\langle c^2\rangle U_i$$

$$+\ \partial_i\langle c^2 u_i\rangle + \partial_i\langle c^2\rangle V_i = -2\langle \varepsilon_c\rangle \tag{2}$$

$$\partial_t\langle cu_i\rangle + \partial_j\langle C\rangle\langle u_i u_j\rangle + \partial_j\langle cu_i\rangle U_j + \partial_j U_i\langle cu_j\rangle$$

$$+\ \partial_j\langle cu_i u_j\rangle + \langle u_i\,\partial_j c\rangle V_j = -\langle c\,\partial_i p\rangle/\rho \tag{3}$$

where ∂_t is a derivative with respect to time, and ∂_i is a derivative with respect to x_i; $\langle\ \rangle$ indicates an ensemble average; and C is the instantaneous value of the particle concentration, while $c = C - \langle C\rangle$. We have already made a few assumptions regarding high Reynolds/Peclet number [16]. Note that we have taken the terminal velocity V_i in an arbitrary direction, presumably alligned with the gravity vector. The exact mechanism for dissipation of particle concentration is a bit mysterious - in addition to Brownian motion, it appears to reflect the fact that it is not possible to have fluctuations in particle concentration on a scale smaller than the interparticle distance. We have introduced a fictitious diffusivity to represent this mechanism.

Realizability and the drift terms

The equation for the particle flux and for the particle concentration variance cannot be modeled independently, since some mechanism must be present to assure that particle concentration variance will never be negative, and the particle fluxes will always obey Schwarz's inequality. We introduce a normalized tensor D_{ij}

$$D_{ij} = (\langle c^2\rangle\langle u_i u_j\rangle - \langle cu_i\rangle\langle cu_j\rangle)/(\langle c^2\rangle\langle q^2\rangle - \langle cu_p\rangle\langle cu_p\rangle) \tag{4}$$

(note that the normalization differs somewhat from that given in [8]. As shown in [8], the non-negativity of the variance, and the satisfaction of Schwarz's inequality by all components, is equivalent to D_{ij} having all non-negative eigenvalues. We may introduce a scalar

$$F_D = 9\mathrm{tr}D^3 - 27\mathrm{tr}D^2/2 + 9/2 \tag{5}$$

In principal axes of D_{ij}, F_D can be written as

$$F_D = 27D_{11}D_{22}D_{33} \tag{6}$$

Hence, F_D will vanish if and only if one of the eigenvalues of D_{ij} vanishes. We may thus satisfy our requirement that the eigenvalues remain non-negative by requiring that F_D remain non-negative. Note that D_{ij} is symmetric, and $\mathrm{tr}D = 1$.

We may implement realizability by requiring that $\partial_t F_D = 0$ when $F_D = 0$; otherwise F_D would be negative at the next instant. We will discuss below how the derivative should vanish - i.e.- should the second derivative also vanish, for example? We can require that the terms on the right hand side of the equation for $\partial_t F_D$ associated with various effects vanish separately: that is, all the terms multiplied by the mean velocity on the right hand side should vanish separately; all those multiplied b the buoyancy vector should vanish separately; and all those multiplied by the terminal velocity should vanish separately. This is because we can imagine a thought experiment in which we arbitrarily change the value or orientation of any of these. If, before the change, we had arranged, by a balance among the terms multiplied by different factors, to have $\partial_t F_D = 0$, after the change it would no longer be so. The fact that, if the system were allowed to run long enough, the values and orientations of all variables would be restored relative to the values and orientations of U_i, V_i, g_i etc. is essentially irrelevant - we must have these inequalities satisfied at all times. This is particularly important when one thinks of recovery of the system of equations from poorly posed initial conditions, for example.

Hence, we must write the equation for F_D, but we will interest ourselves only in the terms multiplied by V_i. We have

$$\partial_t F_D = 27(D_{ij}D_{jk}\,\partial_t D_{ki} - D_{ij}\,\partial_t D_{ji}) \tag{7}$$

and from the definition of D_{ij} we may easily obtain the form of the right hand side. We may write

$$\langle u_i\,\partial_p c\rangle = \langle cu_i\rangle A_p \tag{8}$$

where A_p is an unknown vector we wish to determine. We presume that if there is no correlation between c and u_i there will be no correlation between the gradient of c and u_i. The equation for F_D becomes

$$\partial_t F_D = (54P\langle q^2\rangle\langle c^2\rangle V_p/d_{qq})(A_p - \partial_p\langle c^2\rangle/2\langle c^2\rangle) \tag{9}$$

where $d_{qq} = \langle c^2\rangle\langle q^2\rangle - \langle cu_q\rangle\langle cu_q\rangle$ and P is (see [17])

$$P = b{:}D^2 - b{:}D + \mathrm{tr}D^2 4/3 - 1/3 - \mathrm{tr}D^3 \tag{10}$$

where $b_{ij} = \langle u_iu_j\rangle/\langle q^2\rangle - \delta_{ij}/3$. If an eigenvalue of D_{ij} vanishes, P does not in general vanish unless a very special relationship happens to hold between the eigenvalues of D_{ij} and those of b_{ij}. Hence, we must require that the second parenthesis in eq. (9) vanish when F_D vanishes. That is, that

$$A_p - \partial_p\langle c^2\rangle/2\langle c^2\rangle = 0 \quad \text{if } F_D = 0 \tag{11}$$

What sort of zero is it?

If $F_D = 0$, will it remain there? That is, is this a situation from which the system can extricate itself without external help, or will it remain stuck in this situation until an external disturbance occurs? We can answer this question by considering the exact equations for such a system. Imagine an isotropic, homogeneous turbulence, without mean velocity, in which there is a homogeneous distribution of particles. At $t = 0$ we introduce the artificial initial condition $c \propto u_1$, which results in $D_{11} = 0$, and hence $F_D = 0$. We suppose that initially $\langle cu_2 \rangle = \langle cu_3 \rangle = 0$. Thus, D_{ij} is diagonal. Taking the exact instantaneous equations, and making extensive use of $c \propto u_1$, we can show that $\partial_t \rho = 0$ at $t = 0$, where ρ is the correlation coefficient between c and u_1. This corresponds to our assumption that $\partial_t F_D = 0$, and is in fact required whenever a correlation coefficient $\rho^2 = 1$ (see [17]). If we calculate further, after a great deal of algebra we can obtain the second derivative at $t = 0$

$$\partial_t{}^2 \rho = -2(V^2/\langle u^2 \rangle)\langle \partial_3 u_1 \partial_3 u_1 \rangle \tag{12}$$

where we have taken the terminal velocity V in the 3-direction. Hence, the system will, without external disturbance, draw away from the state of perfect correlation. This is not true of all states of perfect correlation, of course; it appears that if two eigenvalues vanish, the system will remain in that state indefinitely until disturbed. We are indebted to S. B. Pope for pointing this out (private communication).

It is not difficult to show that, if we require that $\partial_t F_D = 0$ when $F_D = 0$, but that $\partial_t{}^2 F_D > 0$, then close to the point where $F_D = 0$ we must have $\partial_t F_D \propto F_D{}^{1/2}$. Of course, when F_D is not close to zero, $\partial_t F_D$ can have a more general dependency. However, we have found it adequate from a computational point of view to take

$$A_p - \partial_p \langle c^2 \rangle / 2\langle c^2 \rangle f(F_D) = V_p(9\langle \varepsilon \rangle / \langle q^2 \rangle^2) F_D{}^{1/2}[a + b(V_i \langle cu_i \rangle)^2 / V_q V_q \langle cu_r \rangle \langle cu_r \rangle] \tag{13}$$

where $f(0) = 1$, and a and b are constants that must be determined by comparison of model predictions with experiment, or with another model. $f(F_D)$ is also an unknown function, which must be determined in the same way. The right hand side must be a vector, and must be operative whether gradients exist in the field or not, and whether or not they are orthogonal to the terminal velocity (since we expect the crossing-trajectories effect to decrease the correlation in any event). Hence, it must be proportional to the terminal velocity vector. We have included a factor to give the proper dimensions, as well as one to give a different diffusivity when the particle flux is colinear with the terminal velocity, from that effective when the flux is orthogonal to the terminal velocity. It appears to be adequate to simply take $f(x) = 1 + cx$, where c is a constant.

Let us imagine a steady, homogeneous situation, with no buoyancy and no mean velocity, and a linear gradient in particle concentration. Then the equation for the particle flux can be written as (where we have introduced a simple model for the pressure correlation - see [8])

$$\partial_j\langle C\rangle\langle u_i u_j\rangle + \langle u_i \partial_j c\rangle V_j = -(\langle\varepsilon\rangle/\langle q^2\rangle)C_c\langle cu_i\rangle \tag{14}$$

Using (8), we can readily solve this equation for the particle flux:

$$\langle cu_i\rangle = -(\langle q^2\rangle/\langle\varepsilon\rangle C_c)\langle u_i u_j\rangle\partial_j\langle C\rangle[1 + A_p V_p\langle q^2\rangle/\langle\varepsilon\rangle C_c]^{-1} \tag{15}$$

Thus, A_pV_p must be a positive quantity, to reduce the particle diffusivity as observed. The factor in square brackets becomes (with (13))

$$[\ldots] = 1 + (3V^2/\langle q^2\rangle)(3F_D{}^{1/2}/C_c)(a + b\alpha) \tag{16}$$

where we have written α for the expression in eq. (13) which takes on the value 0 if the particle flux is orthogonal to the terminal velocity, and 1 if they are colinear.

We can compare this expression with the predictions of Csanady (as described in [12]) for vertical and horizontal dispersion in isotropic turbulence. Briefly, Csanady's model predicts

$$K_H = K/[1 + \beta_H^2 V^2/u'^2]^{1/2}, \quad \beta_H = 4/3$$
$$K_V = K/[1 + \beta_V^2 V^2/u'^2]^{1/2}, \quad \beta_V = 2/3 \tag{17}$$

where K is the diffusivity when the terminal velocity vanishes. Subscript H corresponds to horizontal dispersion while subscript V corresponds to vertical dispersion. All diffusivities are for long times relative to the integral time scale of the turbulence. As discussed in [12], the vertical case is much more reliable; the value of β_H is rather poorly determined. Comparing equations (16) and (17) for the vertical and horizontal cases, with $\alpha = 1$ and 0 respectively, taking V^2/u'^2 small, and using the fact that, with $F_D = (1 - \rho^2)/(1 - \rho^2/3)^3$, if $\rho \approx 0.5$, $F_D \approx 1$, while $C_c \approx 3$, we obtain $a = 8/9$, $b = -2/3$.

Now let us imagine an inhomogeneous situation in which there is a particle flux only colinear with the terminal velocity. We may write the equation for the particle flux correlation coefficient:

$$2\partial_t \rho/\rho = -V_p(\partial_p\langle c^2\rangle/\langle c^2\rangle)cF_D \ldots \tag{18}$$

The omitted terms correspond to the crossing-trajectories effect. We wish to concentrate here on the effect of the term resulting from inhomogeneity. This is an effect which is thought to be new, and has not been previously described. The term in eq. (18), of course, can have either sign, depending on the sign of the gradient of $\langle c^2\rangle$. In many flows, there will be regions of both signs. Consider a situation in which $V_p\partial_p\langle c^2\rangle < 0$; as the concentration field drifts down through the velocity field, we may imagine that the correlation between u_3 and c is dropping (due to the crossing-trajectories effect), but the level of c is rising (because the entire c-field is drifting down, bringing higher and higher values to the measuring point), so that the particle is not dropping asmeasuring point), so that the particle flux is not dropping as fast as it otherwise would. The question is, does this effect make exactly the same contribution to the flux that it makes to the variance, so that the correlation coefficient stays the same (c = 0), or does it overcompensate (c > 0) or undercompensate (c < 0)? If the inhomogeneity is such that it is normalizable, i.e.- if the spacial variation of concentration can be reduced to a homogeneous random variable by normalizing by the local variance and stretching the length scale, then this corresponds to no change in the correlation coefficient. Lacking other information, we have adopted this model, and taken $c = 0$.

The third moments

In [8] a technique is presented for the derivation of forms for the third moments which is based more or less on first principles. Briefly, it is supposed that the turbulence (in the energy containing range) wishes to relax to a Gaussian state in the absence of disturbing conditions such as inhomogeneity, buoyancy, chemical reactions and so forth. This is in accord with experiment. Now it is assumed that the turbulence is fine grained with respect to the scale of the inhomogeneity, and the leading terms are kept.

We wish now to include the effect of the terminal velocity. We have no need to reproduce here the complex expressions from [8]. It is enough to note that the term associated with the terminal velocity in the equation for the instantaneous particle concentration will produce a term in the equation for the moment generating function which is of the same form as the term arising from the substantial derivative. Hence, in the order of magnitude analysis, this term will be no larger than the substantial derivative term as long as the ratio of the terminal velocity to the length scale of the gradient in its direction is no larger than the ratio of the turbulent rms velocity to the length scale of the gradients transverse to the mean velocity. Since, in the worst case, the scales of these gradients are the same, the drift term may be neglected if the terminal velocity is not larger than the turbulent rms fluctuating velocity.

Hence, so long as $V \leq u'$, we may use the same forms for the third moments as are presented in [8], and used in [10].

Computations

Computations were carried out for a simple, non-decaying isotropic turbulence for two cases: in case I, the particles initially uniformly filled the upper half-space; in case II, the particles initially uniformly filled the left half-space. In both cases gravity was directed vertically downward. In respects other than those discussed here, the models were the same as those described in [10]. The computations were carried out for times at least twice, and usually many more times, the integral time scale of the turbulence.

In figure 1 we present results obtained with $a = 8/9$, and $b = -2/3$, for the range of $V \leq u'$. It can be seen, that the model mimics very well the behavior of Csanady's simple model. The results are evaluated at the peak of the $\langle c^2 \rangle$ curve, so that there is no contribution from the inhomogeneity term, nor essentially any from the transport; if the time evolution is taken as self-similar, there is also approximately no contribution from the time term. Thus, this point is roughly a point of quasi-homogeneity, so that equation (15) should apply. As discussed in [12], the value of β_H is much less well determined than that of β_V; a change in this value resulting from comparison with better data would change the values of a and b somewhat. Here we wish only to establish the principle that our model produces results similar to those of Csanady's simple model.

In figure 2 we show the distribution of the horizontal diffusivity (obtained as the ratio of the particle flux and the gradient of the mean particle concentration) for a value of $V/u' = 0.5$. Since in this case the terminal velocity and the gradient of the variance are orthogonal to each other, the first term in eq. (13) has no effect.

In figure 3 we show the distribution of the vertical diffusivity for a value of $V/u' = 0.5$. Here the first term in eq. (13) is active. (Note that the abscissa measures vertical distance in figure 3, but horizontal distance in figure 2). Now, as a particle drifts downward in the upper part of the mixing layer (the left of the figure) it encounters increasing variance, and vice versa in the lower part. Hence, in the lower part the influx of higher variance compensates for the reduction of the particle flux caused by the crossing-trajectories effect, while in the upper part the effect is opposite. Thus, the diffusivity is higher in the lower part, and lower in the upper part. For $z < 33$ and $z > 48$ the values are not particularly reliable since both flux and gradient are very close to zero.

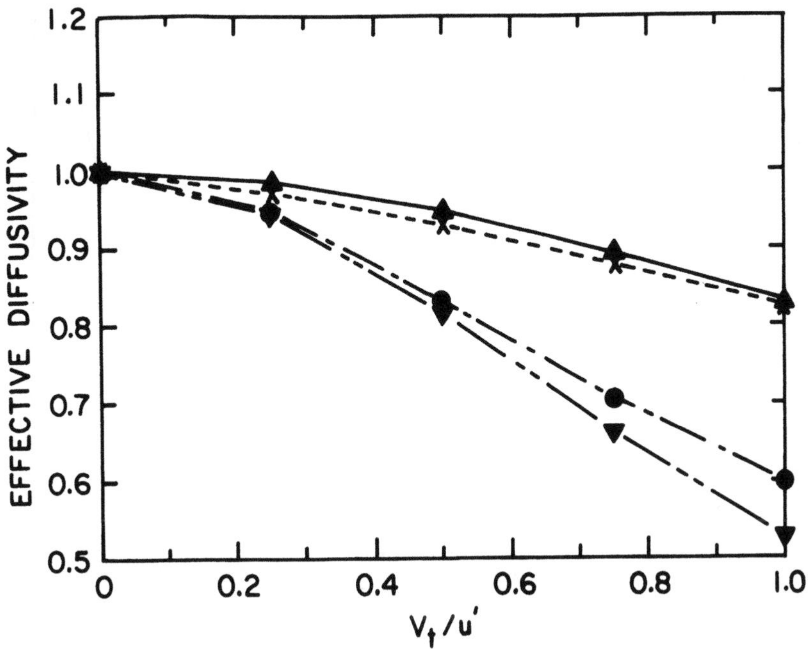

Fig. 1. Relative diffusivities compared with Csanady's predictions. ● - Csanady K_h; ▼ - Second order K_h; ▲ - Csanady K_v; X - Second order K_v.

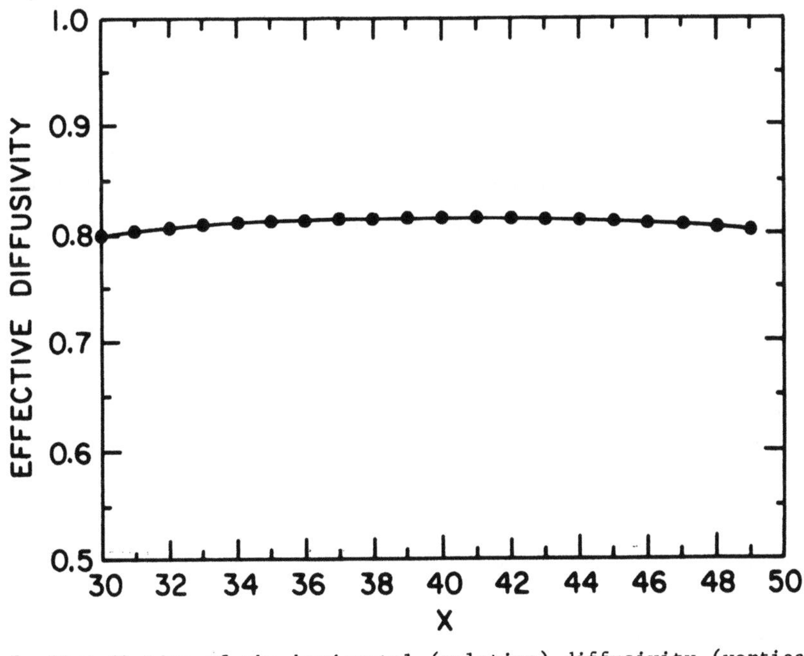

Fig. 2. Distribution of the horizontal (relative) diffusivity (vertically homogeneous).

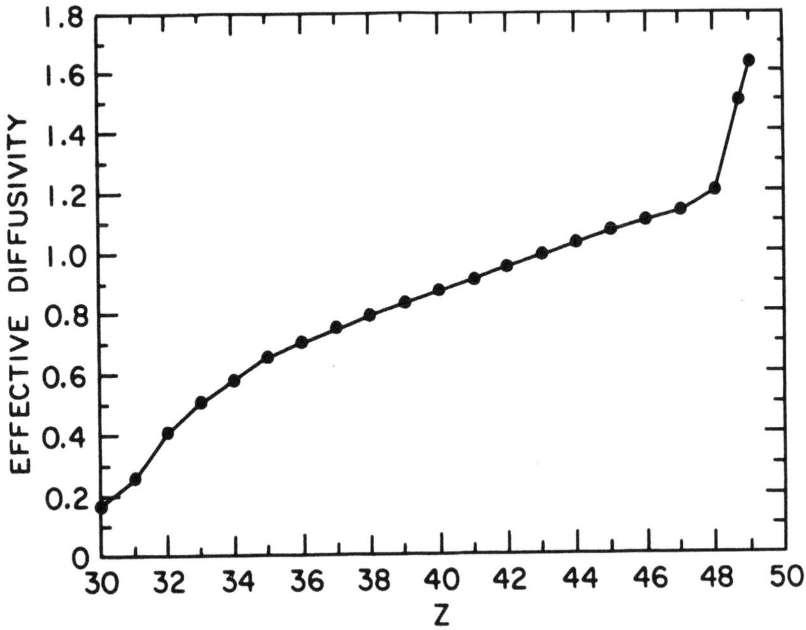

Fig. 3 Distribution of the vertical (relative) diffusivity (horizontally homogeneous).

Particle loading

If the concentration of particles is sufficiently high that they can no longer be considered dynamically passive, it is a relatively simple matter to include this effect in the model. In [12] it is pointed out that heavy, non-interacting inertia-free particles make their presence felt primarily through their downward drag on the fluid. Since (under our circumstances) inertia is negligible, the upward drag on each particle is equal to its weight, and hence the downward force felt by the fluid is equal to the particle weight. Thus, the downward force per unit mass of fluid is simply $g\rho_p/\rho_f$, where ρ_p is the mass of particles per unit volume, and ρ_f is the mass of fluid per unit volume; this downward force must be added to the vertical fluid momentum equation. If the effect of heat and water vapor is described by the Boussinesq approximation, the density anomaly (relative to adiabatic conditions) will also appear in the equation for the vertical momentum, and the two will have the same form. Thus the total term is simply the total relative density anomaly (due to heat, moisture and particles) multiplied by gravity.

From a computational point of view, it is simplist to carry equations for the density anomaly, the moisture mixture fraction and the particle mass per unit volume. In this way, the velocity field and the density anomaly and particle mass may be calculated separately, and the moisture mixture fraction calculated subsequently. If it were not for the drift term in the particle mass equation, this equation also would decouple, and the velocity field and density anomaly could be calculated separately. Fortunately, since the drift terms do not appear in the equations for the third moments, the third moment expressions do decouple. In the expressions for the third moments, it is necessary to retain only terms which are dynamically significant. Thus, if buoyant production of turbulence is significant, the buoyant terms must be retained. From this point of view, the terms in particle density probably could always be neglected in the third moment equations, since particle density anomaly is probably seldom a major source of turbulence production. However, much more is gained in convenience by keeping the particle density, and using total density anomaly as a variable.

As is discussed in [17], second order modeling suffers from a serious drawback: combinations of scalars such as we propose (combining heat, moisture and particles into a density anomaly) create ambiguities regarding the scales. If each primitive scalar has a different length scale of the energy containing eddies, density anomaly will have a very complex spectrum with three different scales, as will the velocity field. In second order modeling, a compromise is reached, and the true spectrum is replaced by a simple spectrum with a single compromise scale, obtained from the individual scales by weighting. So long as we do not attempt to make predictions in situations with extreme disparities of scale, this is probably not too serious.

Bibliography

1. Zeman, O.; Lumley, J. L.: Turbulence and diffusion modeling in buoyancy driven mixed layers. In Proceedings of Third Symposium on Atmospheric Turbulence, Diffusion and Air Quality, Raleigh, NC, (1976) 38-45. Boston, MA: American Meteorological Society.

2. Zeman, O.; Lumley, J. L.: Modeling buoyancy-driven mixed layers. J. Atmos. Sci. 33, 10 (1976) 1974-88.

3. Zeman, O.; Lumley, J. L.: A Second-order model for buoyancy-driven mixed layers. In Proceedings, 9th ICHMT International Seminar

"Turbulent Buoyant Convection", Dubrovnik, pp. 65-76 (1976). Washington, DC: Hemisphere Publishing.

4. Zeman, O.; Lumley, J. L.: Buoyancy effects in entraining turbulent boundary layers: a second-order closure study. In Turbulent Shear Flows I, eds. F. Durst, B. E. Launder, F. W. Schmidt and J. H. Whitelaw, pp. 295-302 (1979). Berlin/Heidelberg: Springer-Verlag.

5. Lumley, J. L.; Zeman, O.; & Siess, J.: The influence of buoyancy on turbulent transport. J. Fluid Mech. 84 (1978) 581-597.

6. Choi, Kwing-So: A Study of the Return to Isotropy of Homogeneous Turbulence. Ph.D. Thesis. Ithaca, NY: Cornell 1982.

7. Shih, T.-H.; Lumley, J. L.: Influence of time scale ratio on scalar flux relaxation: modeling Sirivat and Warhaft's homogeneous passive scalar fluctuations. Sibley School of Mechanical and Aerospace Engineering Report No. FDA-81-06. Ithaca, NY: Cornell 1981.

8. Lumley, J. L.: Computational modeling of turbulent flows. In Advances in Applied Mechanics 18, ed. C.-S. Yih, pp. 123-176. New York: Academic Press 1978.

9. Taulbee, D. B.; Lumley, J. L.: Implications regarding large eddy structures from attempts to model turbulent transport, in Applications of Fluid Mechanics and Heat Transfer to Energy and Environmental Problems, Papailiou, D. D.; Papailiou, D. (eds.), Berlin/Heidelberg/New York: Springer. In Press (1982).

10. Shih, T.-H.; Lumley, J. L.: Modeling heat flux in a thermal mixing layer. In Refined Modeling of Flows, Vol. I. Eds. J. P. Benque, C. J. Chen, R. Glowinski, D. Gosman, B. Mandagaran, W. Rodi, P. L. Viollet, C. S. Yih. pp. 239-250 (1981). Paris: Presses Ponts et Chaussées.

11. Yudine, M. I.: In Advances in Geophysics, Vol.6. (1959) 185.

12. Lumley, J. L.: Two-phase and non-Newtonian flows. In Turbulence (Topics in Applied Physics, Vol. 12), ed. P. Bradshaw. pp. 290-324. Berlin/Heidelberg/New York: Springer 1978.

13. Lottey, J. M.: The Turbulent Transport of Atmospheric Aerosols. M. S. Thesis. Ithaca, NY: Cornell 1979.

14. Lumley, J. L.: Turbulent transport of passive contaminants and particles: fundamentals and advanced methods of numerical modeling. In Pollutant Dispersal, lecture series 1978-7. Rhode-St.-Genese, Belgium: Von Karman Institite for Fluid Dynamics 1978.

15. Csanady, G. T.: Turbulent diffusion of heavy particles in the atmosphere. J. Atmos. Sci. 20 (1963) 201-208.

16. Tennekes, H.; Lumley, J. L.: A First Course in Turbulence. Cambridge, MA: MIT Press 1972.

17. Lumley, J. L.: Turbulence Modeling. J. Applied Mechanics (50th Anniversary Issue). In Press (1983).

Experimental Observations of Entrainment Rates in Dense Gas Dispersion Tests

JAMES A. FAY

Department of Mechanical Engineering
Massachusetts Institute of Technology
Cambridge, Massachusetts

Summary

It is shown how the local entrainment rate and the local Richardson number used in box models of dense gas cloud dispersion can be found from experimental measurements of wind tunnel simulations. These calculated values can then be used to check the common hypothesis of box and other models that the dimensionless entrainment is a universal function of the local Richardson number. It is found that some experiments do not conform to this hypothesis, most likely because of the vigorous initial mixing of the cloud. An empirical mixing law is proposed for this early stage which does not violate energy conservation.

1. Introduction

Numerical models of the dispersion of heavy gas clouds in the atmosphere show a clear distinction between the lateral spreading caused by excess gravity force within the cloud and vertical mixing due to turbulent diffusion. Whether this distinction is a consequence of a priori assumptions (as in the case of box or slab models) or simply a consequence of the conservation equations (as in the case of turbulent transport models), the vertical mixing process dominates the dilution of the heavy gas, at least for most times of interest. Without exception, all current models express the vertical mixing rates in terms of a local parameter, such as the bulk Richardson number (box and slab models) or the gradient Richardson number (transport models). Except for the level of detail of a model, there is no distinction among them with respect to this local mixing hypothesis. In this paper we examine the experimental evidence regarding this assumed sole dependence of the vertical mixing rate on local stratification.

The experimental data which is analyzed in this study consists entirely of wind tunnel tests in which samples of heavy gas are released more-or-less instantaneously under isothermal conditions (thus conserving the negative buoyancy of the heavy gas cloud). The experiments are those of Hall (1979) and Meroney and Lohmeyer (1982), the latter consisting of repeated runs under identical conditions for which ensemble averages were determined. The two data sets are also distinguished by the method of release, the former resulting in much slower mixing during the release process than the latter. Of the substantial amount of data collected, attention in this study is focused solely on measurements of the maximum heavy gas concentration χ and the time t of its occurrence, as measured at various distances downstream from the release point.

In such experiments the initial conditions are characterized by an initial Richardson number (Eq. 2.10), which is approximately the ratio of initial potential energy to the wind turbulent kinetic energy. For these experiments, this ratio is very large (10^2 to 10^4). It is to be expected therefore, that gravitational effects will be very significant.

The common assumption of dense gas models is that the vertical mixing rate depends only upon local variables, most commonly the local Richardson number. To test the validity of a model employing such an assumption, comparisons usually are made between measurements of an experiment and an integral calculation of the model for the corresponding initial conditions. Because of the random variability of wind tunnel and field tests (especially the latter), many tests may be required to discern whether a model reproduces the average behavior of the dense cloud experiments (Fay 1983). But because the comparison is made to a model integral, the detailed assumptions of the model regarding the relationship between vertical mixing rates and local variables are not directly tested in such a comparison.

In this paper a more direct test is proposed and made. It is based upon the recognition that the vertical mixing rate and the local Richardson number can be readily estimated from the test measurements and thereby compared with the assumptions used in various models. It is to be expected that these inferred values will possess statistical variability, so that many test results will have to be averaged to discern the relationships among these local variables. Nevertheless, this more direct test should help to distinguish which entrainment assumptions are in best agreement with test observations.

For simplicity in analyzing the data we make use of a box model in which the dense gas is diluted by entrainment of atmospheric air across its upper surface. As is well known, the radial spreading speed determined by this model is independent of this mixing process (Fay 1980). From the measured values of peak concentration χ and time t at which it occurs, it is relatively straightforward to calculate the entrainment speed u_e and the local Richardson number Ri.

The modelling assumption that the vertical mixing rate depends upon local variables expresses the well-founded notion that turbulence is generated by shear in the wind and shear caused by the horizontal gravitational spread of the dense gas cloud. But because the flow is unsteady and complex, it is not entirely clear how these two simultaneous effects should be combined to determine the vertical mixing rate. The comparisons made in this paper should help to distinguish between these separate processes.

In section 2 we provide an analysis of the variables of interest and express the results in two dimensionless forms, depending upon the expected significance of wind turbulence. In the following section 3 we compare the relationship between entrainment speed and local variables, as inferred from test observations, with those commonly assumed in models. In section 4 the importance of turbulence generated by the dense cloud formation is discussed. Section 5 discusses the limitations imposed on cloud behavior by energy conservation, and proposes a method for accounting for the effects of initial turbulence.

2. Analysis

For the purpose of inferring entrainment rates from wind tunnel experiments, we shall use the simple box model of Fay and Ranck (1983) in which a dispersing cloud is represented by a circular cylinder of radius R and height H. The radial spreading rate of the cloud is determined by:

$$dR/dt = \alpha(g'H)^{1/2} \tag{2.1}$$

in which g' is the reduced gravity,

$$g' \equiv g(\rho/\rho_a - 1) \tag{2.2}$$

ρ and ρ_a being the average cloud density and atmospheric density, respectively, g is the gravitational acceleration and α is an empirical factor close to unity. Entrainment occurs across the top surface of the cloud:

$$d(\pi R^2 H)/dt = \pi R^2 u_e \tag{2.3}$$

(By hypothesis the entrainment speed u_e depends upon local variables, in a manner which in this paper is to be determined by analysis of the experimental observations.) For isothermal experiments, the cloud (negative) buoyancy is conserved:

$$g' \pi R^2 H = g'_v V_v \tag{2.4}$$

in which V_v is the initial cloud volume and g'_v is the value of g' at the time of release. The mean concentration χ of cloud material is related to the cloud volume:

$$\chi = V_v / \pi R^2 H \tag{2.5}$$

The conservation of buoyancy makes it possible to estimate the product $g'H$ quite accurately. Integrating Eqs. (2.1) and (2.4) and substituting the integral in Eq. (2.4),

$$g'H = (g'_v V_v / \pi)^{1/2} / 2\alpha t \tag{2.6}$$

Thus $g'H$ varies inversely with time and is determined only by the time since release and the initial conditions.

Substituting the integral of Eq. (2.1) in Eq. (2.3),

$$u_e = \{(V_v / \pi g'_v)^{1/2} / 2\alpha \chi t^2\}(-d\ell n\ \chi / d\ell n t) \tag{2.7}$$

To estimate the local entrainment speed u_e thus requires not only the local measurement of χ at time t but also determining the time derivative of χ. If the concentration χ is measured at several downwind locations (and hence different times), the final factor in Eq. (2.7) can be determined with reasonable accuracy at each measuring station. Thus $g'H$ and u_e are determinable from the experimental observations.

For dense gas clouds released into a wind field characterized by the friction velocity u_*, the local variable which is expected to affect the value of u_e is the local Richardson number Ri:

$$Ri \equiv g'H / u_*^2 \tag{2.8}$$

For such cases it is also desirable to introduce dimensionless variables (Fay and Ranck 1983):

$$\tilde{f} \equiv u_e/u_*$$

$$\tilde{t} \equiv (g'_v)^{1/2}\ t/Ri_v\ V_v^{1/6}$$

$$\tilde{X} \equiv X\ Ri_v^{3/2} \tag{2.9}$$

in which a quasi-inital Richardson number, Ri_v, is defined as:

$$Ri_v \equiv g'_v V_v^{1/3}/u_*^2 \tag{2.10}$$

In terms of these variables, Eqs. (2.6) and (2.7) take the dimensionless form:

$$Ri = (2\alpha\pi^{1/2}\ \tilde{t})^{-1} \tag{2.11}$$

$$\tilde{f} = (-d\ell n\ \tilde{X}/d\ \ell n\ \tilde{t})/2\alpha\pi^{1/2}\tilde{X}\tilde{t}^2 \tag{2.12}$$

Thus, from measured values of $\tilde{X}$ and $\tilde{t}$, we can find $\tilde{f}$ and Ri and determine whether $\tilde{f}$ is indeed a function of Ri as is postulated in most models.

In the case where there is no wind, or it is so weak as to be inconsequential, a different nondimensionalization can be used (Fay 1980) which does not introduce u_*:

$$t^* \equiv (g'_v/V_v^{1/3})^{1/2} t = Ri_v \tilde{t}$$

$$X^* \equiv X = Ri_v^{-3/2}\ \tilde{X}$$

$$f^* \equiv\ u_e/(g'_v V_v^{1/3})^{1/2} = Ri_v^{-1/2} \tilde{f} \tag{2.13}$$

in which the relationship to Eq. (2.9) is indicated. The corresponding form of Eqs. (2.11-2.12) becomes:

$$Ri^* \equiv g'H/g'_v V_v^{1/3} = Ri_v^{-1}\ Ri \tag{2.14}$$

$$= (2\alpha\pi^{1/2}\ t^*)^{-1} \tag{2.15}$$

$$f^* = (-d\ell n\ X^*/d\ell n t^*)/2\alpha\pi^{1/2}\ X^*(t^*)^2 \tag{2.16}$$

As before, measurements of X and t lead to corresponding values of Ri^* and f^*.

3. Local Entrainment

The common hypothesis of box models is the assumption that the dimensionless entrainment function f depends only upon the local instantaneous

Richardson number Ri. However, the form of this function differs among the various models. Eidsvik (1980) suggests the form:

$$\tilde{f} = (aRi + b)^{3/2}/(Ri+c) \qquad (3.1)$$

in the isothermal case, in which a, b, and c are empirical constants. In the limit of $Ri \to \infty$, u_e would be proportional to $(g'H)^{1/2}$, i.e., the spreading speed dR/dt, while in the limit of $Ri \to 0$ u_e would be proportional to u_*, as would be the case for neutrally buoyant clouds. Meroney and Lohmeyer (1982) use a different form,

$$\tilde{f} = aRi^{1/2} + b/(Ri + c) \qquad (3.2)$$

but which has the same asymptotic limits as that of Eidsvik (1980). In both cases entrainment is influenced by both the shear in the wind field and the shear induced by gravitational spreading.

Fay and Ranck (1983) proposed a form,

$$\tilde{f}^{-2} = a^{-2} + (b/Ri)^{-2} \qquad (3.3)$$

which differs principally in the limit of $Ri \to \infty$, where $\tilde{f}$ varies as Ri^{-1} rather than $Ri^{1/2}$ as in Eqs. (3.1) and (3.2). They argued that in cases where the initial Richardson number Ri_v was very large (which is so for the wind tunnel tests to be discussed), after the cloud has been diluted (say $X \ll 1$), the entrainment rate should not depend upon the spreading speed $(g'H)^{1/2}$ but only upon u_* and Ri.

Irrespective of these different formulations, all models propose a universal relationship between $\tilde{f}$ and Ri, albeit different ones. We propose to test these model formulations against the experimental observations.

The first data set to be examined is that of Hall (1979). In these isothermal wind tunnel tests a sample of test gas was released through a porous hemispherical plug during a finite time interval. The peak concentration and its time of passage at several downwind locations on the tunnel center-line were recorded. No averaging of tests under identical conditions was done. The values of Ri_v ranged from about 400 to 12,000. For each of 14 different Ri_v values, a single value of $d\ell n\,\tilde{X}/d\ell n\,\tilde{t}$ was determined by the best fit to a simple power law, and this was then used to determine $\tilde{f}$ from Eq. (2.12).

Fig. 1 displays the relationship between $\tilde{f}$ and Ri, separate symbols being used for three ranges of Ri_v. All measurements were made at $Ri < 10^2$, and generally the largest Ri for any Ri_v was less than 10^{-2} Ri_v. Considerable dilution had taken place by the time these measurements were made. There is a general downward trend of $\tilde{f}$ with increasing Ri which is detectable despite the larger scatter of the data points. The numerical value of $\tilde{f}$ used by Fay and Ranck (1983) to correlate all isothermal experiments is shown superposed on Fig. 1. Because of the range of Ri shown in Fig. 1, there is no indication of the limiting value of $\tilde{f}$ as $Ri \rightarrow 0$.

More recent isothermal wind tunnel experiments by Meroney and Lohmeyer (1982) differ in several important respects. Gas samples contained in a cylindrical cup were instantaneously released by overturning the cup. Equally important, several runs were made under each test condition and the results were averaged. A wider range of Ri_v was used (450 to 26,000).

The results of these tests are shown in Fig. 2, again segregated by three ranges in Ri_v. The range of Ri is greater, about 1 to 10^3. Compared with the Hall data the experimental scatter is less, as would be expected from the averaging process. In contrast to the Hall data, there is a definite trend for $\tilde{f}$ to increase with Ri, especially for $Ri > 10$, but perhaps more strongly than is suggested by Eqs. (3.1) or (3.2). More importantly, the values of $\tilde{f}$ are higher by a factor of ten than for the Hall data, in the range $10 < Ri < 100$. Also, for any Ri, $\tilde{f}$ seems to decrease with increasing Ri_v, contrary to the assumption that $\tilde{f}$ depends only upon Ri.

There are thus significant differences in entrainment rates between these two sets of wind tunnel tests, especially for large Richardson number. We tentatively ascribe these differences to the effects related to the manner in which the gas sample was introduced into the wind stream. In the experiments of Meroney and Lohmeyer, where an initially compact sample was suddenly introduced, vigorous mixing ensued. In contrast, the Hall samples flowed into the tunnel smoothly over a period of time, reducing the amount of initial turbulent mixing. While the initial conditions which generate turbulent shear flows are generally of little consequence in the far field, this generally accepted rule of thumb may not be true for negatively buoyant clouds and plumes.

4. The Effects of Initial Turbulence

If turbulence generated by the initial motion of a dense gas sample suddenly introduced into a flow is dominant during the period of observation, then atmospheric turbulence, as measured by u_*, should not have any effect on the local mixing rate. The entrainment speed u_e and the excess gravity head g'H should then be scaled with the initial spreading speed and gravity head, as in Eqs. (2.13)-(2.14). We then would expect that the dimensionless mixing rate f^* (Eq. 2.13) would be a function of the hybrid Richardson number Ri^* (Eq. 2.14).

To test this hypothesis, we show in Fig. 3 a plot of f^* versus Ri^* for the data of Meroney and Lohmeyer (1982). Compared with Fig. 2, the data shows somewhat better correlation, less scatter, and less sensitivity to the value of Ri_V (which depends upon u_*). Also shown are the results of additional experiments for no wind flow ($u_* = 0$, $Ri_V = \infty$). Although there are fewer of the latter, it can be seen that in this representation there seems to be little difference between the behavior with and without a wind field, at least for the range of variables at which the observations were made.

It is perhaps not too surprising that the turbulence generated by the initial release should persist for such a long time. The initial potential energy, $g'_V V_V^{1/3}$, is very large compared to the wind turbulent energy, u_*^2, (this ratio, equal to Ri_V, was between 400 and 26,000 in these tests) so that turbulent decay would have to proceed at length for levels to reach u_*^2. Ultimately, of course, gravity effects would become too weak and atmospheric turbulence would dominate. However, these tests do not provide a clear indication of when such a transition would occur.

But the more significant result of the correlation shown in Fig. 3, which is also demonstrated in Fig. 2, is that u_e cannot be a universal function of the local Richardson number Ri only (as most models assume) when initial turbulence dominates the flow. (There is one exception: if $f^* \alpha (Ri^*)^{1/2}$, then $\tilde{f} \alpha Ri^{1/2}$, independent of Ri_V). It would appear that the initial conditions play a role in determining the local entrainment rate in a manner which cannot be represented by the usual entrainment hypothesis.

5. Violating Energy Conservation

In a previous report (Fay 1982), it was pointed out that, in the absence of a wind, an assumption that $u_e \alpha (g'H)^{1/2}$; i.e., $\tilde{f} \alpha (Ri)^{1/2}$ or $f^* \alpha (Ri^*)^{1/2}$, leads to an indefinite growth in the cloud height H and to violation of energy conservation since the potential energy of the cloud would be increasing in the absence of any external energy source. But even in the presence of a wind which is ineffectual in governing the mixing rate (as our analysis suggests is true for the experiments of Meroney and Lohmeyer), there must exist equivalent constraints on the possible relationship between u_e and $(g'H)^{1/2}$. In this section we examine the experimental data from this point of view.

We begin by noting that Eqs. (2.1), (2.3) and (2.4) can be combined to give

$$dH/dt = u_e - H/t \tag{5.1}$$

If the potential energy of the cloud cannot increase, then $dH/dt \leq 0$ and

$$u_e \leq H/t \tag{5.2}$$

Using Eqs. (2.5), (2.7) and the integral of Eq. (2.1), this condition becomes:

$$(-d\ell n\ \chi/d\ell nt) \leq 1 \tag{5.3}$$

As explained previously, the average value of this derivative was determined for each set of measurements corresponding to a common value of Ri_v. The values so determined are plotted in Fig. 4 as a function of Ri_v. Points lying above the horizontal line correspond to positive values of dH/dt, i.e., increasing cloud potential energy.

The experiments of Meroney and Lohmeyer show a clear tendency for dH/dt to approach zero as Ri_v increases. Because of the averaging process, this trend is unmistakable. Noting that the observed range of Ri in these tests is about 10^{-3} to 2×10^{-2} times Ri_v (see Fig. 2), we can conclude that vertical entrainment is energy limited for Ri $\gtrsim$ 5 (i.e., $Ri_v \geq 5 \times 10^3$). In this region,

$$u_e \approx H/t \tag{5.4}$$

This remarkable result indicates that u_e is not governed by local shear in the cloud but by a decaying eddy structure created by the initial cloud release. While H is the local scale length, the scaling time t is the lifetime of the cloud.

It is thus not possible to model such a flow in terms of local variables. From an empirical point of view, we can use Eq. (5.4) to express an entrainment rate due to the initial effects of a release in which the cloud potential energy is of the order of $g'_v V_v^{1/3}$. Since H is constant during the early period of release, we can estimate an empirical value of H from the relation:

$$H/V_v^{1/3} = Ri_v^{1/2}/2\pi^{1/2}\alpha\,\tilde{\chi}\tilde{t} \qquad (5.5)$$

in which only those experiments are used for which $-d\ln\chi/d\ln t$ is approximately unity (say $Ri_v > 5 \times 10^3$). Using the data of Meroney and Lohmeyer, we found an average value of $H/V_v^{1/3}$ of 0.12 (assuming $\alpha = 1$), leading to an empirical entrainment rate

$$u_e \simeq 0.23\ H_v/t \qquad (5.6)$$

in which H_v is the cloud height at the time of release. (For these experiments, $H_v = 0.52V_v^{1/3}$.)

In terms of dimensionless variables, this can be rewritten as:

$$u_e/u_* = \tilde{f} = 0.82\ (H_v/V_v^{1/3})Ri/Ri_v^{1/2} \qquad (5.7)$$

Since Ri decreases inversely with time, simply adding this expression to that of Eq. (3.3) should permit modelling of releases for varying values of $H_v/V_v^{1/3}$.

Comparable values of the derivative of Eq. (5.3) for the Hall experiments are also shown in Fig. 4. Despite the larger amount of scatter, there is clear evidence that growth is not limited by the original cloud energy but is being augmented by wind turbulence, even at the largest values of Ri_v. As we concluded from Fig. 1, u_e seems to be unaffected by the cloud release and depends only upon the local parameters, u_* and Ri.

6. Conclusions

An examination of the vertical mixing rate inferred from measurements made in isothermal dense gas cloud wind tunnel experiments shows that this rate is not always determined by local variables, as has been hypothesized in dense gas models. Instead, evidence suggests that turbulence generated by the formation of the dense gas cloud can dominate mixing during times of interest if the initial cloud Richardson number is sufficiently high. This effect can be absent when the cloud is formed slowly and gently.

In tests for which the initial turbulence is important and while the Richardson number is larger than about 5, the average cloud height is unchanging with time. Based upon test measurements, the entrainment speed under these circumstances is $0.23\,H_v/t$ where H_v is the initial cloud height and t is the time since the cloud was formed.

7. References

Eidsvik K.J. (1980) A model for heavy gas dispersion in the atmosphere. Atmospheric Environment 14, 769-777.

Fay J.A. (1980) Gravitational spread and dilution of heavy vapor clouds. Proc. Second Int. Symp. on Stratified Flows, Vol. 1, 471-444, Tapir, Trondheim.

Fay J.A. (1982) Some unresolved problems of LNG vapor dispersion. MIT-GRI LNG Safety and Research Workshop, Vol. II, Dispersion of Dense Vapors, 72-84, GRI 82/0019.2, Gas Research Institute, Chicago.

Fay J.A. and Ranck D.A. (1983) Comparison of experiments on dense gas cloud dispersion. Atmospheric Environment 17, 239-248.

Hall D.J. (1979) Further experiments on a model of an escape of heavy gas. Report LR(312)AP, Warren Spring Laboratory, Stevenage.

Meroney R.N. and Lohmeyer A. (1982) Gravity spreading and dispersion of dense gas clouds released suddenly into a turbulent boundary layer. GRI-81/0025, Gas Research Institute, Chicago.

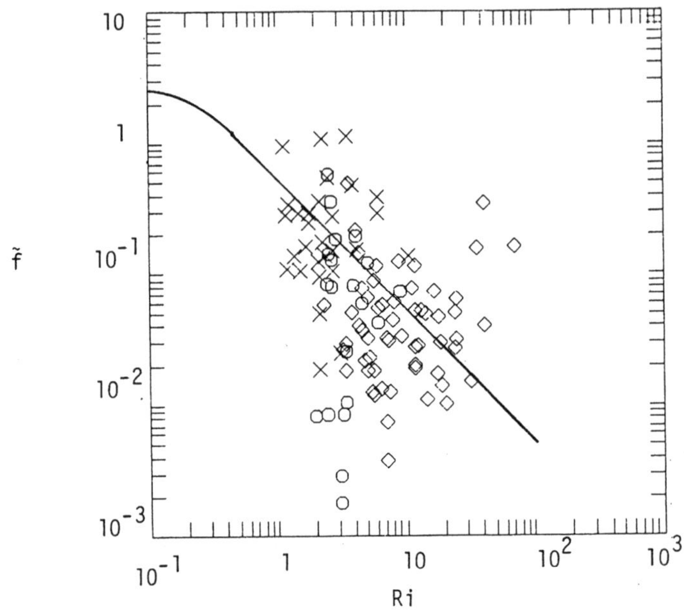

Fig. 1. Dimensionless entrainment speed $\tilde{f}$ as a function of Richardson number Ri, inferred from measurements of Hall (1979). The solid line is the relationship used by Fay and Ranck (1983). Range of Ri_v: x, $3.8\text{-}8.2 \times 10^2$; ○ $1.2\text{-}2.6 \times 10^3$; ◇, $3.7\text{-}11.9 \times 10^3$.

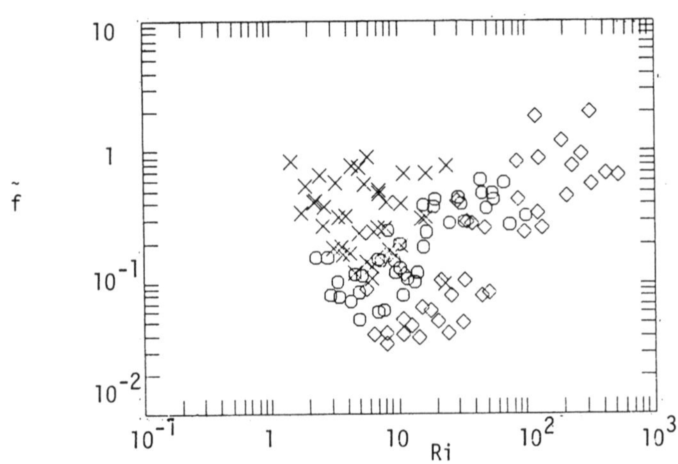

Fig. 2. Dimensionless entrainment speed $\tilde{f}$ as a function of Richardson number Ri, inferred from measurements of Meroney and Lohmeyer (1982). Range of Ri_v: x, $4.5\text{-}12.4 \times 10^2$; ○, $2.1\text{-}4.6 \times 10^3$; ◇, $6.5\text{-}26 \times 10^3$.

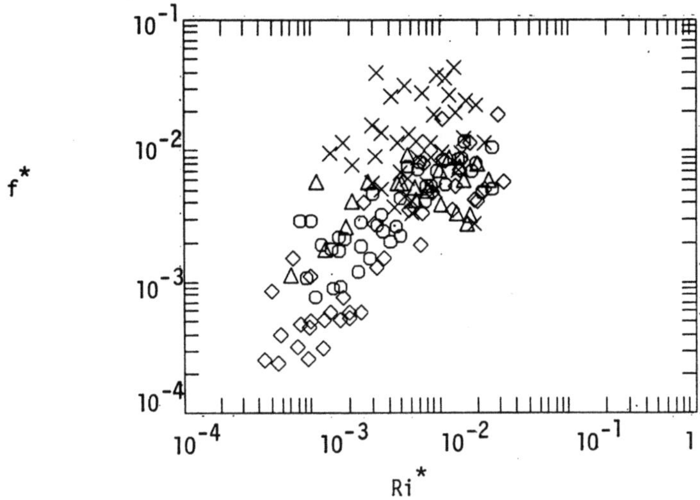

Fig. 3. Dimensionless entrainment speed f^* as a function of the hybrid Richardson number Ri^*, inferred from measurements of Meroney and Lohmeyer (1982). Range of Ri_V: x, 4.5-12.4x10^2; ○, 2.1-4.6x10^3; ◇,6.5-26x10^3; △,∞.

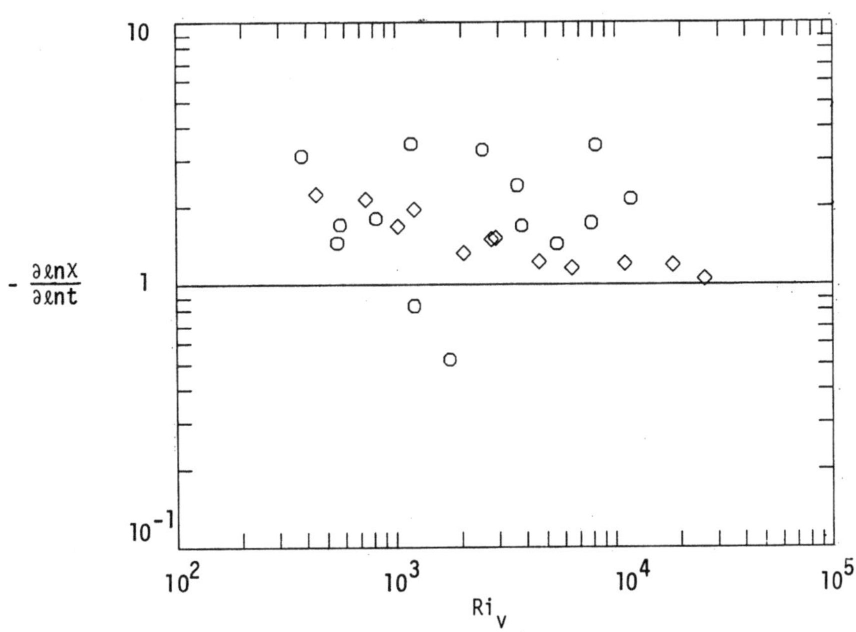

Fig. 4. Measured value of ($-\partial\ell n/\partial\ell nt$) as a function of Ri_V for the experiments of Meroney and Lohmeyer (◇) and Hall (○).

Entrainment in Gravity Currents

OTTO ZEMAN*

The Johns Hopkins University
Baltimore, MD 21218, U.S.A.

Summary

In this paper the following topics associated with the entrainment modeling of natural and manmade gravity currents are discussed: reexamination of the definition and observability of entrainment in unsteady gravity currents, applicability of the empirical entrainment laws to gravity currents, the concept of entrainment energy conversion, and of two-way entrainment.

I. The Concept of Entrainment in Turbulent Flows

The most crucial problem in the modeling of heavy gas release flows is undoubtedly the parameterization of turbulent mixing processes which in the slab model approximation are represented by an entrainment velocity (Zeman [1]). Apart from the definition in the Oxford English dictionary, we understand entrainment as a process by which a finite turbulent region grows into a surrouunding (nonturbulent) fluid; the entrainment velocity (w_e) specifies the rate of the entrainment. As pointed out by Hunt (in these proceedings) there exists a variety of definitions of w_e; we shall adhere to the definition appropriate for boundary layer flows, i.e.

$$\partial H/\partial t + \nabla_h \cdot \int_0^H \underline{U}\, dz = w_e \tag{1}$$

where H(x,y,t) is a depth of the turbulent flow, $\underline{U}$(x,y,z,t) is the horizontal velocity vector (parallel to the boundary). Both H and $\underline{U}$ are statistical averages suitably defined so that the vertical entrainment velocity w_e has a proper physical meaning (see e.g. Zeman & Tennekes [2]). In one-dimensional or steady flows (in the mean) the value of w_e can be inferred with reasonable accuracy from experiments (for example in wakes, jets, in flows with density interface). However, in unsteady gravity

*present address 205 E. Marshall St., Ithaca, NY 14850.

currents (such as due to heavy gas releases) measurement of local entrainment velocities presents great difficulties. To demonstrate this, consider a two-dimensional progressing gravity current as depicted in Fig. 1 at two successive times; (1) reduces to

$$\partial H(x,t)/\partial t + \partial/\partial x\,([U]H) = w_e(x,t) \quad , \tag{2}$$

where $[U] = 1/H \int_0^H U dz$ is the layer-average value within the current (hereafter indicated by square brackets). It is evident that due to spatial and temporal variability of the flow, the local value of w_e cannot be determined with reasonable accuracy. Furthermore, the relationship (2) breaks down near the front of the current where the slope $\partial H/\partial x$ is large. A partial remedy to the problem is to average a number of experiments under identical conditions to approximate ensemble averages $\langle H \rangle$ and $\langle U \rangle$. In a natural environment this is not possible and one has to resort to some observable appoximations to the true local value of $\langle H \rangle$ and $\langle U \rangle$. Conveniently, the manmade heavy gas releases contain scalar constituents and the entrainment velocities can be inferred from the dilution rates. If the scalar S is uniformly distributed in the vertical so that $S \simeq [S]$, then in analogy with (2), the species conservation equation gives

$$\partial/\partial t([S]H) + \partial/\partial x\,([SU]H) = 0 \quad \text{or}$$
$$w_{es} = -H/[S]\left(\partial[S]/\partial t + [U]\,\partial[S]/\partial x\right) \tag{3}$$

In real flows, S is not uniformly distributed within the layer ($[SU] \neq [S][U]$) and the effective depth H for the scalar S may differ from that for the momentum. Thus in general w_{es} is not identical with w_e as defined in (2). Understandably there exist no data on the local values of w_e or w_{es} for natural or manmade gravity currents. Entrainment velocities reported in published literature on heavy gas releases usually refer to some global entrainment velocity (as used, for example, in the so-called pancake models which assume uniform properties within the pancake). Nevertheless, the understanding of entrainment in the local sense, i.e. as a process depending on local flow parameters is of importance because w_e varies significantly in the horizontal direction. Slab models that resolve the horizontal variation of the properties of the heavy gas flow require a local value of w_e as an input parameter. Therefore in the rest of this paper we shall discuss the dynamics of local entrainment and its pecularities in gravity driven flows.

2. Empirical Entrainment Laws: Applicability to Gravity Currents

The flow we are dealing with may be classified as a stratified boundary layer flow with shear or buoyancy driven mixing or entrainment. Let us summarize some more important experiments designed to find empirical laws for entrainment in stratified flows:

1) Convective entrainment in buoyancy driven flows (see e.g. Deardorff at al [3]); the scaling convective velocity is $w_* = (-g/\rho_0 \; [\rho w]_0 H)^{1/3}$, where $[\rho w]_0$ is the surface (turbulent) density flux.

2) Shear-stress induced entrainment (Kato & Phillips [4], Kantha et al. [5], Deardorff and Willis [6]); the scaling velocity is u_*, the friction velocity based on the applied stress at the boundary.

3) Entrainment due to an interfacial shear between two layers of different densities (Moore & Long [7]); the scaling velocity is ΔU the mean velocity difference between layers.

4) Entrainment in stratified boundary layers (Lofquist, [8]; Piat and Hopfinger [9], and others); both u_* and ΔU are present.

5) Entrainment in steady gravity currents (no density flux at the boundary); notably experiments by Ellison and Turner [10]. Again $\Delta U, u_*$ are scaling velocities; for free surface boundary $u_* = 0$.

Now, the question is to what extent the above experiments can be utilized to provide some rational entrainment laws for heavy gas spills. Obviously the experiments investigate singular entrainment mechanisms under simplified boundary conditions, either of horizontal homogeniety or steadiness. On the other hand a gravity flow, such as one resulting from a finite LNG spill into the atmosphere, changes rapidly in time and space and interactions of a variety of mechanisms may conceivably contribute to the ultimate entrainment rate. An example of possible physical structures that may arise in a spreading LNG cloud is illustrated in Fig. 2: at the arrested upwind edge of the flow the entrainment is dominated by the interfacial shear ΔU between the ambient and gravity flows, while near the downwind edge the turbulence which is produced by the surface-flow interaction is important (i.e. the friction velocity u_*). In the cloud core, where the spreading velocity is small, a significant contribution to entrainment may come from the convective turbulence due to surface heating of the cold cloud. As the internal turbulence decays, the ambient turbulence begins to dominate mixing and a new set of scaling parameters must be introduced.

Hence to answer the question of applicability of the laboratory experiments to the actual gravity currents, these experiments may, in principal, provide information only on certain asympotic forms of the entrainment laws. Despite these limitations, the slab models which typically use an interpolative entrainment formula (e.g. combining laws from the above laboratory experiments) have been relatively successful in predicting important features of heavy gas flows (see, for example, Morgan et al. in these proceedings). This is perhaps because a majority of the empirical entrainment laws obey certain energy conversion principles which are common to different classes of entraining flows. We suggest that this unifying principle be utilized to formulate entrainment laws for more complex flows, and we shall discuss this in the next section.

3. Entrainment from the Energy Viewpoint

If one inspects different classes of entrainment experiments one finds that within a certain range of input parameters the entrainment velocity obeys a general law

$$w_e/u_c \propto (\Delta g' H/u_c^2)^{-1} = (R_{ic})^{-1} , \qquad (4)$$

where u_c is a characteristic velocity of turbulence in a particular experiment; $\Delta g' = g([\rho]/\rho_0-1)$ is the reduced gravity corresponding to the mean density difference between the turbulent layer (of depth H) and the ambient fluid, and R_{ic} is the characteristic Richardson number. As is well known, the inverse law (4) can be inferred from the postulate that the rate of increase of the potential energy $PE \simeq g\int_0^H(\rho/\rho_0-1)zdz$ due to entrainment is a constant fraction (χ) of the energy input that is converted to turbulence, i.e. of the rate of turbulence production P_t, or,

$$DPE/Dt = \chi P_t \ . \qquad (5)$$

It is of interest that this postulate yields the sought-for difference in entrainment between the rotating screen experiments of Kato and Phillips [4] and Kantha et al. [5] (hereafter KP and KPA). In both experiments $PE \approx 1/2 \ \Delta g' H^2$ and $P_t = (U_s-[U])u_*^2$ where U_s is the fluid speed at the screen; $(U_s-[U])/u_*$ is expected to be constant, say C_D and (5) becomes

$$1/2\ \partial/\partial t(\Delta g'H^2) = \chi C_D u_*^3 \tag{6}$$

Now in the KP experiment $\Delta g' = 1/2\ \gamma H$ where $-dg'/dz = \gamma$ is the initial buoyancy stratification. In the KPA experiment $\gamma = 0$ and $\Delta g'H$ is constant. Substituting for $\Delta g'$ in (6) we obtain

$$w_e/u_* = 2/3\ \chi C_D R_i^{*-1}$$

for the KP experiments, while for the KPA experiment the normalized w_e is larger by the factor of 3. This difference has been observed for moderate values of $R_i^* = \Delta g'H/u_*^2$. At least two other theoretical arguments have been offered to explain the KP-KPA disparity (see the discussion of the subject by Deardorff and Willis [6]).

4. The Principle of Entrainment Energy in Gravity Currents

A special feature of the gravity currents propagating into the fluid at rest is that they are gravity-driven and purely baroclinic, i.e., the (hydrostatic) mean pressure gradient increases from zero at the layer top to a maximum value at the surface. This has a profound effect on distribution of momentum, density and Reynolds stresses in the current and inevitably on the entrainment as well. To examine plausible forms of the generalized entrainment law for unsteady gravity currents, let us use the principle outlined in the preceding section. To simplify the problem let us assume two-dimensionality and, similarity of profiles of U and $g' = g(\rho/\rho_0-1)$ as proposed by Keulegan [11]

$$U = \Delta U(x,t)e(\eta) \quad \text{and}$$

$$g' = \Delta g'(x,t)f(\eta) \quad \text{where} \quad \eta = z/H(x,t) .$$

Then, according to (5)

$$DPE/Dt = D/Dt(1/2\,\Delta g' H^2 \int_0^1 f\,\eta d\eta) = \chi P_t =$$

$$\chi[-\Delta U\ \partial/\partial x(\Delta g'\ H^2/2)\int_0^1 e(\eta)\int_\eta^1 f(\eta')d\eta' d\eta - D/Dt(\Delta U^2 \int_0^1 e^2 d\eta)]. \tag{7}$$

With further assumptions and simplications, (7) can be reduced to a basic functional dependence of the following form:

$$w_e/\Delta U = (\chi/S_1)[aS_2 R_{ib} R_{io}^{-1} + bR_i^{*-1}]f\{R_{io}, S_1, S_2\} \tag{8}$$

where $R_{io} = \Delta g'H/\Delta U^2$, $R_i{*} = \Delta g'H/u_*{}^2$ and S_1, S_2 are shape factors associated with the similarity profiles e,f (indicated in (7)). The new unfamiliar parameter in (8) is $R_{ib} = -\partial/\partial x(\Delta g'H^2)/\Delta U^2$ which compares the drivng (motive) force to inertial forces. In the theories of turbulent boundary layers with pressure gradients there exists a similar parameter $-(\partial P/\partial x)H/\rho_0 u_*{}^2$ (Clauser [12]). This parameter enters empirical laws for various shape factors and the growth rate of the boundary layer. Hence, it can be argued that R_{ib} is a fundamental parameter in gravity currents (with no ambient motion) and that the local parameters such as R_{io}, the shape factors S_1, S_2 and ultimately the entrainment rate depend in some way on the nonlocal parameter R_{ib} (preliminary numerical experiments by this author indicate that it is so). Since R_{ib} is an observable quantity it would be fairly easy to verify this hypothesis by experiments.

In more complex flow situations (e.g. with ambient wind and surface heating) the idea of energy conversion embodied in (4) is equally valid, however, the concept of a slab model with entrainment on top is perhaps too crude to realistically represent the physics of the flow. Van Ulden (in these proceedings) used the idea of the energy conversion described here to estimate the global entrainment rate in his new model. This is probably the best approach to estimating the global entrainment.

5. Two-Way Entrainment

As the final topic we outline the concept of two-way entrainment. This concept is intended to deal in an uncomplicated way with turbulent mixing, and entrainment across a density interface when turbulence of comparable intensity exists on both sides of the interface. In such flows the entrainment can proceed either up or down depending on the relative turbulence levels in the two layers.

The gravity currents released in the atmosphere or oceans always encounter a certain level of ambient turbulence. It is conceivable that if the self-generated turbulence within the current layer subsides, the fluid is entrained from the layer upwards and the layer thickness may in fact decrease without noticeable dilution. As the layer thickness decreases the energy of ambient turbulence becomes comparable to the potential energy of the dense layer, and the layer will eventually break up. There exists some visual evidence for the reverse entrainment (or detrainment) but no quantitative one.

To eliminate the ambiguity of which way the entrainment proceeds, we postulate (positive) entrainment velocities w_{e1}, w_{e2} pertaining to their respective layers below (1), and above (2), the density interface. Assuming that $H_1 + H_2$, the total depth of the two layers is constant, that the flow is horizontally homogeneous, and that the mean buoyancy $g' = [g']$ is uniformly distributed in either layer we obtain the following set of relationships to describe the kinematics of the flow:

$$\partial H_1/\partial t = -\partial H_2/\partial t = w_{e1} - w_{e2} \quad \text{and} \tag{9}$$

$$\partial[g_1']/\partial t = w_{e1}\,([g_2']-[g_1'])/H_1 \;, \quad \partial g_2'/\partial t = w_{e2}\,([g_1']-[g_2']/H_2$$

The subscripts 1, 2 refer to the respective layers (1) and (2). An extention of (9) to the equations for a general flow is straightforward.

In the practical case the lower layer is the gravity current and the layer above the current is the ambient (atmospheric) flow. The crucial question is, of course, how to represent w_{e1} and w_{e2} in terms of the turbulence characteristics of the current and atmosphere. The answers can be found only from experiments.

6. Conclusions

In conclusion, a few words concerning the application of one layer (slab) models for heavy gas flow predictions. Entraining, one-layer models have been widely used in geophysical flows such as atmospheric and upper ocean mixed layers (Zeman and Tennekes [2]; Zeman [13]). In heavy gas flow applications the one-layer models appear to be competitive, as far as the predictive capabilities, with the far more complex eddy viscosity models. However, on physical grounds, the application of the one-layer model cannot be justified at the gravity head of the gas flow, where most of the model assumptions are violated. For example, the model cannot represent the dynamics of the current-ambient flow interactions, also the turbulent mixing processes at the gravity head are unlike the entrainment process.

To remove these limitations of the one-layer models one should perhaps consider the idea of composite models as suggested for example by van Ulden (in these proceedings). Different, dynamically coupled segments of the flow would be treated by different modeling techniques to achieve the most optimal results.

Supported in part by the U.S. Office of Naval Research under the following programs: Physical Oceanography (Code 422PO), Power (Code 473); in part by the U.S. National Science Foundation under grant no. ATM 79-22006; and in part by the U.S. Air Force Geophysics Laboratory.

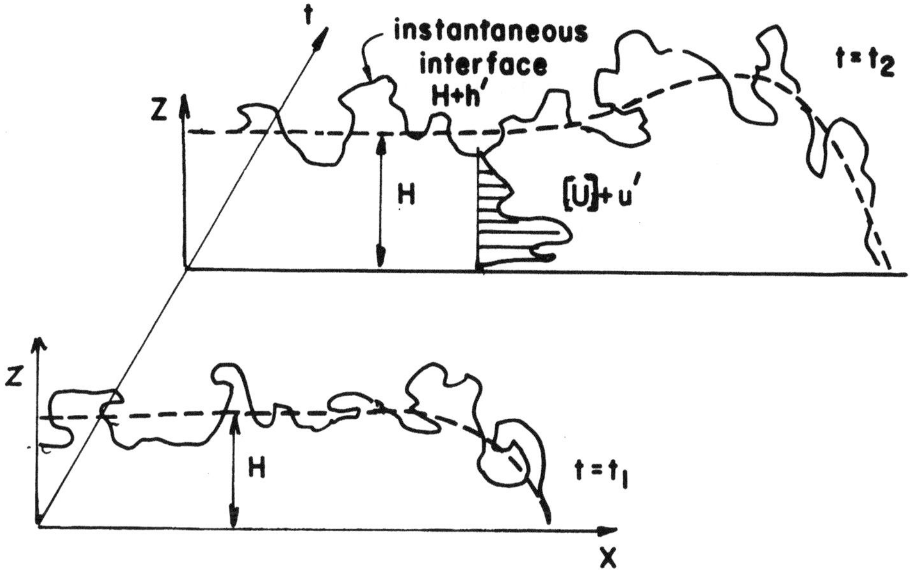

Fig. 1. Instantaneous interface H+h' between the gravity current and ambient fluid (at rest) at two successive times t_1 and t_2. [U]+u' represents an instantaneous horizontal velocity.

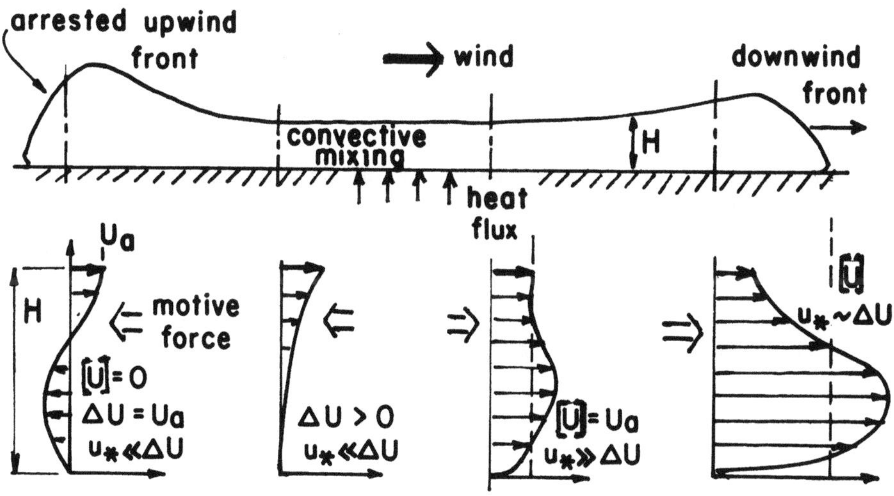

Fig. 2. Possible mean velocity profiles across a spreading heavy gas cloud. The velocity deficit $\Delta U = U_a - [U]$.

References

1. Zeman, O.: The dynamics and modeling of heavier-than-air cold gas releases. Atmos. Environ., 16 (1982) 741-751.

2. Zeman, O.; Tennekes, H.: Parameterization of the turbulent energy budget at the top of the daytime atmospheric boundary layer. J. Atmos. Sci., 34 (1976) 111-123.

3. Deardorff, J. M.; Willis, G. E.; Stockton, B. H.: Laboratory studies of the entrainment zone of a convectively mixed layer. J. Fluid Mech., 100 (1980) 41-64.

4. Kato, H.; Phillips, O. M.: On the penetration of a turbulent layer into stratified fluid. J. Fluid Mech., 37 (1969) 643-655.

5. Kantha. L. H.; Phillips, O. M.; Afzad, R. S.: On turbulent entrainment at a stable density interface. J. Fluid Mech., 79 (1977) 753-768.

6. Deardorff, J. W.; Willis, G. E.: Dependence of mixed-layer entrainment on shear stress and velocity jump. J. Fluid Mech., 115 (1982) 123-149.

7. Moore, M. J.; Long, R. R.: An experimental investigation of turbulent statified shearing flow. J. Fluid Mech., 49 (1971) 635-655.

8. Lofquist, K.: Flow and stress near an interface between stratified fluids. Phys. Fluids., 3 (1960) 158-175.

9. Piaf, J. F.; Hopfinger, E. J.: A boundary layer topped by a density interface. J. Fluid Mech. 113 (1981) 411-432.

10. Ellison, T. H.; Turner, J. S.: Turbulent entrainment in stratified flows. J. Fluid Mech., 8 (1959) 514-544.

11. Keulegan, G. H.: The motion of saline fronts in still water. Nat. Bar. Stan. Rep. 8531 (1958).

12. Clauser, F. H.: Turbulent boundary layers in adverse pressure gradients. J. Aeronaut. Sci., 21 (1954) 91-108.

13. Zeman, O.: Parameterization of the dynamics of stable boundary layers and nocturnal jets. J. Atmos. Sci. 36 (1979) 1974-88.

The Incorporation of Wind Shear Effects into Box Models of Heavy Gas Dispersion

P. C. CHATWIN

Department of Applied Mathematics and Theoretical Physics, The University of Liverpool, P.O. Box 147, Liverpool L69 3BX, England

Summary

Box models are widely used in predicting the potential hazards associated with the accidental release of heavy gas clouds, and this paper discusses whether, and how, such models should be amended to incorporate the effects of ambient wind shear. As is desirable, the proposed changes do not materially alter the simplicity of box models, which is their most important practical advantage.

Introduction

This paper summarizes work done for the U.K. Health and Safety Executive (HSE). The work is described in more detail in Chatwin [4]. Since 1976 HSE has led a research programme to clarify the behaviour of massive releases of toxic or flammable gases. Discussion of this programme is given by McQuaid [10]. It includes several series of experiments (the latest on Thorney Island in 1982-1983 comprising 10-15 trials of 2000 m^3 releases) and, simultaneously, the investigation and development of predictive models.

One class of such models is dealt with here. These models are known as box models, and are described in a recent comprehensive survey by Webber [15]. Here there is need (and space) only to emphasize the particular features of box models that distinguish them from other classes of models. Above all, box models have the single practical purpose of predicting, to acceptable accuracy, the location (in space and time) of hazards associated with accidental releases. Box models are intended for easy and rapid use, and the only test which can legitimately be applied to them is how well they succeed in predicting hazards. They do not aim, or claim, to model accurately the details of all the physical processes affecting the dispersion, but only to describe these processes sufficiently well for their stated purpose.

The only situation considered in this paper is the instantaneous release (by sudden catastrophic loss of containment) of a finite volume of heavy gas of uniform density ρ_0 into an atmosphere with uniform density ρ_a. A basic assumption of box models is that for all time t, the dispersing gas cloud has the shape of a circular cylinder with a vertical axis perpendicular to the ground. As shown in Fig.1, the horizontal displacement of the cloud

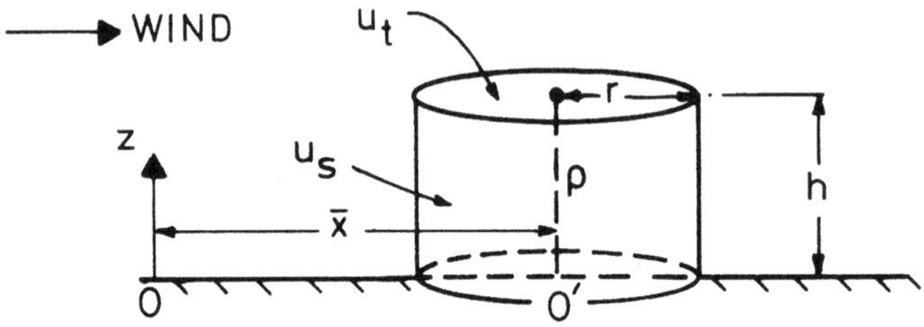

Fig.1. Sketch for discussion of existing box models

OO' at time t since release will be denoted by $\bar{x}$, the radius and height of the cylinder by r and h, and the gas density by ρ. It is further assumed that ρ, and the gas concentration C, are spatially uniform. Thus, by mass conservation,

$$C/C_0 = v_0/v = (r_0^2 h_0)/(r^2 h), \tag{1}$$

where $v = \pi r^2 h$, and zero subscripts denote initial values. The basic aim of every box model is to predict how $\bar{x}$ and v vary with t, and hence, using (1), to predict the location of hazards.

Under isothermal conditions, the total negative buoyancy of the cloud is conserved, so that

$$vg(\rho - \rho_a)/\rho_a = vg' = \pi b_0, \tag{2}$$

where b_0 is a constant of dimensions $L^4 T^{-2}$. All box models include an equation for the rate of spreading of the cloud about its axis. The relationship used in most models is

$$dr/dt = \alpha (g'h)^{1/2}, \tag{3}$$

where α is a constant of order unity. It follows (Picknett [11]) from (2) and (3) that

$$r^2 = r_0^2 (1 + t/t_0) \tag{4}$$

where

$$t_0 = r_0^2/(2\alpha b_0^{1/2}) \tag{5}$$

is a characteristic time scale of order $0.3 - 0.5\,s$ for both the Thorney

Island trials, and the Porton trials described by Picknett.

Box models assume (see Fig.1) that mixing of the ambient atmosphere with the heavy gas takes place over the surface of the box with entrainment velocities u_t and u_s over the top and side surfaces of the cloud respectively. Both u_t and u_s are assumed not to vary with position on the appropriate surface. These assumptions lead to the following entrainment equation for $d\upsilon/dt$:

$$d\upsilon/dt = \pi r^2 u_t + 2\pi r h u_s. \tag{6}$$

Existing box models differ widely in their prescriptions of u_t and u_s, and one or two examples will be given later. However, given these prescriptions, use of (4) and $\upsilon = \pi r^2 h$ enable (6) to be integrated, often analytically (Webber [15]).

The final ingredient of box models is a formula for $d\bar{x}/dt$, the rate of advection of the cloud as a whole.

The effect of wind shear on side entrainment in the early stages of dispersion

Despite the great variety of prescriptions of the entrainment velocities in (6), not one takes ambient wind shear explicitly into account. This section examines whether this omission is justified in the case of u_s. As indicated in Fig.2, the term involving u_s in (6) is the net effect of

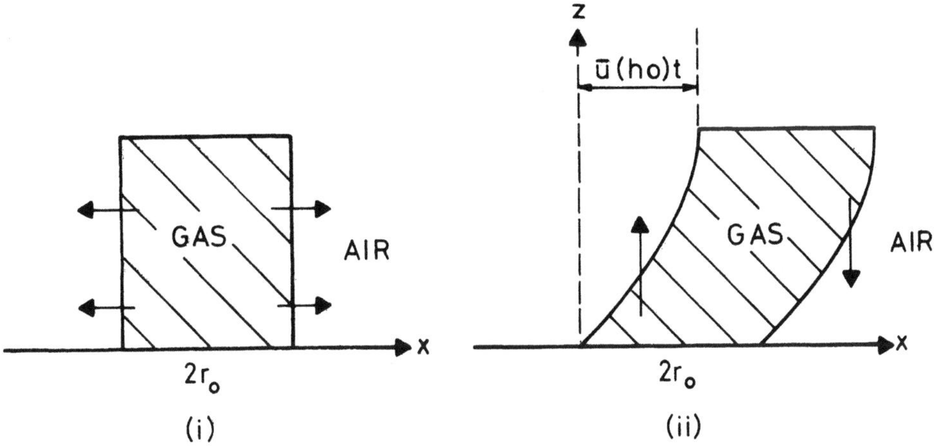

Fig.2. Mechanisms causing horizontal entrainment. (i) Direct horizontal diffusion. (ii) Direct effects of wind shear reduced by vertical diffusion, i.e. shear dispersion.

two distinct mechanisms causing mixing of the heavy gas with air. One such mechanism is direct horizontal diffusion, indicated in Fig.2(i); this needs no explanation here. The other mechanism is shear dispersion, indicated in Fig.2(ii), first described in a different context by Taylor [13]. Acting by itself the mean wind $\overline{u}(z)$ would shear the cloud at a rate proportional to $\overline{u}(h_0)$, thereby creating vertical gradients of concentration. Shear dispersion is the net effect of the resulting vertical diffusion, and shearing due to the wind.

For the dispersion of a passive cloud in an unbounded neutrally stable atmosphere, the value of u_S for each of these mechanisms is proportional to the shear velocity u_* (Batchelor [1], Chatwin [2], Yaglom [16]). The situation for heavy gas clouds is more complicated, since the negative buoyancy of the cloud generates turbulence additional to that already present in the ambient atmosphere. On dimensional grounds, typical velocities associated with this buoyancy generated turbulence are of order $(g'h)^{1/2}$. Therefore a reasonable proposal for the direct horizontal diffusion contribution to u_S is (on dimensional grounds)

$$u_S = (g'h)^{1/2} f(Ri), \tag{7}$$

where Ri is a Richardson number defined by

$$Ri = g'h/u_*^2. \tag{8}$$

In most cases of interest, the initial values of Ri are large, of order $10^2 - 10^3$ for both the Porton and Thorney Island trials. The value of u_S is then inevitably dominated by gravity effects so that

$$Ri \gg 1 \Rightarrow f(Ri) \approx \text{constant}. \tag{9}$$

Consider now the effect of shear dispersion. Note first that wind shear itself alters the shape, but not the volume, of the cloud. Thus (as expected) vertical diffusion must be considered in assessing the effects of wind shear on the term involving u_S in equation (6) for dV/dt. Consider high values of Ri, when, as shown in Fig.3, shear tends to cause the upstream edge of the cloud to become stably stratified with consequent inhibition of vertical mixing. Conversely vertical mixing will be vigorous on the downstream edge. However the speed of free fall on this edge is of order $(g'h)^{1/2}$, no greater than the direct horizontal diffusion contribution to u_S, given by (7) and (9). This suggests the following two conclusions, supported by somewhat more detailed arguments in Chatwin [4]:

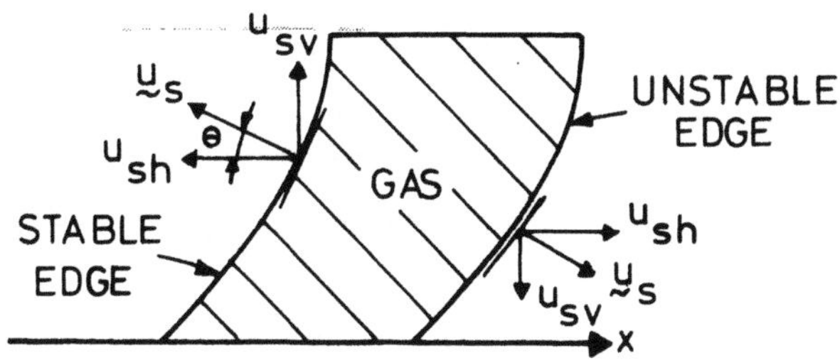

Fig.3. Shear dispersion for heavy gas clouds

(i) for $Ri \gg 1$, the contribution of shear dispersion to the rate of entrainment over a small surface element on the side of the cloud is at most of the same order as that due to direct horizontal diffusion;

(ii) for $Ri \gg 1$, the rate of entrainment over a small side surface element of area δA is of order $(g'h)^{1/2}\delta A$, i.e. u_s in (6) is of order $(g'h)^{1/2}$.

These conclusions, however, relate only to the local rate of entrainment. There are other effects of wind shear. In particular, wind shear will, acting by itself, increase the surface area of the cloud without altering its volume (for given radius and vertical height). Therefore wind shear will tend to increase the total rate of entrainment. This increase will be greatest in the early stages of dispersion, when the height to width ratio of the cloud is largest. This conclusion is in qualitative agreement with observations by van Ulden [14] and Picknett [11] of intense mixing immediately after release. It therefore seems worthwhile to attempt, in a simple way, to model this tendency by means of a small amendment.

The geometrical ideas of the amendment are shown schematically in Fig.4. The vertical cylinder of existing box models is replaced by a cylinder of identical circular cross-section and vertical height (and hence identical volume), but with an axis inclined at an angle $\phi = \phi(t)$ to the vertical. To adequate approximation for practical purposes, the side surface area of the cloud is $2\pi r h \sec\phi$ (Chatwin [4]), so that the entrainment equation (7) is replaced by

$$dv/dt = \pi r^2 . u_t + 2\pi r h \sec\phi . u_s . \quad (10)$$

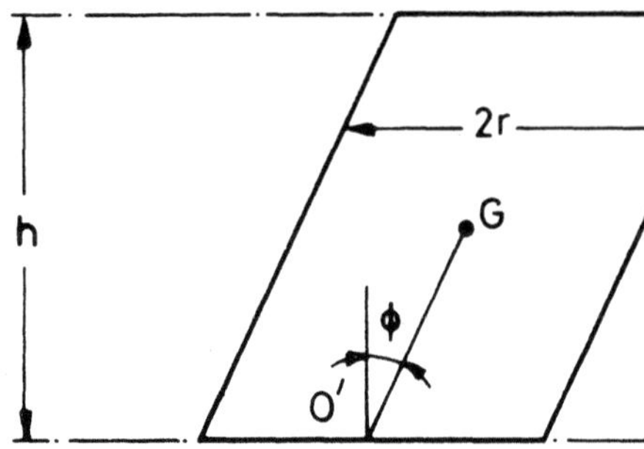

Fig.4. The tilted cylinder of the amended box model

It follows that box models with $u_s \equiv 0$ are not affected by the amendment. Fay and Ranck [7] note that their model with $u_s \equiv 0$ may therefore "not be accurate at the earliest times", when the effect being modelled here is most important.

An equation for $\phi(t)$ was developed in Chatwin [5], and is

$$\tan\phi = \delta(u_* t/h)\ln(h/ez_0), \tag{11}$$

where $e \approx 2.72$ is the base of natural logarithms, and δ is a constant which, theoretically, is of order 5, but will be determined from data. It was shown in Chatwin [5] that, with $\delta = 5$, equation (11) was consistent, at least approximately, with data in Hall, Hollis and Ishaq [9].

Equations (4), (10) and (11), together with $v = \pi r^2 h$ and prescriptions of u_t and u_s, form a closed system. Solutions of this system, with the prescriptions of u_t and u_s in three existing models, are given in Chatwin [4]. They show, as anticipated, far greater entrainment than in the models without the amendment described above. Because of pressure on space, and for reasons that will become apparent below, no further discussion of these solutions is given here.

Other aspects of entrainment

There is no obvious physical reason why wind shear should affect the modelling of u_t, representing entrainment over the top surface of the cloud. Accordingly, formulae used in existing models can be retained. Two such formulae are

$$u_t = \beta\beta_1 u_* / \{\beta + \beta_1(Ri)\}, \tag{12}$$

(Eidsvik [6]) and

$$u_t = \beta\beta_1 u_* / \{\beta^2 + \beta_1^2 (Ri)^2\}^{1/2} \tag{13}$$

(Fay and Ranck [7]). In both cases β and β_1 are constants to be determined empirically, and Ri is a Richardson number. These formulae are very similar in their likely effect on the box model solution. For high values of Ri (i.e. soon after release when the dispersion is gravity dominated) both formulae, like those in many other models, give u_t proportional to $(Ri)^{-1}$. As Ri decreases to zero (representing the increasing influence of ambient turbulence on the dispersion as the cloud becomes more like a passive marker), both formulae evolve gradually into a situation where u_t is proportional to u_*.

This gradual transition from buoyancy dominated behaviour to passive behaviour is physically realistic and practically desirable. While the prescription of u_t is probably the most sensitive ingredient of box models, certainly in the early stages, there is no reason why gradual transition, like that incorporated in equations (12) and (13), should not be built into the other ingredients, such as the prescription of u_s. Lagrangian similarity (Batchelor [1]) requires that u_s, like u_t, is proportional to u_* when the gas cloud is behaving passively. Thus a prescription of u_s is required which gives u_s proportional to $(g'h)^{1/2} \sec\phi$ for large values of Ri. (This is consistent with the earlier discussion provided equation (7) is retained as the basic entrainment equation.) Also, u_s must be proportional to u_* when Ri is small. A formula which has both of these properties, and also has no abnormal behaviour for values of ϕ near 90° (unlike the simple amendment discussed above), is (Chatwin [4])

$$u_s = (g'h)^{1/2} \{\gamma (Ri) + \gamma_1 \gamma_*\} / \{\cos\phi \, (Ri) + \gamma_* (Ri)^{1/2}\}, \tag{14}$$

where γ, γ_1 and γ_* are constants to be determined empirically. The role of γ_* (which should perhaps be set at unity in a first attempt at validating this proposal) is to allow tuning of the transition from buoyancy dominated behaviour (when $(Ri)^{1/2} \gg \gamma_* \sec\phi$) to passive behaviour (when $(Ri)^{1/2} \ll \gamma_* \sec\phi$).

Equations (4), (7) and (11), together with $v = \pi r^2 h$, equation (12), or (13), for u_t and equation (14) for u_s, form a closed system whose solutions are being investigated in conjunction with the use of available data.

The advection of the cloud

The prescription of the advection of the cloud as a whole, i.e. a prescription of $d\bar{x}/dt$, where $\bar{x}$ is the horizontal displacement of the cloud centroid, is an essential ingredient of all box models, an ingredient separate from the processes of spreading and entrainment discussed in previous sections. Some models take $d\bar{x}/dt$ to be a constant fraction of the mean wind speed $\bar{U}$. This procedure is somewhat unsatisfactory from a dimensional point of view, since $\bar{U}$ is the value of $\bar{u}(z)$ when z is a specified number of metres. This prescription is therefore independent of the cloud height, which cannot be correct. A more satisfactory prescription is to take $d\bar{x}/dt$ as a weighted average of $\bar{u}(z)$ over the cloud height, or, more simply, to take

$$d\bar{x}/dt = \bar{u}(\epsilon h), \tag{15}$$

where ϵ is a constant. This prescription is adopted in, for example, van Ulden [14] with $\epsilon = e^{-1} \approx 0.37$, Fryer and Kaiser [8] with $\epsilon = 0.5$, and Fay and Ranck [7] with $\epsilon = 0.4$. Chatwin [4] proposed taking $\epsilon = 0.32$ as a compromise between the different values appropriate for buoyancy dominated behaviour and passive behaviour. It is important to note that, since $\bar{u}(z)$ varies logarithmically with z , the precise choice of ϵ is relatively unimportant. Thus, with $\bar{u}(z) = (u_* / \kappa) \ln(z/z_0)$, the difference between $d\bar{x}/dt$ when $\epsilon = 0.5$ and when $\epsilon = 0.32$ is about $1.1 u_*$, which is invariably small.

Concluding remarks

The amendment presented in this note, and in more detail in Chatwin [4,5], to take account of wind shear is, of course, not a true representation of the physics. The real physics is far more complicated, and aspects of it are currently being investigated as described, for example, by Rottman and Simpson [12]. However, as stressed in the Introduction to this paper, the correct representation of the physics is not a primary purpose of box models. Should further investigation reveal a better understanding of the effects of wind shear, the results of such investigations will be worth incorporating in box models if, but only if, the consequent amendments do not cause the model to become too intricate for practical use. It is of course possible that it will eventually be decided that box models are not capable of predicting with sufficient accuracy some, or all, of the hazards associated with dispersing heavy gas clouds. Such a decision will be justified only after detailed comparison of many box models with the increasing quantity of reliable data now becoming available. These comparisons are in progress but no firm conclusions seem imminent.

In the event, unlikely in my opinion, that box models eventually prove inadequate, it will be necessary for practical prediction either to use another (more complicated) class of models (e.g. slab models, or solutions of three-dimensional partial differential equations), or to develop stochastic models (Chatwin [3]). The latter alternative has the advantage of being more physically realistic.

Acknowledgement

The work described in this paper was supported by the U.K. Health and Safety Executive under Contract No. 1189.1/01.01. I am grateful to Mr R. Fitzpatrick of HSE for his efficient computing, and to Dr J. McQuaid of HSE for his encouragement and practical wisdom. I was also helped by many other people; Dr P.W. Brighton, Dr D.M. Webber and Dr C.J. Wheatley of SRD, Dr D.J. Hall of Warren Spring Laboratory and Dr J.S. Puttock of Shell.

References

1. Batchelor, G.K.: Diffusion from sources in a turbulent boundary layer. Arch. Mech. Stosowanej 16 (1964) 661-670.

2. Chatwin, P.C.: The dispersion of a puff of passive contaminant in the constant stress region. Q.J.R. Met. Soc. 94 (1968) 350-360.

3. Chatwin, P.C.: The use of statistics in describing and predicting the effects of dispersing gas clouds. J. Haz. Mat. 6 (1982) 213-230.

4. Chatwin, P.C.: Towards a box model of all stages of heavy gas cloud dispersion. In Turbulence and Diffusion in Stable Flows (J.C.R. Hunt, ed.). Oxford University Press 1984.

5. Chatwin, P.C.: The advection and tilting of gas clouds. Submitted to Atm. Envir. (1983).

6. Eidsvik, K.J.: A model for heavy gas dispersion in the atmosphere. Atm. Envir. 14 (1980) 769-777.

7. Fay, J.A. and Ranck, D.: Scale effects in liquified fuel vapor dispersion. DOE/EP-0032, U.S. Dept. of Energy, Washington, D.C., December 1981.

8. Fryer, L.S. and Kaiser, G.D.: DENZ - A computer program for the calculation of the dispersion of dense toxic or explosive gases in the atmosphere. SRD R152, Safety and Reliability Directorate, UKAEA, Culcheth, Warrington, England. 1979.

9. Hall, D.J., Hollis, E.J. and Ishaq, H.: A wind tunnel model of the Porton dense gas spill field trials. LR 394 (AP), Warren Spring Labor-

atory, Stevenage, England. 1982.

10. McQuaid, J.: Future directions of dense-gas dispersion research. J. Haz. Mat. 6 (1982) 231-247.

11. Picknett, R.G.: Field experiments on the behaviour of heavy gas clouds. Ptn. IL 1154/78/1, Chemical Defence Establishment, Porton Down, Salisbury, England. 1978.

12. Rottman, J.W. and Simpson, J.E.: The initial development of gravity currents from fixed-volume releases of heavy fluids. In Proc. of IUTAM Symposium on the Atmospheric Dispersion of Heavy Gases and Small Particles, The Hague. Berlin, Springer (1984).

13. Taylor, G.I.: The dispersion of matter in turbulent flow through a pipe. Proc. Roy. Soc. A 223 (1954) 446-468.

14. van Ulden, A.P.: On the spreading of a heavy gas released near the ground. In Proc. of 1st Int. Symposium on Loss Prevention and Safety Promotion in the Process Industries, Delft. (1974) 221-226.

15. Webber, D.M.: The physics of heavy gas cloud dispersal. SRD R 243, Safety and Reliability Directorate, UKAEA, Culcheth, Warrington, England. 1983.

16. Yaglom, A.M.: Diffusion of an impurity emanating from an instantaneous point source in a turbulent boundary layer. Fluid Mech. - Soviet Research 5 (1976) 73-87.

Gravity Spreading and Turbulent Dispersion of Pressurized Releases Containing Aerosols

K. S. Mudan

Technological Risk Assessment Unit
Arthur D. Little, Inc.
Cambridge, MA

Summary

Sudden release of a pressurized chemical results in instantaneous flashing of fraction of the chemical into vapor. If the pressure inside the tank or a pipeline falls very rapidly, much of the liquid fraction may be thrown into the vapor cloud as a result of vigorous boiling process. A gravity spreading and a turbulent dispersion model has been developed to determine the dispersion hazards of pressurized releases containing liquid aerosols. The gravity spreading model determines the initial lateral spreading of the vapor cloud. A thermodynamic equilibrium model is used to determine the evaporation of liquid aerosols, the mean density and temperature of the cloud. Finally, when the effects of gravity are small, the transition from the gravity spreading model to a turbulent diffusion model takes place. Results are presented for pressurized propane releases at various exit velocities and different initial aerosol mass fraction.

Introduction

In the event of a tank or a pipeline failure resulting in a sudden release, a certain portion of the pressurized chemical stored in the tank will vaporize instantly. This sudden "flash off" is a function of the initial storage pressure and can be calculated using thermodynamic considerations. The remainder of the contents is in the form of liquid at its boiling point.

The density of vapor will determine the spreading characteristics of the vapor cloud. A buoyant puff (as in the case of ammonia) will rise in the atmosphere and then disperse. However, even in the case of buoyant vapors, if the pressure inside the storage tank falls very rapidly, much of the liquid fraction may be thrown into the air as a consequence of the vigorous boiling process (the champagne effect). The liquid fraction is in the form of fine droplets with negligible settling velocity and therefore is suspended in the vapor cloud. This liquid fraction determines the initial density of the mixture and hence the spreading and dispersion characteristics. These aerosols evaporate gradually to form more vapor and the density of the mixture is continuously changing with time. Whenever the average density of the mixture is greater than the ambient air density, the spreading of the vapor cloud is dominated by the effects of gravity.

The spreading of the cloud is further enhanced by the entrainment of air. Air entrainment is caused by the shear layer formed between the spreading cloud and the air above it. The vigor with which the vapor escapes into the atmosphere is such that the shear layer formed by it and the surrounding air is turbulent. The entrained air increases the heat content of the mixture and evaporates the liquid droplets.

Finally, when the effects of gravity are small, the dispersion of the vapor cloud will take place. The dispersion is essentially due to the turbulence in the atmosphere. The atmospheric dispersion can be estimated by assuming the vapor cloud to be passive (neutrally buoyant) and using the well-established theories of Pasquill and Gifford.

Gravity Spreading Model

The initial pressure in the storage vessel or pipeline will give the velocity of release of a pressurized chemical. If the chemical is a liquefied gas under pressure, then a fraction of the released liquid will vaporize instantaneously. The fraction that flashes may be determined by considering isenthalpic mass balance for the initial and final pressures. If f_v is the flashing mass fraction, the liquid mass fraction in the vapor cloud will have a maximum value of $(1 - f_v)$. The actual mass fraction of liquid can be anywhere between zero and the theoretical maximum. At present, there are no analytical techniques or experimental data to determine the exact mass fraction of liquid in a vapor cloud. Several high pressure pipeline accident investigations indicate that high pressure releases do not result in any liquid accumulation on ground. This implies that the vigorous boiling of the chemical throws the entire liquid fraction into the vapor cloud in the form of fine aerosols. The vapor/liquid two phase jet that will result in the event of a pressurized release will be at a velocity much greater than that of ambient air. Because of this high velocity jet, substantial amounts of air will be entrained in the "jet phase" of the plume. The mean density of the two-phase mixture will be large compared to that of air and small compared to that of liquid. Because of this density difference, the vapor cloud moves laterally. The lateral spreading velocity, initially, is small compared to the axial jet velocity. Also, the amount of air entrained due to lateral spreading will be small compared to that entrained by the turbulent jet. As the ambient air is entrained, the density of the mixture decreases. If the amount of air entrained an any downwind location is known, the spreading characteristics of the vapor cloud can be calculated. The assumptions made in developing the gravity spreading model are as follows:

- The cross section of the plume is rectangular on the ground surface.
- The entrainment of air is only from the top (the entrainment from sides is neglected by assuming that the height of the cloud is small compared to width).
- The concentration of vapor/aerosol at each cross section is uniform. The axial velocity is also assumed to be uniform.
- The longitudinal and lateral entrainment at the density interface is inversely proportional to the local Richardson number.

A schematic of the geometry of the negatively buoyant plume is shown in Figure 1. The mass and momentum balance equations for a slice of cloud are derived in the original version of the paper (Mudan, 1983).

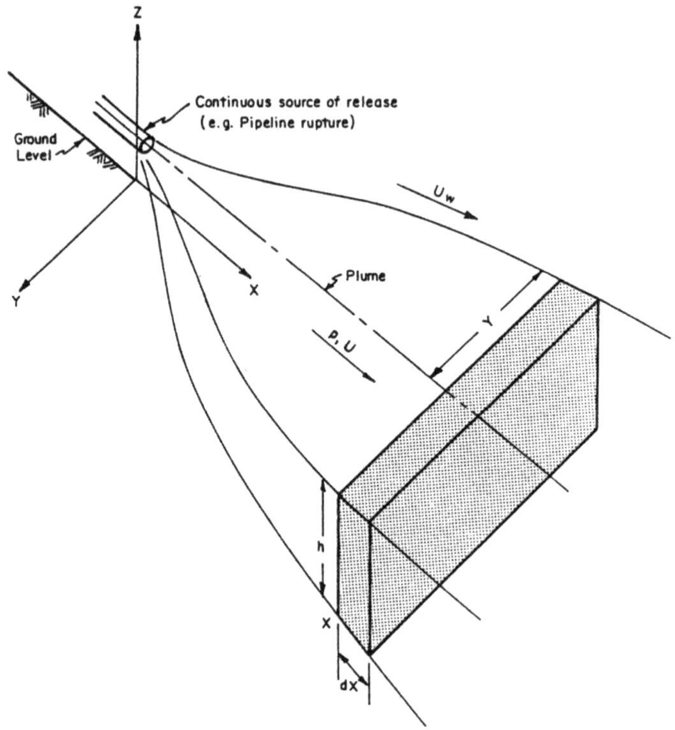

Figure 1

Schematic Diagram of a Negatively Buoyant Plume

Thermodynamic State of a Mixture of Air-Vapor and Liquid Aerosol

When a small quantity of air is mixed with a vapor cloud containing liquid aerosols, the aerosols will begin to evaporate owing to the reduction in the partial pressure of the surrounding vapor and the heat provided by the air. In early stages of dilution, the temperature of the cloud and the partial pressure of the vapor will follow the saturated vapor pressure-temperature curve of the substance until all the droplets are evaporated. Subsequent dilution will raise the temperature of the mixture. Details of the thermodynamic analysis are given in Raj and Aravamudan (1980).

Turbulent Dispersion of a Negatively Buoyant Vapor Cloud

In the "gravity spreading" region, the concentration of the vapor at any downwind distance is assumed to have a "top hat" profile, i.e., the concentration within the cloud is uniform. The spreading (i.e., the increase in the lateral dimension) of the cloud is essentially due to the effects of gravity. As the cloud spreads, the average density of the vapor in the cloud steadily decreases and asymptotically, the mean density of the cloud will reach that of the ambient air. When the density differences are small, the interaction between ambient turbulence and the vapor cloud is likely to dominate the spreading of the vapor cloud. The classical dispersion theory assumes that the concentration profile is Gaussian in shape. Once a criterion for transition from a top hat profile to a Gaussian profile is established, the further dispersion of the cloud due to atmospheric turbulence can be determined. The lateral spread velocity is equated to the ambient lateral turbulent velocity at the cloud height to determine the transition from gravity spreading to atmopsheric dispersion.

Development of a Numberical Algorithm

A closed-form solution to the problem of gravity spreading of a denser than air cloud can be found only under certain simplifying assumptions (Raj and Aravamudan, 1980). The presence of liquid aerosols and their subsequent evaporation makes the problem more complicated. A numerical solution was sought to determine the extent of gravity spreading of a negatively buoyant vapor cloud with entrained liquid droplets. The amount of liquid aerosols initially released into the atmosphere is specified by the user. For the purposes of hazard analysis it is recommended that the maximum value of the liquid fraction be used. In Figure 2 is shown the width of a propane vapor cloud as a function of down wind distance. The source release rate is 100 kg/s and the initial vapor fraction and liquid fraction are 0.35 and 0.65 respectively. The chosen atmospheric conditions are D-Stability and a wind speed of 5 m/s. Shown in Figure 2 are the footprints of the vapor cloud for release velocities of 400 m/s and 100 m/s. As

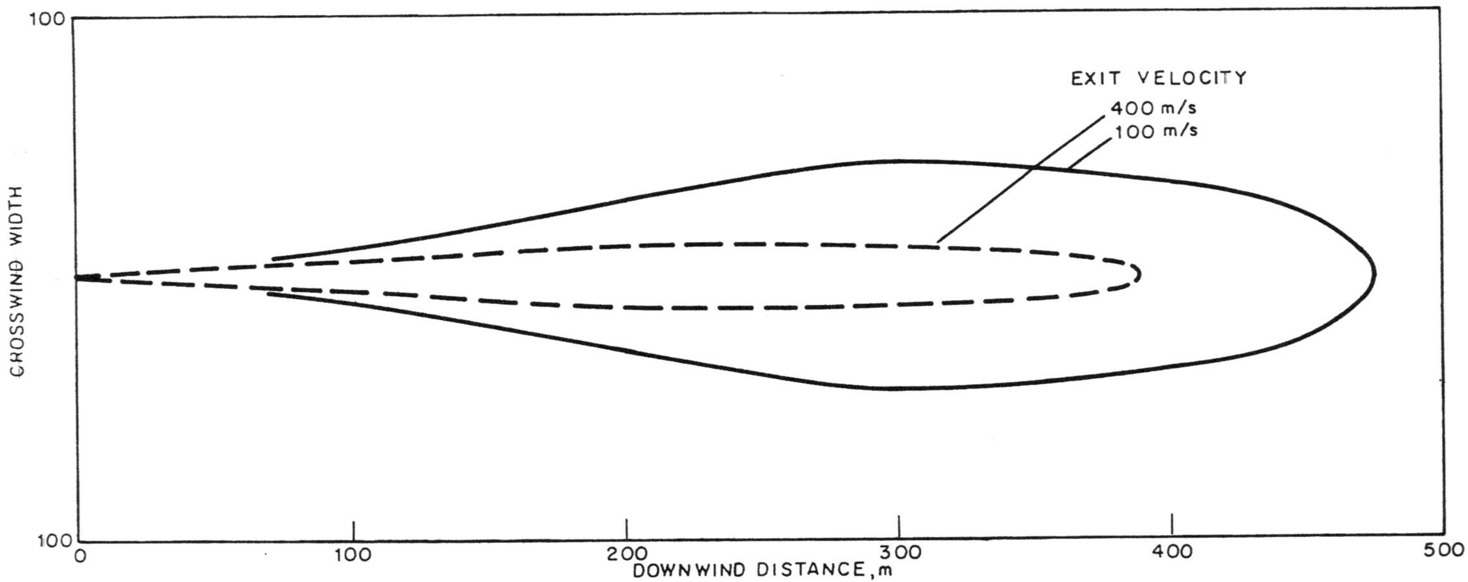

Figure 2

VARIATION OF PLUME WIDTH WITH DOWNWIND DISTANCE

can be seen from Figure 2, the downwind distance increases slightly because of a decrease in axial velocity. But the more profound effect of decrease in axial velocity is in the increase of the width of the vapor cloud.

In Figure 3 is shown the fraction of the liquid aerosols contained in the vapor cloud for various exit velocities. As the exit velocity is increased, the mass of ambient air entrained into the vapor-aerosol system is also increased and the liquid aerosols evaporate faster. In Figure 4 is shown the mean temperature of the mixture as a function of downwind distance. The mixture temperature drops from the saturation temperature until all the aerosols are evaporated. Then, the temperature of the mixture increases and asympotically reaches that of the ambient air. The effect of liquid aerosols on the distance where transition from gravity spreading to atmospheric dispersion is shown in Figure 5. Also shown in Figure 5 are the cross wind widths at transition location.

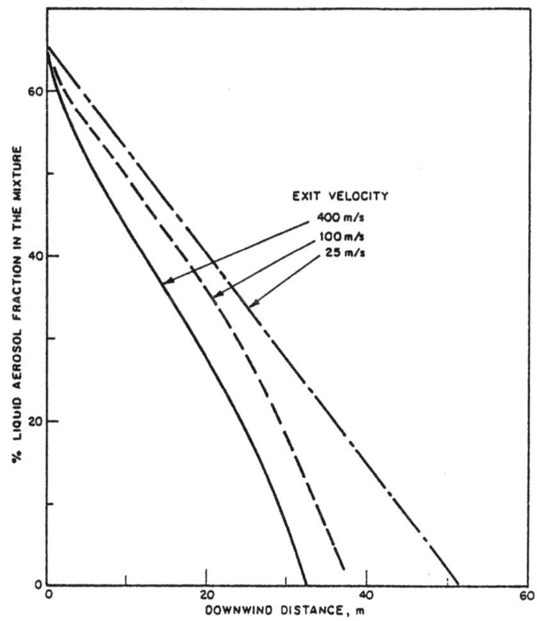

Figure 3

Liquid Aerosol Fraction in the Vapor Cloud is a Function of Downwind Distance for Various Exit Velocities

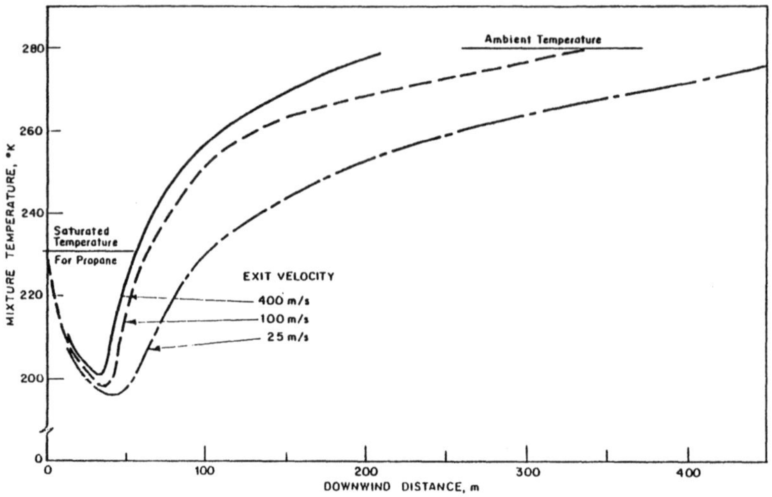

Figure 4

Variation in Mixture Temperature with Exit Velocity

These two distances are normalized with respect to downwind and crosswind distances in the absence of liquid aerosols (i.e. 100% propane vapor). As can be seen from Figure 5, the pressence of liquid aerosols modifies the behavior of the dense gas cloud. Increase of liquid fraction (for the same mass release rate) increases both downwind transition distance and the width of the vapor cloud.

Conclusions

Based on the analysis of the results, certain key parameters affecting the spreading and dispersion of negatively buoyant vapor clouds resulting from pressurized releases may be identified. The more obvious parameters affecting the dispersion distances are the source release rate, wind speed and the atmospheric stability. Additional parameters that affect the dispersion of negatively buoyant vapor cloud are as follows:

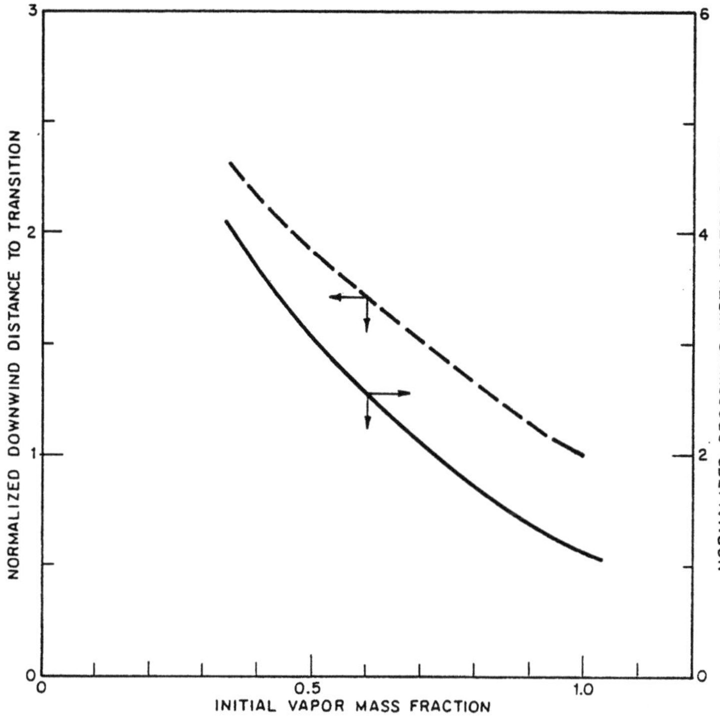

Figure 5

Normalized Downwind Distance to Transition and Crosswind Width at Transition for Various Liquid and Vapor Fractions

- Release Velocity: The release velocity appears to affect the downwind dispersion distances inversely. In general, the larger the velocity, the larger is the amount of air entrained and the shorter is the downwind dispersion distance. The crosswind dimensions increase rapidly with decreasing release velocity.

- Liquid Fraction: For a given release rate, increase of liquid fraction increases both downwind dispersion distance and the maximum crosswind dimensions of the vapor cloud.

- Wind Speed: The wind speed affects the dispersion distances inversely in the turbulent dispersion region. For high speed jet releases, the wind speed has very little influence on both downwind and crosswind dispersion distances.

References

Mudan, K.S. (1983) "Gravity Spreading and Turbulent Dispersion of Pressurized Releases Containing Aerosols," Paper presented at the IUTAM Conference on Atmospheric Dispersion of Heavy Gases and Particles, August 29 - September 2, 1983, The Hague, The Netherlands.

Raj, P.K. and Aravamudan, K.S. (1980) "Theoretical models supporting the design of ammonia spill experiments," Report to the Fertilizer Institute, Washington, D.C., Arthur D. Little Report Reference 82315.

Simulations and Parameter Variation Studies of Heavy Gas Dispersion Using the Slab Model – Condensed*

D.L. MORGAN, JR.; E.J. KANSA; L.K. MORRIS

Liquefied Gaseous Fuels Program

Lawrence Livermore National Laboratory

Livermore, California

Summary

We are employing the SLAB model in ongoing studies of the atmospheric dispersion of heavy gases. SLAB computer simulations of four of the Burro series large-scale 40-m^3 liquefied natural gas (LNG) spill experiments at China Lake, California [1] have been successful in predicting distances to the lower flammability limit (LFL) [2]. We have used this model in simulations of three of the Coyote series of experiments [3] as well as in parameter variation and sensitivity studies [4] and improved simulations of some of the Burro tests. The parameters studied include source rate, wind speed, atmospheric stability, type of source gas, and source duration, as well as the parameters important to certain physics submodels.

The SLAB Model

The SLAB model is a one-dimensional model describing diffusion and gravity flow of a heavy gas released into the atmosphere [2, 4-6]. The properties of the air-gas cloud are treated explicitly in their dependence on downwind distance (x) and time. The properties are slab-averaged in the horizontal (y) and vertical (z) crosswind directions.

Five coupled, partial differential equations (PDEs) of the model express the conservation of air and gas masses, downwind and horizontal momenta, and thermal energy. They are derived by averaging the Navier-Stokes conservation equations over y and z within the limits of the cloud. With the use of the hydrostatic approximation for pressure, these equations relate cloud motion, density, and temperature to the forces that affect them: gravity, the mixing in of air, heat flow from the ground, ground friction, air resistance, and the source gas. Another PDE defines the cloud

*Work performed under the auspices of the USDOE by LLNL under contract number W-7405-ENG-48 and the Gas Research Institute.

width by stating that the downwind-Lagrangian speed of the cloud edge in the y-direction is the material speed (v_g) plus the horizontal air-entrainment speed (v_e). Together with the ideal gas law this equation provides the additional information necessary for defining the size and shape of the cloud. Thus, the SLAB model is quasi-three-dimensional. Algebraic submodels are employed to calculate turbulent diffusion (using entrainment), heat flow, friction, height-dependent wind speed, and crosswind gas concentration. In the averaging, it is assumed that concentration and temperature are independent of y and z, while the downwind cloud speed is independent of y and has a prescribed power law dependence on z. The speed in the y-direction is assumed to be proportional to $|y|$. Since the model cloud is symmetric about $y = 0$, it is only necessary to consider one half of the cloud. This is done in defining the dependent variables of the SLAB equations which are (per unit downwind distance for all but B):

- m = total mass of air and heavy gas,
- m_1 = mass of heavy gas,
- ε = thermal energy,
- P_x = downwind component of momentum,
- P_y = mean horizontal crosswind component of momentum, and
- B = width of the half-cloud.

The variables, height (h), density (ρ), temperature (T), cloud speed (u), and volume concentration (C_o), as well as v_g can be related to the above variables, e.g.,

$$C_o = \frac{m_1}{m_1 + (m - m_1)M_s/M_a}, \quad u = P_x/m, \quad v_g = 2\,P_y/m,$$

where M_s and M_a are the molecular weights of the source gas and air. A formula based on experimental and theoretical information is employed to obtain the concentration distribution in a crosswind plane:

$$C = C_o\, e^{-z/h\, -\, \pi y^2/4B^2},$$

where it is assumed that the maximum concentration in the crosswind plane is equal to C_o.

Turbulent mass diffusion is modeled by entrainment of air into the cloud surface. The entrainment rate depends on the air-cloud density and velocity differences and on the friction and convection velocities of the cloud. As the cloud becomes dilute, the entrainment rate approaches ambient. Our formulae for the vertical and horizontal entrainment speeds, w_e and v_e, are fitted to experimental data and were derived by Morgan, Ermak, and Zeman [4-6]. The submodel formulae for w_e and v_e are:

$$w_e = \frac{0.4\ w}{D}, \qquad v_e = (1.8)^2\,\frac{h}{B}\,w_e\ ,$$

where

$$w = (u_*^2 + 0.02\ (\delta u)^2 + 0.27\ w_*^2)^{1/2},$$

$$D = \begin{cases} 1 + 0.28\ Ri & \text{for } Ri \geq 0\ , \\ (1 - 0.90\ Ri)^{-1/4} & \text{for } Ri < 0\ , \end{cases}$$

where u_* is the cloud friction velocity, δu is a density-adjusted cloud-air velocity difference, and w_* is the convection scale velocity. Ri is the Richardson number for the cloud, $g(1-\rho_a/\rho)\ h/w^2$, where ρ_a is the air density, but its value is adjusted to approach the ambient value as the cloud becomes dilute.

We use an empirical formula for heat flux from the ground, based on measurements made during the Burro series [1]:

$$j \simeq 0.0125\ C_p\ \rho(T_a - T),$$

where C_p is the specific heat of the cloud and T_a the air temperature. The formulae for the fluxes of horizontal momentum from the cloud into the ground (the effect of the ground friction) follow from atmospheric surface boundary layer theory [5]:

$$\tau_x = \rho u_*^{\,2} u/(u^2 + v^2)^{1/2}, \quad \tau_y = \rho u_*^{\,2}\ v/(u^2 + v^2)^{1/2},$$

where $v = P_y/m$ is the mean horizontal crosswind cloud speed.

The SLAB code employs a height-dependent wind speed u_a. The speed of the air entrained into the cloud is assumed to be equal to the ambient wind speed at $z = h$:

$$u_a = u_{a2}(h/h_2)^n,$$

where u_{a2} is the measured average wind speed at $h_2 = 2$ m. The exponent is chosen to match the variation of wind speed with z for the conditions of interest.

The theory of the SLAB model was initially developed by Zeman [5]. The current theoretical form was derived by Ermak and Morgan [2,6]. The SLAB computer model was developed by Morgan and Morris [6] and is described in detail in [4] and [6].

Simulations

We have conducted simulations of seven of the Burro and Coyote tests. The principal spill and meteorological parameters describing these experiments are given in the following table.

LNG Spill Tests

	B3	B7	B8	B9	C3	C5	C6
LNG Volume (m^3)	34.0	39.4	28.4	24.2	14.6	28.0	22.8
Spill Rate (m^3/min)	12.2	13.6	16.0	18.4	13.5	17.1	16.6
Wind Speed (m/s)	5.4	8.4	1.8	5.7	6.0	9.7	4.6
Ambient Richardson No. at 2m	-0.22	-0.02	+0.12	-0.01	-0.32	-0.08	+0.03

Figure 1 shows the maximum observed values of 10-s averaged LNG vapor concentration (by volume) observed at various downwind distances from the source for six of the seven tests. From such information the maximum distances to the LFL (C = 5%) can be determined (least squares linear fit on a log-log plot, emphasizing points near the LFL). These distances are shown in Fig. 2, where they are compared to SLAB predictions. The comparisons show good agreement, with Burro 8 being the weakest which we believe is due to the relatively high ambient stability, low wind speed, and high spill rate resulting in terrain-influenced gravity flow and cloud bifurcation[7]. The SLAB model cannot predict such effects. It is also possible that the entrainment and heat flow submodels may not be sufficiently accurate for Burro 8.

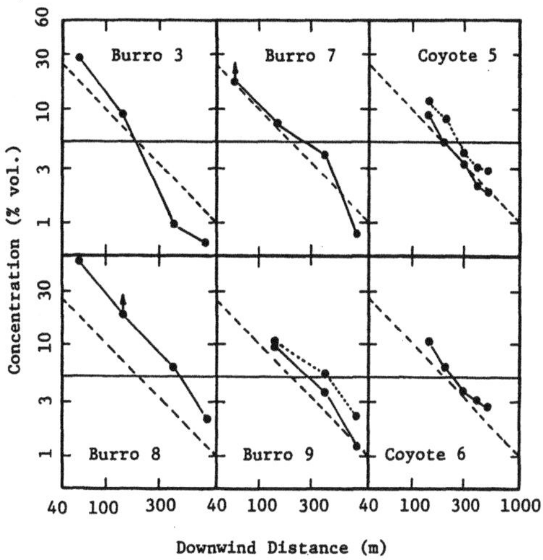

Fig. 1. Maximum values of 10-s averaged data. Dotted lines indicate data in which RPT effects have not been removed. The solid horizontal line is the 5% LFL. The dashed line is for visual reference (C[%] = 1000/x[m]). Arrows indicate instrument saturation. The results for Coyote 3 are simular to those for Burro 3.

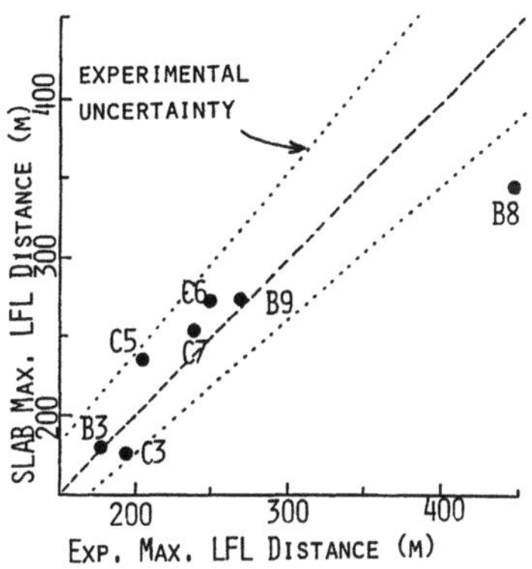

Fig. 2. SLAB vs experiment LFL distance comparison.

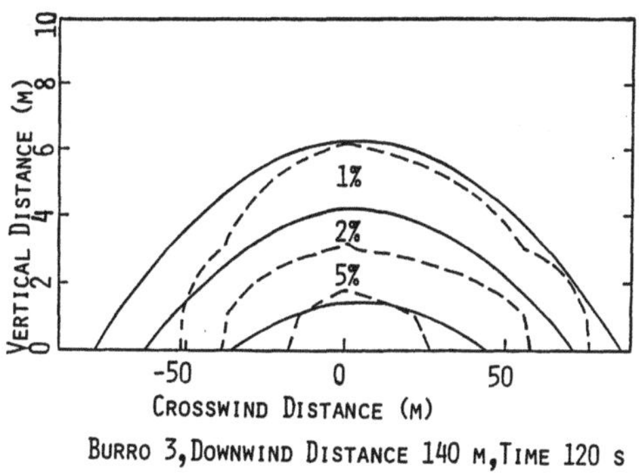

Fig. 3. SLAB (solid) vs experiment (dashed) ccrosswind concentration comparison.

Figure 3 is an example of a comparison between crosswind concentration contours calculated by SLAB and from experimental data. The good agreement is typical of most comparisons excluding the bifurcated cloud of Burro 8.

Figure 4 shows typical examples of SLAB vs experiment, time-history comparisons for 1 m above the surface. Two comparisons at 3 m (Coyote 5) are included. In each case, the SLAB center-line (y = 0) concentration for the indicated height above ground and downwind distance is compared to experimentally determined concentrations for all sensors in a horizontal crosswind row at the same height and distance. Such a comparison is made due to

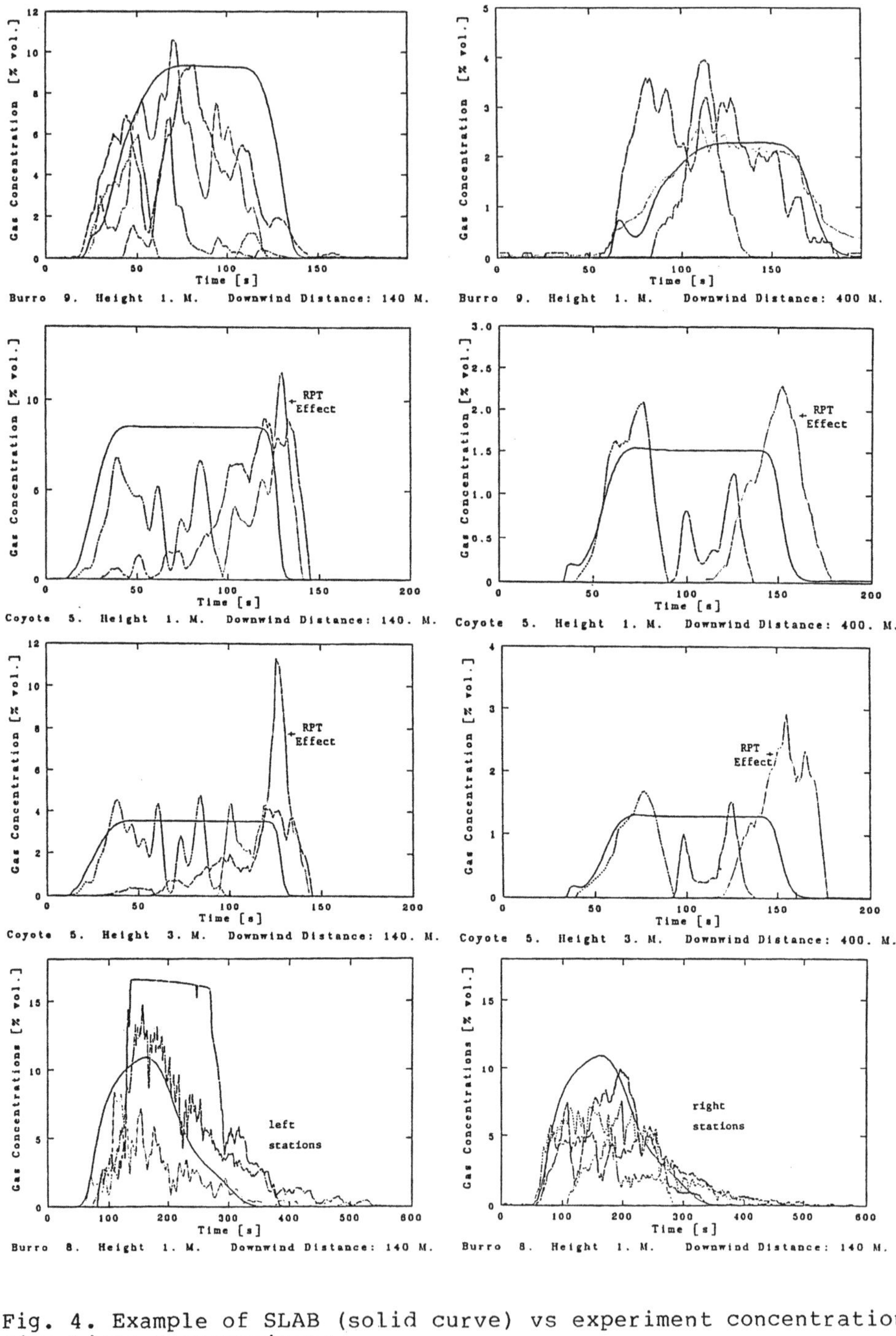

Fig. 4. Example of SLAB (solid curve) vs experiment concentration time history comparisons.

meander of the cloud centerline. When the centerline passes over a sensor, the resulting concentration peak is the mean centerline concentration except for the presence of relatively smaller turbulent variations in the 10-s averaged data. Thus a correct model result would probably fall slightly below the highest concentration peaks.

In Burro 9 and Coyote 5, RPT explosions [8] released puffs of LNG vapor that momentarily increased concentrations downwind. Such phenomena were not modeled by SLAB. Their presence is noted in the Coyote 5 comparisons where the effect was substantial.

The SLAB time-history curves compare well with experimental data for Coyote 5 and Burro 9. The same is true for Burro 8 except for the very high concentration measured by the G11 sensor at 1 m (the nearly flat top of this curve is due to instrument saturation) and the fact that the cloud tended to arrive and depart somewhat later than predicted by SLAB. Both of these differences are probably due to terrain-influenced gravity flow. The G11 station was 4m lower than the average elevation of the 140 m row.

Parameter Study

Results of the parameter study using SLAB show the effects of individual variations of five parameters about values that define an LNG base case: rate 135 kg/sec; wind speed 5.7 m/s; neutral atmospheric stability. The base case is similar to Burro 9 except a value of 0.05 is used for the friction coefficient c_f, instead of 0.08 as subsequently recommended by Zeman [private communication] and the source remains on until steady state is reached (c_f affects the entrainment rate since u_* is taken to be proportional to it).

Results for variations in wind speed, stability, and in source rate, type, and duration are in Ref. [4]. Except for possibly wind speed, they agree with physical expectation. The effects on concentration, height, and width of the cloud, due to resulting variations in gravity flow, turbulent mixing, and cloud heating, are seen.

Fig. 5. SLAB predicted effects of wind speed variation on LNG vapor dispersion (3m/s dashed; 5.7 m/s solid; 15 m/s dotted). Ambient stability, source rate, and other parameters are held constant. The horizontal line in the concentration plot is the 5% LFL. The effects of wind speed variation are markedly different from the wind speed effects on the dispersion of trace pollutants, where cloud height and width are independent of wind speed and concentration is inversely proportional to wind speed.

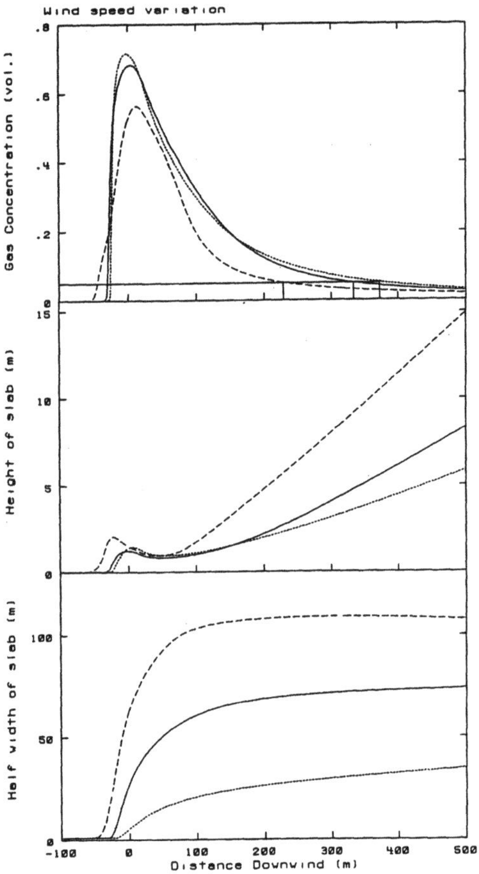

The results for wind speed variation (stability is held constant) are shown in Fig. 5. Increasing u_{a2} reduces the height and width of the cloud (they are constant for trace pollutants), since the air-cloud density difference leads to entrainment speeds that are much less than proportional to u_{a2}. This decreased dependence of entrainment on wind speed is sufficient, in the SLAB model, to lead to the higher values of downwind concentration for higher wind speed shown in the figure. This result is in contrast to trace pollutant dispersion where concentration is inversely proportional to wind speed. In light of the approximate nature of SLAB's entrainment formulation, the exact magnitude of this difference in wind speed dependence is uncertain, but it is clear that such a difference exists. Our experimental data may indicate a slight inverse dependence of concentration on wind speed for constant source rate and ambient stability, but they are definately inconsistent with the inverse proportionality for trace pollutants.

We have also tested the sensitivity of our results to the friction coefficient value c_f, the choice made for the heat flow submodel, and to inclusion of the retarding force of surface friction on motion of the cloud. Details of these tests are given in Ref. [4]. A base case similar to the above was employed. We find that the recommended value of $C_f = 0.08$ gives good agreement with experiment but values of 0.04 and 0.12 give poor agreement. Employing a theoretical ground heat flow model in place of our empirical model also significantly reduces agreement [4]. In contrast, taking the surface friction to be zero has no signifcant effect on our results. Thus, at least for the conditions of the base case, accurate submodels for entrainment and heat flow are important for accurate modeling of LNG vapor dispersion, but the retarding force of surface friction is not.

Conclusions

The SLAB model has achieved good agreement with experimental data in predicting gas concentration levels measured in the seven Burro and Coyote vapor dispersion experiments it simulated. In the most difficult case (Burro 8), SLAB did reasonably well in predicting those features for which it was designed.

The dependence of SLAB model results on source rate, atmospheric stability, source type, and source duration is physically reasonable. The dependence of downwind concentration on wind speed indicates less of a dependence in heavy gas dispersion than the inverse proportionality in trace pollutant dispersion. This alteration is significant and worthy of further investigation.

The choice of model for heat flow from the ground into the cloud is found to be quite significant to the dispersion of the cold

LNG vapor. The level of air entrainment, which is somewhat uncertain, is also quite significant, and it is therefore important to formulate accurate entrainment/turbulence models for heavy gas dispersion. For the cases studied, the retarding effects of ground friction on the cloud appear to be of much less significance.

References

1. Koopman, R.P.; Cederwall, R.T.; Ermak, D.L.; Goldwire, H.C. Jr.; Hogan, W.J.; McClure, J.W.; McRae, T.G.; Morgan, D.L.; Rodean, H.C.; Shinn, J.H.: Analysis of Burro series 40-m^3 LNG spill experiments. J. Haz. Mat. 6 (1982) 43-83.

2. Ermak, D.L.; Chan, S.T.; Morgan, D.L.; Morris, L.K.: A comparison of dense gas dispersion simulations with the Burro series LNG spill test results. J. Haz. Mat. 6 (1982) 129-160.

3. Goldwire, H.C., Jr.; Rodean, H.C.; Cederwall, R.T.; Kansa, E.J.; Koopman, R.P.; McClure, J.W.; McRae, T.G.; Morris, L.K.; Kamppinen, L.M.; Kiefer, R.D.; Urtiew, P.A.; Lind, C.D.: Coyote series data report, LLNL/NWC 1981 LNG spill tests, dispersion, vapor burn, and rapid phase transition. (in press), LLNL, Livermore, Calif. (1983).

4. Morgan, D.L., Jr; Kansa, E.J.; Morris, L.K.: Simulations and parameter variation studies of heavy gas dispersion using the SLAB model. UCRL-88516, LLNL, Livermore, Calif. (expanded version of this report).

5. Zeman, O.: The dynamics and modeling of heavier-than-air cold gas releases. Atmos. Environ. 16 (1982) 741-751.

6. Morgan, D.L. Jr.; Morris, L.K.; Ermak, D.L.: SLAB: a time-dependent computer model for the dispersion of heavy gases released in the atmosphere. UCRL-53383, LLNL, Livermore, Calif. (1983).

7. Chan, S.T.; Rodean, H.C; Ermak, D.L.: Numerical Simulations of Atmospheric Releases of Heavy Gases Over Variable Terrain. UCRL-87256, LLNL, Livermore, Calif, (1982)

8. McRae, T.G.: Preliminary analysis of RPT explosions observed in the LLNL/NWC LNG spill tests. UCRL-87564, LLNL, Livermore, Calif. (1982).

0129t/3t

Application of Advanced Turbulence Models in Determining the Structure and Dispersion of Heavy Gas Clouds

D M Deaves

Atkins Research and Development Epsom, Surrey, UK

Summary

The present paper reviews the use of a particular form of higher order turbulence modelling in heavy gas dispersion, and demonstrates its potential for use in complex situations. It is also shown how such models can be used to provide more general information on the detailed turbulence structure within a heavy gas cloud. A currently available 3D heavy gas dispersion model is also described, and examples given of typical applications.

1. INTRODUCTION

The current interest in heavy gas dispersion, which is likely to increase as the Seveso Directive(1) takes effect, has resulted in the development of a number of mathematical models. A useful assessment of currently available models has recently been given by Blackmore et al(2). A realisation of the limitations of the simple 'box' type of model has provoked increasing interest in alternative K-theory models, whose main advantage is the ability to include the effects of complex terrain, buildings etc.

It has recently been pointed out(3), however, that all the models currently under consideration predict mean concentrations, where the average is taken over some time period. This raises difficulties in predicting a phenomenon which is basically non-deterministic, and comparing with non-repeatable field or model tests. These problems can only be overcome by consideration of probability density functions, or, at the very least, computation of rms concentration fluctuations.

2. MODELLING OF HEAVY GAS DISPERSION

2.1 Flow Effects

The relative density difference of the gas compared with that of air will tend to affect the mean motion of both the cloud and the atmosphere in its neighbourhood. In the early stages after release, the cloud will tend to retain some of its inertia, and provide an effective obstruction to the flow. As the gas is accelerated towards ambient windspeed the negative buoyancy will induce a downward velocity, which will also draw air downwards and thus modify the ambient flow field.

Whilst the mean flow effects are fairly obvious, and reasonably easy to visualise, those associated with turbulence are less apparent, but have greater significance to the dispersion properties of heavy gases. An analogous situation to that provided by a heavy gas cloud is found in natural atmospheric conditions when a temperature inversion exists, and the stratification is extremely stable. The reduction of the diffusion coefficient to zero at the inversion height in these cases implies that a certain amount of turbulent mixing is still taking place below the inversion level, but almost no mixing occurs across this interface.

The downwind dispersion of a heavy gas cloud in a steady wind is also characterised by a region in which the negative density gradient suppresses turbulent mixing, and this stratification not only affects the dispersion characteristics of the cloud, but also tends to modify the structure of the atmospheric turbulence. Figure 1, from Gibson & Launder(4) shows the variation of the normal stress ratio with stability. They also discuss the importance of modelling the wall effects correctly; the dashed line in the figure shows that $\overline{w^2}/\overline{u^2}$ decreases with stability in a free shear layer, whereas the solid line, along with atmospheric data, displays the opposite behaviour. This phenomenon occurs because of the combined effects of stability and wall proximity; increasing stability gradually de-couples the eddy size scaling from its proportionality to distance from the wall.

2.2 K-theory Models

These models set up transport equations for momentum, concentration, temperature etc. The coupling between these equations enables the effect of the heavy gas on the flow patterns to be modelled, but necessitates an iterative solution. The turbulence is modelled using K-theory, which requires an eddy viscosity (K) in order to close the system of equations. For undisturbed atmospheric flows, it is straightforward to specify K algebraically and this type of model can be used with confidence. However, the presence of the heavy gas will affect the turbulence in a way which depends on the solution, and will actually reduce the diffusion coefficient,

K, to zero in certain circumstances. Similarly, specification of K in the presence of obstructions is unlikely to be good enough to enable the complex dispersion effects to be computed accurately.

An alternative method of specifying K is to relate it to local values of turbulence energy (k) and dissipation (ε), which in turn are found as solutions to transport equations. This k-ε model is an improvement over an algebraic specification of K since, with the appropriate turbulence suppression term due to buoyancy in the equations, regions of high density gradient result in low values of k, and hence very small values of K ($K \alpha k^2/\varepsilon$). The model has also been used extensively in many engineering flows in which regions of recirculation are present(5). It therefore opens up a wider class of problems, namely those in which obstructions play a significant part, and these are the subject of the current phase of the Thorney Island Heavy Gas Dispersion Trails(6).

2.3 Algebraic Stress Modelling

The eddy diffusivity which is used in the k-ε model is assumed isotropic, whereas, as indicated in Figure 1, stability is known to affect the different stress components in different ways. This could result in a differential suppression of turbulence, which cannot be reproduced in a k-ε model, and a more realistic representation requires a consideration of the equations for the Reynolds stresses. When concentration fluctuations are included, a large set of transport equations(7) results, all of which would need to be solved by finite differences (or some other suitable numerical technique). This makes such methods economically unattractive, and their present state of development does not really justify their application to the complexities of heavy gas dispersion.

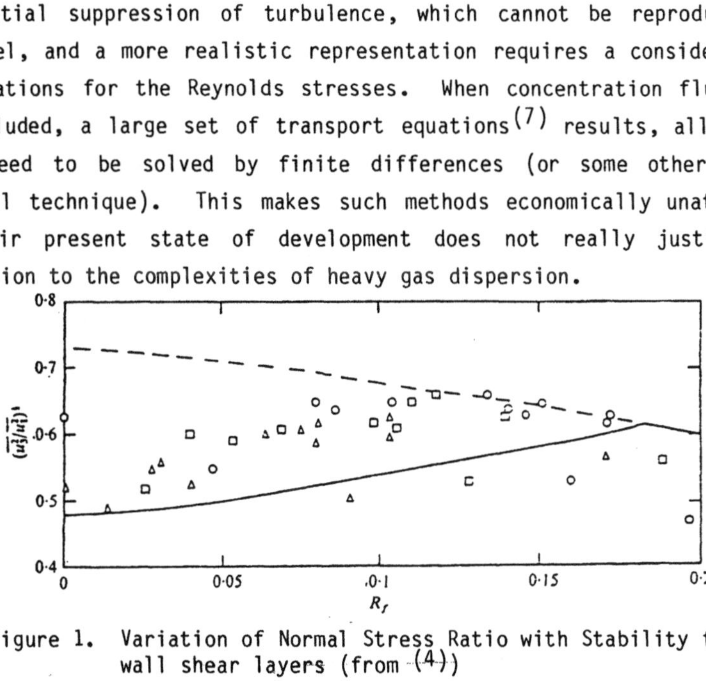

Figure 1. Variation of Normal Stress Ratio with Stability for wall shear layers (from (4))

However, some fairly simple modelling assumptions enable the transport equations to be reduced to a set of algebraic equations which can then be solved in conjunction with a modified k-ε model. This is known as algebraic stress modelling (ASM)[8], and has been successfully applied to other dispersion problems in the atmosphere [9]. Smith & Takhar[10] applied this technique to a study of saline intrusion in tidal channels and estuaries, while Meroney[11] applied the method to stably stratified atmospheric flows. He concluded that the level of sophistication afforded by algebraic stress models was both necessary and also sufficient for predicting the detail of such stratified flows.

It therefore appears that algebraic stress modelling, which requires only a little more computational effort than the standard k-ε model, could provide significant improvements in the understanding of heavy gas cloud structure, and in the prediction of the dispersal of such clouds.

3. APPLICATION OF ASM TO HEAVY GAS DISPERSION

3.1 Objectives

It was outside the scope of this paper to formulate and program a completely general form of algebraic stress model for heavy gas dispersion. The objective was therefore to identify an appropriate form of the model, simplify it to an equilibrium boundary layer flow, with a layer of heavy gas, and deduce the behaviour of various turbulence quantities as the stability (i.e. density of the gas layer) is increased.

3.2 Formulation for Equilibrium Wall Layers

It has already been indicated that different stress components vary with stability in different ways. It is also evident from Figure 1 (and expanded in greater detail by Gibson and Launder[4]) that the presence of a wall also modifies the way in which stability affects the stresses. Since a heavy gas will always disperse from a cloud which hugs the ground, it was decided to use the Gibson-Launder formulation, which includes both wall proximity and stratification effects.

Full details of this scheme are given in[4], but the results are outlined here. The first important parameter is the flux Richardson number, R_f, defined as the ratio of buoyancy to shear turbulence production.

Secondly, $f = \frac{(-\overline{uw})^{3/2}}{\varepsilon \kappa z}$ (1)

defines the wall length-scale function, where: $-\overline{uw}$ is wall shear stress ($= u_*^2$), κ is von-Karman's constant (= 0.4) and z is distance from the wall.

For a full application of ASM, $-\overline{uw}$ and ε would both be computed for use in equation (1). However, for the present simplified study, the empirical variation of the dimensionless mean shear is used to relate f to R_f:

$$f = 1 - 4.5\, R_f \qquad (2)$$

This is not an integral feature of the model, but is used to illustrate stability dependence, and enable simple numerical results to be obtained.

The resulting equations for an equilibrium ($P + G = \varepsilon$) wall shear layer are given by Gibson and Launder and relate the quantities $\overline{u^2}$, $\overline{w^2}$, $\overline{uw}$, $\overline{u\rho'}$, $\overline{w\rho'}$ and $\overline{\rho'^2}$ to k, ε, $\partial U/\partial z$ and $\partial \rho/\partial z$. For given values of velocity and density gradients, the various assumptions made can be applied to obtain consistent values of ε, u_* and k, and hence the correlations $\overline{w^2}$, $\overline{w\rho'}$ etc. More general application of ASM would require an iterative solution of a complete set of simultaneous equations of the form given above.

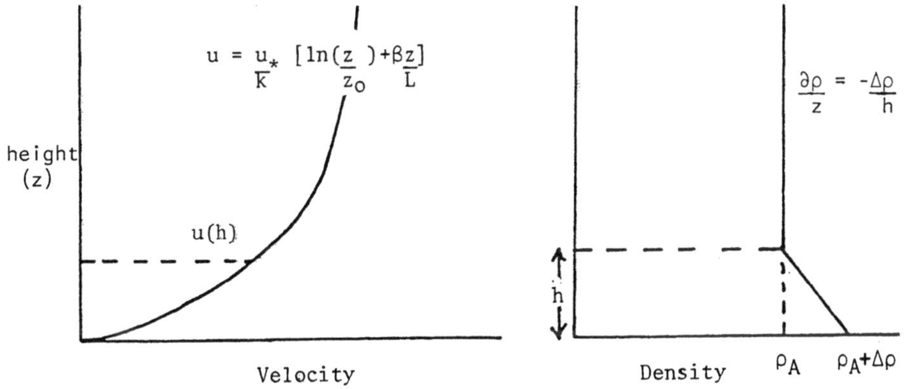

Figure 2. Conditions for ASM Test Case

3.3 ASM Results for a Simple Test Problem

A simple equilibrium flow has been chosen in which a layer of dense gas h metres thick is being transported and dispersed by a wind of speed U(h) at the height of the cloud top, as shown in Figure 2. The density difference between the top and bottom of the cloud is $\Delta\rho$, implying a density gradient of $\Delta\rho/h$. If it is assumed that changes to the flow variables occur much more slowly in the streamwise direction than in the vertical, the simplified formulation given above can be used.

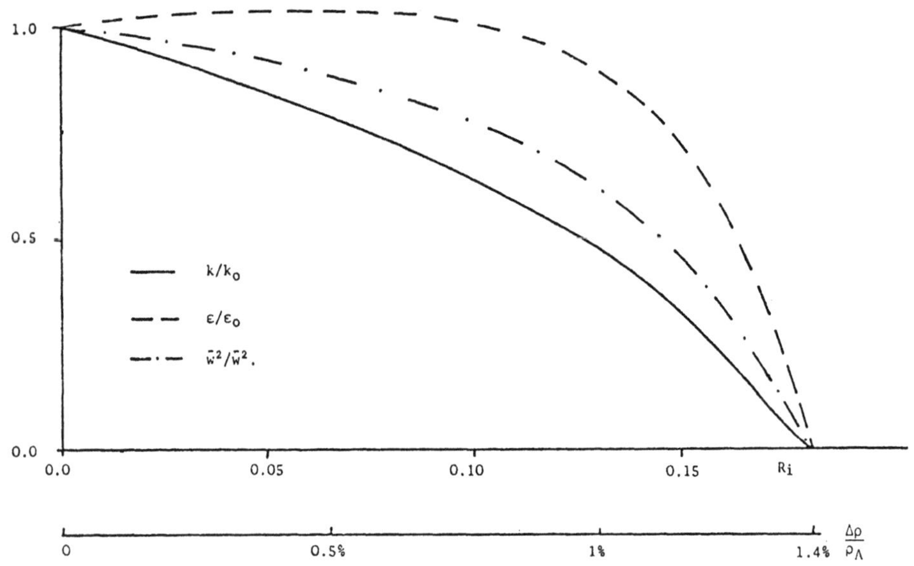

Figure 3. Variation of Normalised Turbulence Quantities with Cloud Density (h=1m u(h)=4m/s)

Figure 3 shows the variation of k and ε with stability for increasing Richardson number R_i. The assumptions concerning the length scale function f imply a critical R_f of 0.22, giving a critical R_i of 0.18, which is therefore dependent on the empirical results used in equation (2). In any more general application of ASM, the appropriate buoyancy suppression terms in the k and ε equations would determine the local conditions under which turbulence was suppressed. The results of this Figure demonstrate the expected behaviour, and imply that the effective diffusivity, proportional to k^2/ε, decreases almost linearly to zero at the critical value of R_i. The curves shown are not universal, but depend weakly on both h and U(h); their general shapes, however, remain the same.

Figure 4 shows the corresponding variation of the correlations $\overline{u\rho'}$, $\overline{w\rho'}$, and $\overline{\rho'^2}$; the normalisations used in this figure result in universal profiles, valid for any h and U(h). Since concentration fluctuations can be obtained from $\rho' = \Delta\rho C'$ (for $\Delta\rho/\rho_A$ small), correlations involving concentrations, rather than densities, will be as shown in Figure 4. This interpretation of that figure illustrates more clearly the reduction in turbulent mixing due to stability.

The results given in Figures 3-4, although dependent upon empirical specification of the length scale function, do at least indicate the

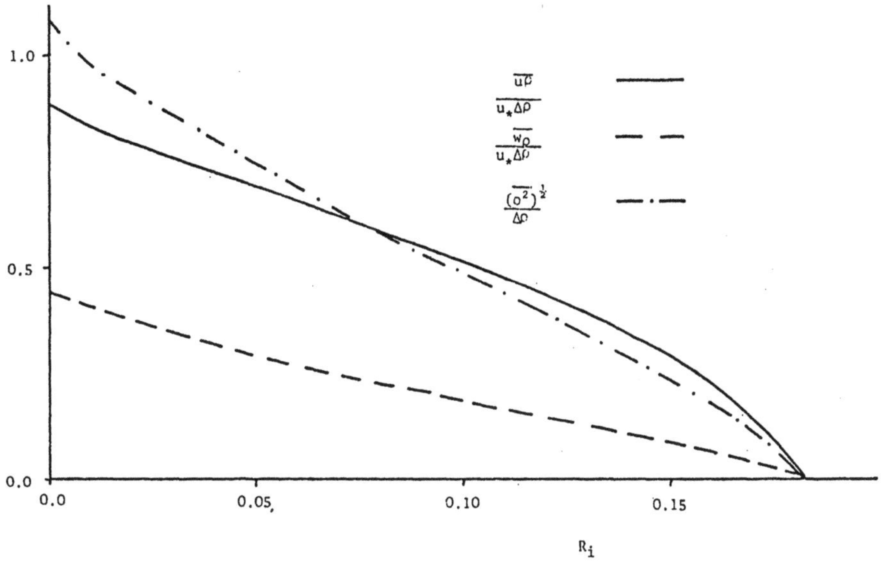

Figure 4. Universal Variation of Normalised Correlations with Stability Parameter

potential of ASM in heavy gas modelling. In principle, it should be possible to feed back the stresses and correlations into the mean flow equations, thus obviating the need for the specification of isotropic diffusivities. The only other requirement is a consistent k-ε model to compute these quantities for insertion into the algebraic equations.

4. APPLICATIONS OF THE COMPUTER PROGRAM HEAVYGAS

4.1 Turbulence Modelling Used

The Atkins R&D computer program HEAVYGAS has been developed to provide a 3D prediction method for complex heavy gas dispersion problems in which such effects as obstructions, terrain irregularities etc, are present. The current version of the program incorporates a modified form of the k-ε turbulence model, as described in 2.2. The main effects of the heavy gas on the equations of motion are to put an additional body force in the mean momentum equation, and turbulence suppression terms, proportional to $\frac{\partial\rho}{\partial z}$, in the k and ε equations. The particular form of the modelling of these suppression terms is that given by Rodi[8].

4.2 Gas Dispersion by Water Spray

This work has recently been reported in more detail elsewhere[14]. In order to show the the agreement between experimental and predicted results, one of the three configurations considered will be presented here in detail.

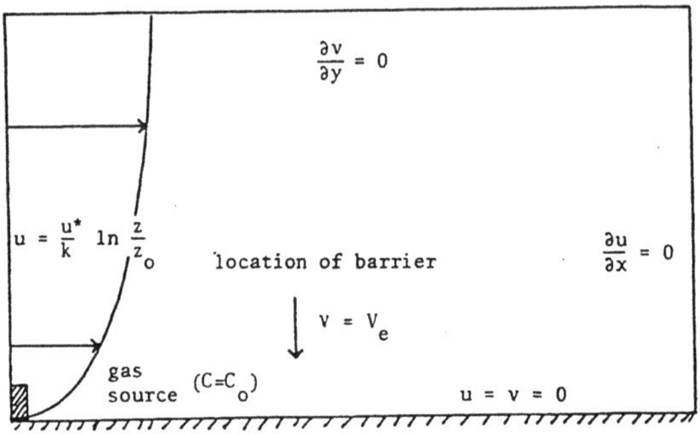

Figure 5. Typical Flow Domain

The experimental set-up consisted of a water spray barrier pointing 45° forwards into the wind and entraining a large volume of air. A section through the plane of symmetry of the computational domain is shown in Figure 5. The spray is only treated as a source of momentum, and the velocity specified at the point at which the spray is effectively acting is that computed by Moodie et al(15), who undertook the measurements. 1.4 Kg/s of CO_2 was released from a 'point source' 15m upwind of the barrier. The 10m windspeed was 2.7 m/s, and the entrainment velocity of 3.6 m/s acted over the barrier width of approximately 30m. Unfortunately, only 3 gas sensors, each at 0.5m above the ground, were operational.

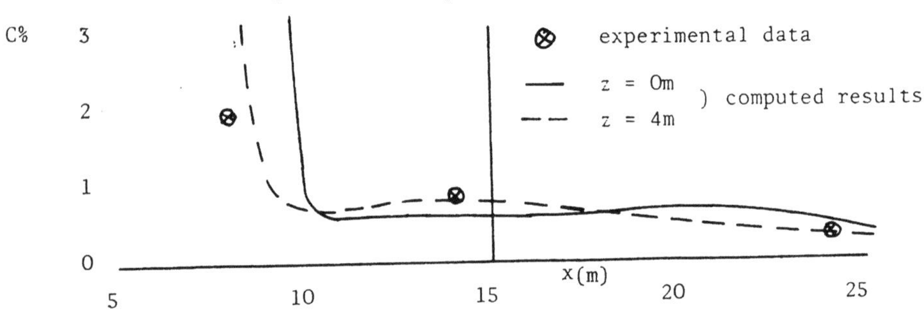

Figure 6. Concentration Profiles at 0.5m Height (Run 19)

The comparison between computed and experimental results is shown in Figure 6, in which z represents lateral distance from the centre-line. The agreement on this figure is good, and the method of dispersal can be clearly seen in Figure 7; the entrained air forms an upwind recirculation which both lifts and mixes the gas, which had previously been hugging the ground. Velocity vectors are shown since streamlines are not defined in 3D flows.

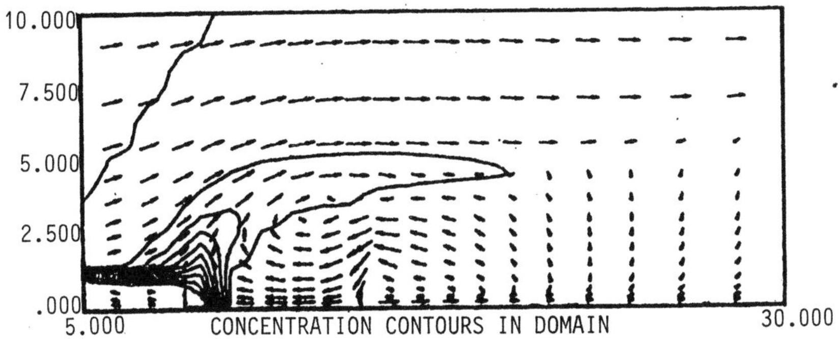

Figure 7. Computer Output (1% Contours)

4.3 Predictions for Thorney Island Phase 2

The second phase of the Thorney Island trials(6) is concerned with dispersion of heavy gases in the presence of buildings and fences. Although the primary objective is to obtain data against which to validate wind tunnel tests, the data will also be particularly useful for validation of the program HEAVYGAS. Indeed, the results presented here were produced in response to the HSE's request for model predictions.

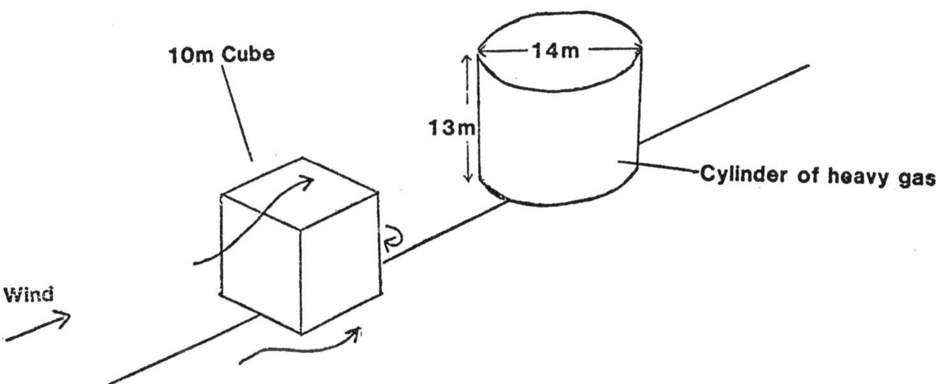

Figure 8. Test Case from Thorney Island Phase 2 Trials

The configuration under consideration is shown in Figure 8; the wind speed is 2 m/s, the edge of the building is 20m upwind of the edge of the gas tent, and the relative density of the gas released is 2. Since the release is instantaneous, it is anticipated that the wakes, not only of the building, but also of the tent, will be significant in the early stages of cloud development. A steady state wind flow solution was therefore obtained with the tent in place before releasing the gas into a transient computation which enabled the progress of the cloud to be tracked.

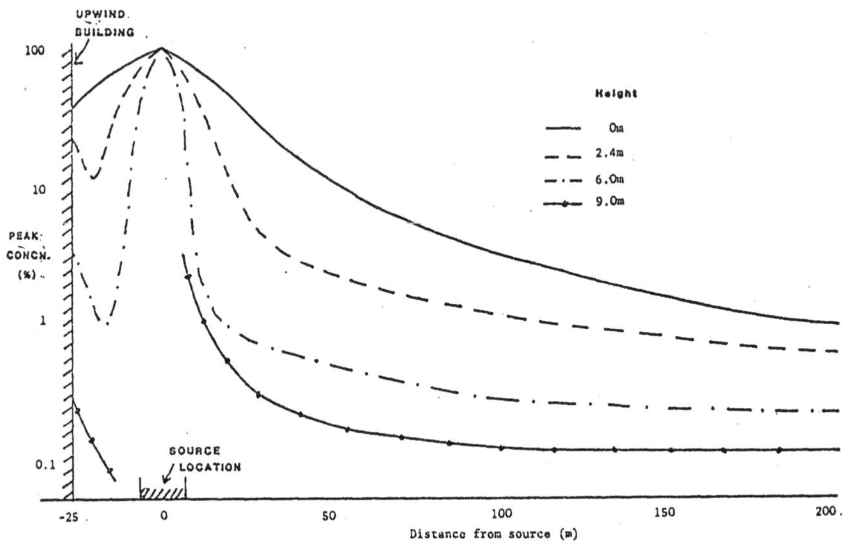

Figure 9. Thorney Island Phase 2 Predictions

The results are presented in Figure 9 in the form requested by the HSE. This was obtained by plotting a time history of the results at each grid node, then picking off the maximum concentration. The effects of the upwind building can be clearly seen and could not be predicted by simpler box-type models. The building causes not only a blocking effect, but also tends to entrain the gas upwards within its wake. This implies that concentrations will remain relatively high in this region for some time, and the entrained gas will be slowly picked up by the ambient wind to feed more of the heavy gas into the cloud which has developed downwind.

5. CONCLUSIONS

a) It appears that algebraic stress modelling is necessary in order to provide both more realistic modelling, and also better information on the structure of dispersing heavy gas clouds.

b) HEAVYGAS in its current form predicts, at least qualitatively, the effects of flow modifications on the dispersion of heavy gases. Quantitative validation will be more feasible when results emerge from Phase 2 of the Thorney Island trials.

c) Future development of HEAVYGAS will be directed towards the incorporation of ASM, and the development of an efficient algorithm for solution of the modified system of equations.

REFERENCES

1) 'Proposal for a Council Directive on the Major Accident Hazards of Certain Industrial Activities' EEC Official Journal L230, 1982.

2) Blackmore, D R, Herman, M N and Woodward, J L 'Heavy Gas Dispersion Models' J Hazardous Materials, Vol 6, pp 107-128, 1982.

3) Chatwin, P C 'The Use of Statistics in Describing and Predicting the Effects of Dispersing Gas Clouds' J Hazardous Materials, Vol 6, pp 213-230, 1982.

4) Gibson, M M and Launder, B E 'Ground Effects on Pressure Fluctuations in the Atmospheric Boundary Layer' J Fluid Mech, Vol 86, p 491, 1978.

5) Moult, A and Dean R B 'CAFE a Computer Program to Calculate the Flow Environment' CAD80, 4th Int Conf on Computers in Design Engineering, Brighton, 1980.

6) McQuaid, J 'Future Directions of Dense-gas Dispersion Research' J Hazardous Materials, Vol 6, pp 231-247, 1982.

7) Launder, B E, Reece, G J Rodi, W 'Progress in the Development of a Reynolds Stress Turbulence Closure' J Fluid Mech 68, pp 537-566, 1975.

8) Rodi, W 'Turbulence Models and Their Application in Hydrualics' State-of-the-art paper, IAHR, 1980.

9) El Tahry, S and Gosman, A D 'The two- and three-dimensional Dispersal of a Passive Scalar in a Turbulent Boundary Layer' Int J Heat & Mass Transfer, Vol 24, pp 35-46, 1981.

10) Smith, T J and Takhar, H S 'Stratification Effects on the Turbulent Transport of Mass and Momentum' Report, Simon Engineering Labs, Univ of Manchester, Sept 1976.

11) Meroney, R N 'An algebraic Stress Model for Stratified Turbulent Shear Flows' Computers and Fluids, Vol 4, pp 93-107, 1976.

12) Deaves, D M 'Application of a Turbulence Flow Model to Heavy Gas Dispersion in Complex Situations' 2nd Battelle Symposium on Heavy Gas and Risk Assessment, Frankfurt, 1982.

13) Van Ulden, A P 'On the Spreading of a Heavy Gas Released Near the Ground' 1st Int Loss Symposium, The Hague, 1974.

14) Deaves, D M 'Experimental and Computational Assessment of Full-scale Water Spray Barriers for Dispersing Dense Gases' 4th International Symposium on Loss Prevention in the Process Industries, September 1983.

15) Moodie, K, Taylor, G and Beckett, H HSE, Research and Laboratory Services Division, Internal Report IR/L/FM/81/3, 1981.

Recent Results in Simulating LNG Vapor Dispersion over Variable Terrain*

Stevens T. Chan and Donald L. Ermak

Lawrence Livermore National Laboratory
University of California
Livermore, California 94550, U.S.A.

Summary

The performance of a three-dimensional, conservation equation model (FEM3) is assessed herein via simulating three distinctly different, approximately 30 m^3 LNG spill experiments. In general, good agreement between model predictions and field measurements was observed and many important features associated with heavy gas dispersion were successfully reproduced.

Introduction

During the past several years, the Lawrence Livermore National Laboratory (LLNL), under the sponsorship of the U.S. Department of Energy, has been conducting safety research related to the possible consequences of liquefied natural gas (LNG) spills. Under this program, LLNL and the Naval Weapons Center (NWC) jointly conducted the Burro and Coyote series of LNG spill experiments at China Lake, California during 1980 and 1981 respectively (see [1] and [2]). As part of the program, LLNL has also been developing numerical models for predicting the vapor dispersion from LNG spills. FEM3 is one of such recently developed models.

An early version of the FEM3 model was applied to simulate several of the Burro series experiments and yielded results that correlated quite well with field data [3]. In particular, the model successfully predicted the bifurcated structure of the vapor cloud in Burro 8, which was conducted under calm wind conditions. Since then FEM3 has been extended to treat variable terrain, together with an improved K-theory turbulence submodel to account for density stratification and ground heat transfer effects. It was demonstrated subsequently that, even with the gentle terrain at the test site, its effects on the dispersing vapor cloud were quite profound, especially in the gravity-flow dominated regimes such as that of Burro 8 [4].

In the present paper, we further assess the performance of the FEM3 model, using data from both the Coyote and Burro series. The tests selected for such purpose include Burro 8, Burro 9, and Coyote 5, which were conducted under very

*This work was performed under the auspices of the U. S. Department of Energy by the Lawrence Livermore National Laboratory under contract No. W-7405-Eng-48.

different atmospheric conditions. The two Burro tests are reexamined here for two reasons: (1) some of the field data had been updated recently, and (2) some additional, and perhaps more appropriate means for comparing model predictions and data are considered. Model validations using the Coyote data are being conducted for the first time but, for brevity, only the results of Coyote 5 are presented in this report.

The Numerical Model

The present model is based on solving the three-dimensional, time-dependent conservation equations, of which a generalized anelastic approximation is employed to accommodate large density variations in both space and time and yet to preclude sound waves. These governing equations are:

$$\frac{\partial(\rho\underline{u})}{\partial t} + \rho\underline{u} \cdot \nabla\underline{u} = -\nabla p + \nabla \cdot (\rho\underline{\underline{K}}^m \cdot \nabla\underline{u}) + (\rho - \rho_h)\underline{g}, \tag{1}$$

$$\nabla \cdot (\rho\underline{u}) = 0, \tag{2}$$

$$\frac{\partial\theta}{\partial t} + \underline{u} \cdot \nabla\theta = \frac{1}{\rho C_P} \nabla \cdot (\rho C_P \underline{\underline{K}}^\theta \cdot \nabla\theta) + \frac{C_{PN} - C_{PA}}{C_P} (\underline{\underline{K}}^\omega \cdot \nabla\omega) \cdot \nabla\theta, \tag{3}$$

$$\frac{\partial\omega}{\partial t} + \underline{u} \cdot \nabla\omega = \frac{1}{\rho} \nabla \cdot (\rho\underline{\underline{K}}^\omega \cdot \nabla\omega), \tag{4}$$

and

$$\rho = \frac{PM}{RT} = \frac{P}{RT\left(\frac{\omega}{M_N} + \frac{1-\omega}{M_A}\right)}, \tag{5}$$

where $\underline{u} = (u, v, w)$ is the velocity, ρ is the density of the mixture, p is the pressure deviation from an adiabatic atmosphere at rest, with corresponding density ρ_h, $\underline{g}$ is the acceleration due to gravity, θ is the potential temperature deviation from an adiabatic atmosphere at θ_o, ω is the mass fraction of NG vapor, and $\underline{\underline{K}}^m$, $\underline{\underline{K}}^\theta$, and $\underline{\underline{K}}^\omega$ are the eddy diffusion tensors for momentum, energy, and NG vapor, C_{PN}, C_{PA}, and $C_P = \omega\, C_{PN} + (1-\omega)\, C_{PA}$ are the specific heats for NG vapor, air, and the mixture, respectively. In the equation of state, P is the absolute pressure, R is the universal gas constant, M_N, M_A are the molecular weights of NG and air, and T is the absolute temperature.

The above set of equations, together with appropriate initial and boundary conditions, are solved via a modified finite element method. Detailed description of the model and its applications can be found in Chan, et al. [4] and in Chan [5]; the related numerical techniques were described in Gresho, et al. [6].

Model-Data Comparisons

In this section, we compare model predictions with field data for the three selected spill tests, for which the spill and meteorological conditions are summarized in Table 1. Attention is given here to the downwind distance to the lower flammability limit (LFL), i.e., 5% volume fraction, the time variation of concentration at some specific locations, the size and shape of the concentration contours, and the maximum concentration as a function of downwind distance. More thorough comparisons will be given in Morgan et al. [7].

Table 1. Spill and Meteorological Parameters

Parameter	Burro 8	Burro 9	Coyote 5
Spill Volume (m^3)	28.4	24.2	28.0
Spill Rate (m^3/min)	16.0	18.4	17.1
Wind Speed at 2 m (m/s)	1.8±0.3	5.7±0.7	9.7±1.3
Friction Velocity, Diabatically Adjusted (m/s)	0.074	0.252	0.437
Momentum Diffusivity at 2 m, Diabatically Adjusted (m^2/s)	0.037	0.212	0.426
Richardson Number at 2 m	+0.121	-0.014	-0.076
Monin-Obukhov Length Scale (m)	+16.5	-140	-26.5
Atmospheric Stability	Slightly stable	Neutral	Slightly unstable

Perhaps one of the more basic comparisons that can be made between model simulations and experiments is the time history data of concentration at individual sensor locations. Time history data, however, are generally quite sensitive to the location within the cloud due to the effects of cloud meander, statistical variations, and vapor source generation fluctuations, all of which are not treated in the present model. In order to minimize these effects, the predicted time-history curves of gas concentration at a specified height and downwind distance on the center plane are compared with the time history data of all the operating sensors on a crosswind arc for the same elevation and downwind distance.

The concentration contours on horizontal and crosswind planes provide a more global view of the dispersing vapor cloud, its time-varying size and shape, and particularly the extent of the flammable zone. However, due to the sparseness of field measurements, some uncertainties obviously exist in the generated experimental contours. For instance, the uncertainty in the distance to the LFL was estimated to be approximately ±15% for all the Burro and Coyote tests [7].

Another type of useful comparison is the maximum concentration as a function of downwind distance. In taking the maxima, a particular value of downwind distance is chosen, while time and location on the crosswind plane (for convenience, the crosswind plane is taken to be a cylindrical surface in the experiments and a plane normal or nearly normal to the mean wind direction in the numerical simulations) are varied to find a maximum value of concentration. Such comparisons are useful because cloud meandering and certain statistical variations are taken into account.

Coyote 5

Coyote 5 was conducted under high wind and slightly unstable atmospheric conditions with significant wind meandering (±7°) during the spill test. Near the end of the spill, a relatively large rapid phase transition (RPT) explosion was recorded. This test was simulated with a flat terrain assumption.

In Fig. 1 the concentration contours on three crosswind sections are compared. Taking into account the effects of cloud meandering, the present model apparently predicts the overall cloud shape and sizes rather well. The greatest discrepancy appears to be at x = 140 m (in maximum concentration) and is probably attributable to RPTs and other effects such as vapor source generation fluctuations and wind meandering, which are not numerically modelled.

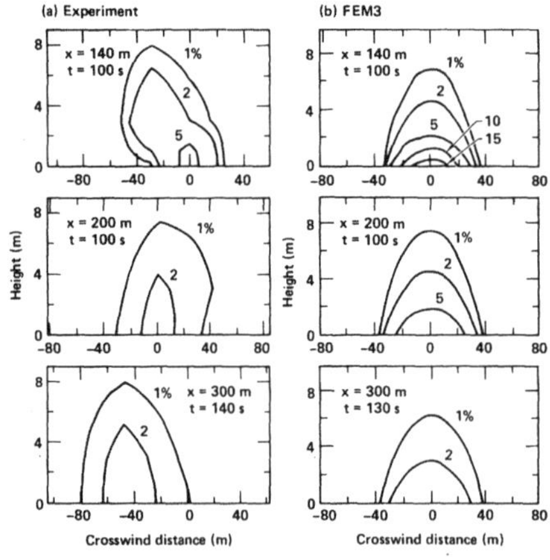

Fig. 1. Coyote 5 crosswind contour plots of concentration at various downwind locations and times.

The maximum concentrations versus downwind distance are compared in Fig. 2. Two experimental curves were plotted, one with data from all sensors and the other constructed only with data unaffected by RPTs. The agreement is generally quite good. The agreement on the downwind distance to the LFL, 260 m numerically vs. 210 m experimentally, is also reasonable.

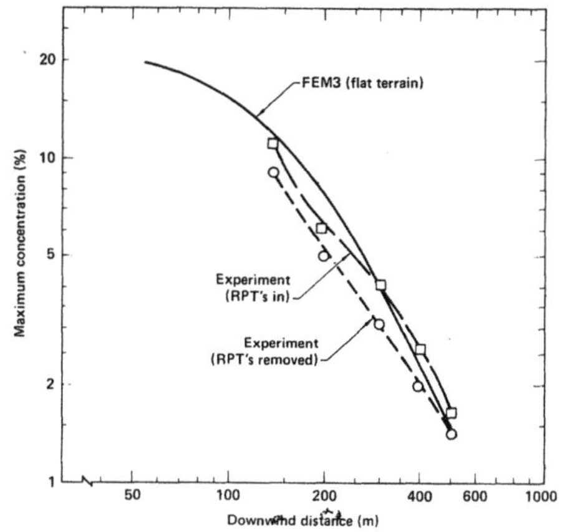

Fig. 2. Coyote 5 maximum concentrations versus downwind distance.

Burro 9

The Burro 9 test conditions include a high spill rate, moderate and fairly steady wind speed, and neutral atmospheric stability. However, a series of RPT explosions occurred and rendered some of the data unreliable. This test was simulated with both a flat terrain and a simulated variable terrain of the test site.

In Fig. 3, FEM3 results for two specific locations (x=140 m and 400 m, respectively; both at z=1 m) on the cloud centerline are compared with the respective time history data of gas sensors. Since the vapor cloud of Burro 9 did not meander very much, it is

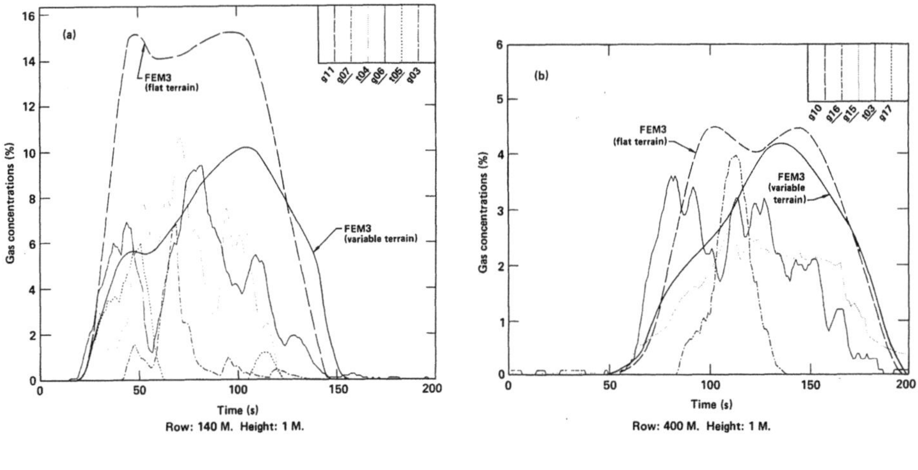

Fig. 3. Burro 9 time history data of gas concentration with FEM3 results superimposed.

appropriate in this case to compare the predicted time history curves with data from the corresponding gas sensors, G06 and T03 respectively. In general, the agreement between variable terrain results and field data is very good; the results from the case of flat terrain are satisfactory except the peak value for x = 140 m was overpredicted by approximately 5% in volume fraction.

In Fig. 4, the variable terrain results of concentration contours 1 m above ground are compared with field data, at the times when the downwind distance to the LFL reaches its maximum value. Despite the difference in times (80 s vs. 120 s), the agreement between model predictions and field measurements is remarkably well regarding the cloud size, shape, and the slight deflection of the vapor cloud toward the lower terrain. The discrepancy in timing is probably due to two main reasons: (1) the variations of the ambient wind speed (± 0.7 m/s) during the spill test, and (2) the fluctuations in vapor source generation, both of which are not treated in the present model.

The comparison of maximum concentrations as a function of downwind distance is shown in Fig. 5, with the 57-m row data omitted because of RPT explosions. Overall the agreement is very good, with the variable terrain results being closer to the experimental curve with RPTs removed. The maximum downwind distance to the LFL is approximately 325 m at t ≈ 80 s in the experiment; the calculated results are approximately 350 m at t ≈ 90 s for the case with flat terrain, and 330 m at t ≈ 120 s for the case with variable terrain.

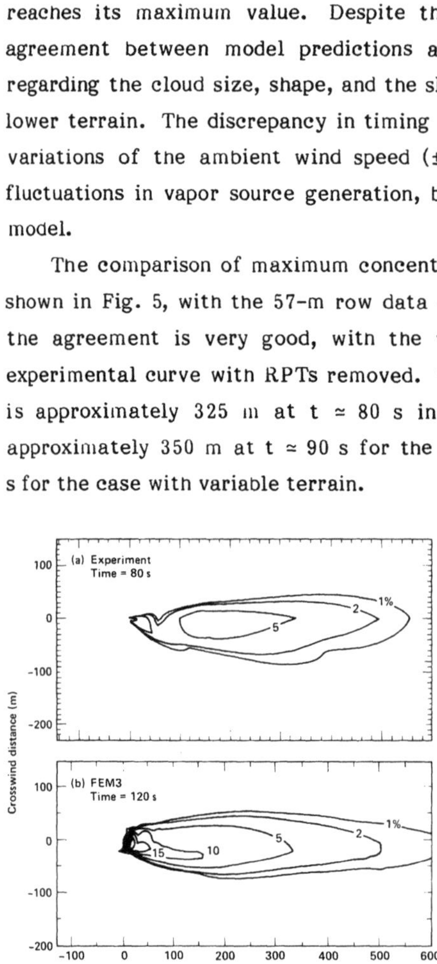

Fig. 4. Burro 9 contour plots of concentration 1 m above ground at equilibrium state.

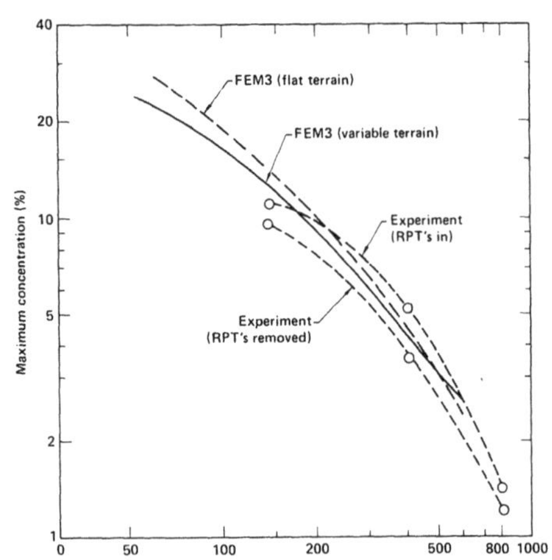

Fig. 5. Burro 9 maximum concentrations versus downwind distance.

Burro 8

Burro 8 was conducted under calm wind and slightly stable atmospheric conditions. The Burro 8 test is especially interesting because the very low wind speed permitted the gravity flow of the cold, dense gas to become so dominant that the flow within the cloud was almost "decoupled" from the ambient atmospheric boundary layer. The cloud spread in all directions including upwind, developed a very distinct bifurcated structure, and lingered over the source region for more than 100 seconds after the spill was terminated.

In Fig. 6, the time histories of gas concentration for sensors G07 and G11 are compared. These gas sensors are located within the larger lobe of the vapor cloud. The agreement between model predictions and data for G07 is excellent except, for late times (t>250 s), the vapor cloud appears to linger at least 100 seconds longer than predicted. The agreement with G11 is quite good for cloud arrival time and for concentrations less than 15%; the agreement for maximum concentration is probably reasonable. Unfortunately, this sensor was saturated at about 16% and the actual peak value was not available. The discrepancies in cloud departure times are probably caused by a combination of the following factors: (1) the ambient wind speed was decreasing in a fairly steady fashion, by about 30%, over the entire duration of the test, which indeed was not modelled, (2) the actual vapor generation rate probably fluctuated considerably during the test (as opposed to a uniform rate assumed in the model) and the actual duration of vapor generation might be somewhat longer than assumed in the simulation, and (3) the present turbulence and ground heat transfer submodels might not be sufficiently adequate for the flow regime being simulated.

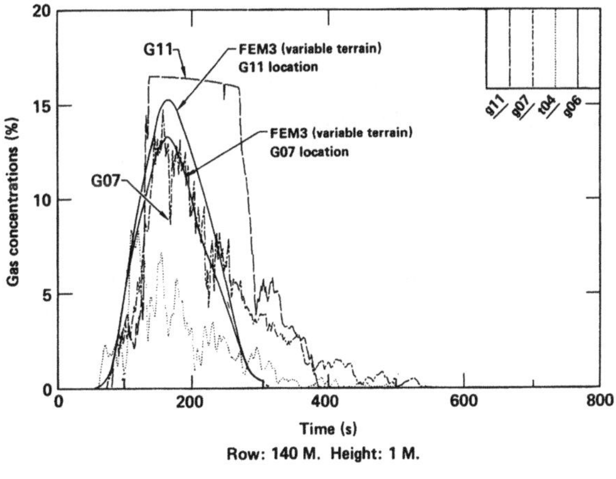

Fig. 6. Burro 8 time history data of gas concentration for sensors G07 and G11 with FEM3 results superimposed.

The concentration contours at a height 1 m above the ground and t = 180 s are compared in Fig. 7, where the dashed lines indicate the edges of the instrument array. The numerical results are in fair agreement with the data with respect to the downwind extent of the concentration contours, although the shapes of some of the contours differ significantly. There is considerable bifurcation in the experimental contours. In the numerical simulation, while the 5% contour shows only a hint of bifurcation, the higher concentration contours are definitely bifurcated and seem to agree rather well with the data. The difference between the numerical results and the data is believed partly due to the reasons stated above and partly due to the lack of data between the 140-m and 400-m rows.

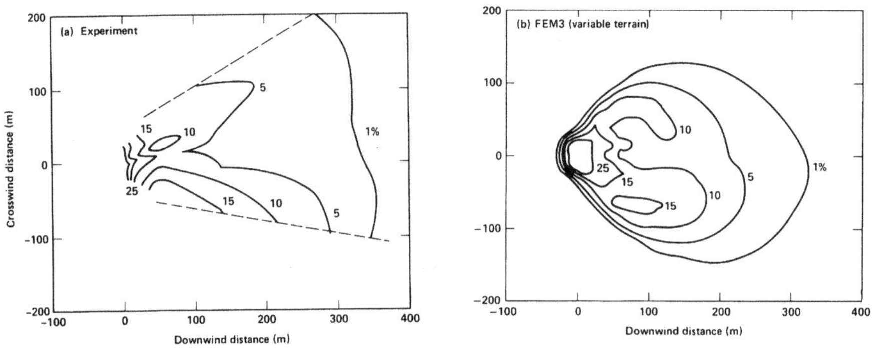

Fig. 7. Burro 8 contour plots of concentration 1 m above ground at t = 180 s.

The crosswind concentration contours for the same time at a downwind distance of 140 m are compared in Fig. 8. Overall, both numerical results are in fairly good agreement with the data regarding cloud height and cloud bifurcation. As expected, the variable terrain results do have better detailed agreement, namely, a higher left lobe and the absence of a 15% contour in the right lobe.

In Fig. 9 the maximum concentrations versus downwind distance are compared. In general the agreement is good, with the variable terrain results being somewhat lower than the other two curves. Besides the aforementioned reasons, some of the discrepancies are probably attributable to the statistical variations (in time) of the concentration data, for which the experimental curve represents an upper bound. The inconsistency adopted in defining the crosswind plane in the experiment and the numerical simulations might also have some effects, especially for a highly bifurcated vapor cloud such as Burro 8.

The predicted maximum downwind distance to the LFL is approximately 470 m at t ≃ 310 s from the flat terrain simulation, and is about 400 m at t ≃ 300 s from the variable terrain simulation. In the experiment, based on the concentration contour plots (which are somewhat incomplete because the vapor cloud extended well beyond the edges of the sensor array), the best estimate appears to be about 440 m at t ≃ 400 s. In view of the uncertainties associated with interpolating the field data and various numerical simplifications employed in the numerical model, the agreement is apparently satisfactory.

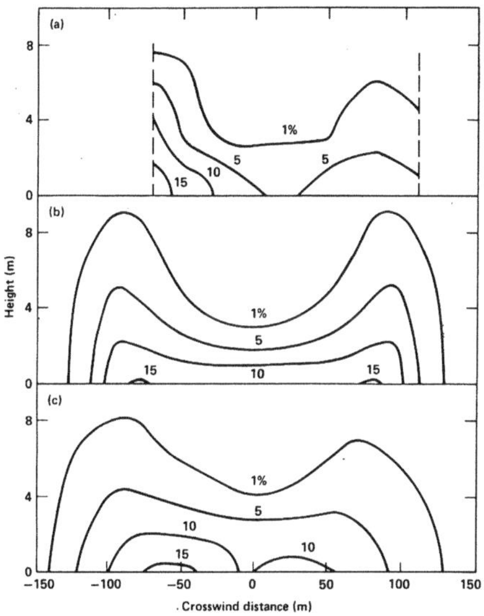

Fig. 8. Burro 8 crosswind contour plots of concentration 140 m downwind at t = 180 s: (a) experiment; (b) flat terrain simulation; (c) variable terrain simulation.

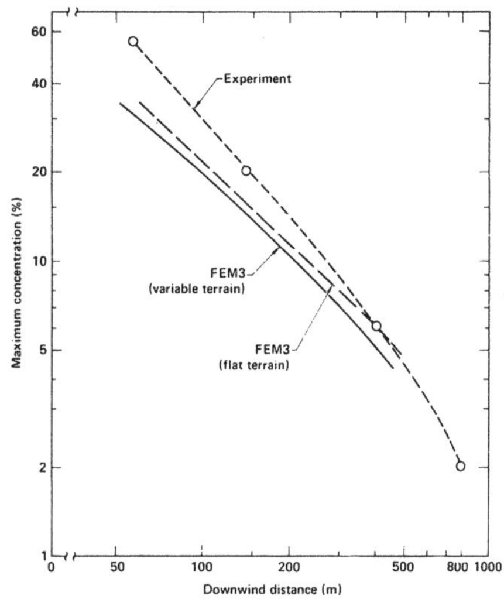

Fig. 9. Burro 8 maximum concentrations versus downwind distance.

Concluding Remarks

In this paper, the FEM model has been further assessed via simulating three distinctly different LNG spill experiments. In general, good agreement between model predictions and field measurements was observed. In particular, the overall results obtained in variable terrain simulations were shown to correlate much better with the field data. Many important features of the dispersing cloud observed under light wind conditions, including spreading of the vapor cloud in all directions, cloud bifurcation, and cloud deflection due to sloping terrain, were all successfully reproduced.

Thus far, predictions of the present model have been compared only with field tests representing a limited range of spill scenarios. Additional comparisons with well-instrumented, larger scale dispersion experiments are needed before the model can be applied with confidence to simulate the atmospheric dispersion of a wide variety of heavy gas spills of practical interest.

References

1. Koopman, R. P., R. T. Cederwall, D. L. Ermak, H. C. Goldwire, Jr., J. W. McClure, T. G. McRae, D. L. Morgan, H. C. Rodean, and J. W. Shinn, "Analysis of Burro Series 40-m^3 LNG Spill Experiments, Dense Gas Dispersion, Elsevier Scientific Publishing Co., Amsterdam, 1982, pp. 43-84.

2. Goldwire, H. C., Jr., Rodean, H. C., Cederwall, R. T., Kansa, E. J., Koopman, R. P., McClure, J. W., McRae, T. G., Morris, L. K., Kamppinen, L., Kiefer, R. D., Lind, C. D., and Urtiew, P. A., "Coyote Series Data Report, LLNL/NWC 1981 LNG Spill Tests: Dispersion, Vapor Burn, and Rapid Phase Transition", Lawrence Livermore National Laboratory, Livermore, CA, UCID-(to be published), 1983.

3. Ermak, D. L., S. T. Chan, D. L. Morgan, and L. K. Morris, "A Comparison of Dense Gas Dispersion Model Simulations with Burro Series LNG Spill Test Results," Dense Gas Dispersion, Elsevier Scientific Publishing Co., Amsterdam, 1982, pp. 129-160.

4. Chan, S. T., H. C. Rodean, and D. L. Ermak, "Numerical Simulations of Atmospheric Releases of Heavy Gases Over Variable Terrain," Lawrence Livermore National Laboratory, Livermore, CA, UCRL-87256, 1982; also in Proc. NATO/CCMS 13th International Technical Meeting on Air Pollution Modeling and Application, Ile d'Embiez (Toulon), France, 1982.

5. Chan, S. T., "FEM3 - A Finite Element Model for the Simulation of Heavy Gas Dispersion and Incompressible Flow: User's Manual," Lawrence Livermore National Laboratory, Livermore, CA, UCRL-53397, 1983.

6. Gresho, P. M., S. T. Chan, C. D. Upson, and R. L. Lee, "A Modified Finite Element Method for Solving the Time-Dependent, Incompressible Navier-Stokes Equations," Lawrence Livermore National Laboratory, Livermore, CA, UCRL-88937, 1983; to appear in Int. J. Num. Meth. Fluids.

7. Morgan Jr., D. L., S. T. Chan, L. K. Morris, et al., "Data Analysis and Model-Data Comparison of Heavy Gas Dispersion in the Burro and Coyote 40-m^3 LNG Spill Experiments," Lawrence Livermore National Laboratory, Livermore, CA, UCRL-(to be published), 1983.

On the Dilution of a Dense Gas Plume: Investigation of the Effect of Surface Mounted Obstacles

N.O. JENSEN

Meteorology Section
Physics Department
Risø National Laboratory,
DK-4000 Roskilde, Denmark

Summary
The various stages and mechanics in the formation of a dense plume resulting from a continous flashing of a pressurized liquid gas is described. Further some simple considerations is offered regarding the effect of a simple porous two-dimensional obstacle on such a plume. Finally a limited set of full scale experimental results, in which CO_2 is used as a model gas, is presented.

Introduction

In the assessment of potential hazards from inadvertent releases of e.g. chlorine, a number of uncertainties exists. Examples from the extreme ends of the course of an incident are: the uncertainty related to the effective gas release rate associated with various leakage geometry (blow-down, two-phase flow, aerosol formation): and the uncertainty related to physiological effect-dose relationships. These uncertainties are normally dealt with by making conservative assumptions. For the phases in between the two, the dispersion phase, formulae are available for calculation of dispersion over flat even terrain. In presence of obstacles (which in practice is the usual case) little guidance can presently be given. Furthermore it is not possible even in principle to take a conservative point of view, since obstructions of the order of the plume dimensions can either enhance or diminish the dosage locally. To work towards a solution of this problem, a project has been initiated, in which the obstruction used is a usual shelter fence placed perpendicular to the plume axis, and two-dimensional in the sense that it is wider that the cross wind dimension of the plume. The advantage of starting out with an obstacle of this type is that its aerodynamic characteristics are well documented: for an early account see Jensen [1], and results from recent experiments are given by Bradley and Mulhearn [2].

Plume Stages and Mechanics

Wanting the tests to be as realistic as possible dictated that the heavy plume should be obtained through blow-down from the liquid phase of pressurized gas. Fig. 1 shows the various stages which may be identified in such a release.

Immediately outside the point of release a fraction: typically ~ 20% of the liquid flashes. The heat of vapourization is delivered by the remaining liquid the temperature of which is brought down to the boiling point at atmospheric pressure, At the same time the vigour of the boiling process causes the remaining liquid to be shattered into fine droplets. The diameter of these droplets is supposed to be quite small; Thus according to Resplandy [3] typically 10 μm. The sudden evaporation causes the jet cross sectional area to widen substantially, typically 100 times (≈ ρ_{liq} × (molar volume/molecular weight) × fraction of liquid flashed) corresponding to a 10 times increase in jet diameter in the outflow zone. The phenomenon is clearly visible on the close-up photo (fig. 2) of the release conditions in the present experiments.

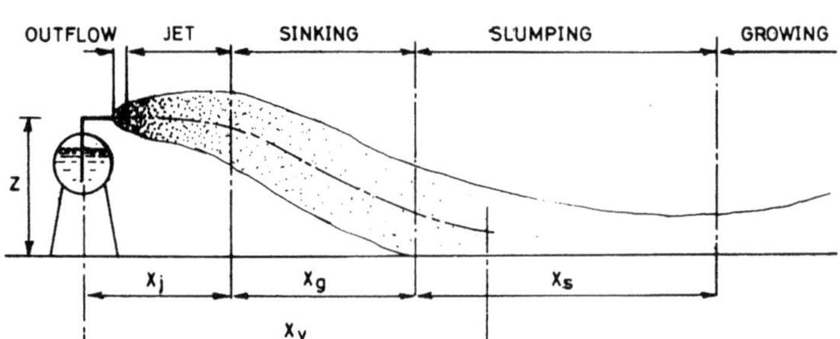

Fig. 1 Blow-down from the liquid phase of a pressurized gas, with ensuing vapour/aerosol jet. Various stages in the development of the resulting heavy gas plume is identified. Even if the gas itself is lighter than air, evaporative cooling of entrained air will ensure that the density of the plume will remain larger than the ambient air density for some time.

Fig. 2. The close-up shows the 1/2" release pipe and the rapid expansion of the jet diameter in the outflow region. The photo also gives an impression of the complete obscuration of the jet due to the CO_2-aerosol.

The next phase, which might be identified is one in which the jet momentum is dominating over the momentum of the ambient flow. Here we will only consider the case in which the jet extends in the downwind direction. A suitable criterion for the length of this jet stage is

$$x_j = \left(\frac{q\, v_o}{\pi\, \rho_{air}}\right)^{1/2} (\alpha\bar{u})^{-1} \qquad (1)$$

in which q is the release rate (mass/unit time) of liquid, v_o is the release velocity (discussed below), $\bar{u}$ is the ambient wind velocity and α is the jet entrainment coefficient (jet radius = αx). A fuller treatment of the jet stage would naturally have to recognize the mixture of jet and ambient turbulence, the translational effect of the ambient wind ($x \rightarrow x-\bar{u}t$) and that entrainment causes drop evaporation with further expansion of the jet as a result.

Depending on the height of the release point z above the ground the jet might impact on the surface before the above development is completed or it might be high enough above the ground not to have made contact at $x = x_j$. In the latter case its centerline will start a descend due to gravity. The distance from x_j to the point of tuch down can be estimated from

$$x_g = \left(\frac{x^3\bar{u}^3}{gq/\rho_{air}}\right)^{1/2} \qquad (2)$$

where the r.h.s. symbols have the same meaning as defined above. After this point gravity causes a gradual downwind flattening of the plume during which stage the plume widens laterally. If entrainment during this stage is neglected, height h times width y is preserved whereby it may be derived [4] that

$$y \simeq 2(gq/\rho_{air})^{1/3}\, x^{2/3}/\bar{u} \qquad (3)$$

where x is the distance from the tuch down point. At further downwind distances the decrease in h due to gravity slumping is gradually off-set by entrainment of air through the top of the plume. This process is governed by the Richardson number $Ri = g'h/u_*^2$ where g' is defined as $g(\rho-\rho_{air}/\rho)$ where ρ is the local plume density. Assuming that the entrainment velocity is given by an expression of the form u_*Ri^{-1}, balance between slumping and entrainment exists when u_*Ri^{-1} equals $dh/dt \sim (h/y)$ dy/dt, which by use of (3) is seen to correspond to a distance

$$x_s \sim q^{1/7} \left(\frac{q}{\rho_{air}}\right)^{4/7} \left(\frac{v_o}{u^3}\right)^{3/7} . \qquad (4)$$

The main point of eq. (4) is that it shows that x_s is rather insensitive to the ambient conditions. This is in contrast to the other lengths discussed above.

The last length to be discussed in this section is the visible length of the plume, x_v. The problem here is a little different as it involves thermodynamic properties. If

$$k = \left(\frac{fL}{\Delta T c_p} \frac{\bar{u}}{v_o}\right)^{1/2} < 1, \qquad (5)$$

where f is the initial fraction that remains in the liquid phase; L is the heat of vapourization; ΔT is the difference between the ambient temperature and the boiling temperature of the liquid; and c_p is the heat capacity of the entrained air - the aerosol will evaporate on a distance of the order $x_v = kx_j$. If the conditions are such that $\bar{u}$ is too large or the ambient temperature too small for (5) to be fulfilled the visible length of the plume may be substantially longer than x_j. This is because, when the plume has lost its excess momentum the rate of spread becomes less, whereby the required diameter for complete evaporation (assuming similar ambient temperature conditions) occurs at a longer distance downwind (see fig. 3). Differences in the length of the visible plume are illustrated in figs. (6) and (7).

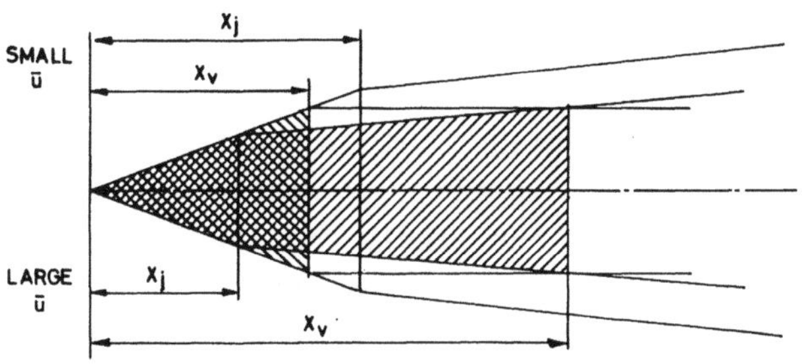

Fig. 3. For small values of u the aerosol may have a chance to evaporate over a distance $x_v < x_j$. For thermodynamic properties fixed, a larger windspeed will result in a shorter distance over which the jet dominates the ambient momentum, the entrainment over a given distance less, and therefore, x_v larger than before.

The Theoretical Effect of Simple Obstacles

The primary influence of an obstacle on the flow may be taken as the streamline distortion it causes. However, what we are interested in here is its effect on enhanced diffusion across streamlines. The situation we want to consider is depicted in fig. 4 which also gives some essential definitions.

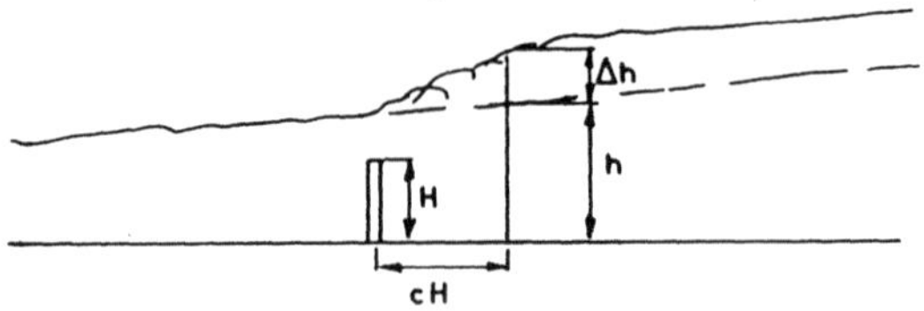

Fig. 4. Over a distance of O(H) the depth of the plume increases as a result of extra turbulence introduced by the fence.

The basic idea is that the fence introduces extra turbulence, which in turn causes a change in the vertical entrainment velocity through its influence on the Richardson number.

Assuming this effect is large enough to locally make Ri (defined earlier) less than unity, the diffusion becomes passive: $dh/dx \sim u_*/\bar{u}$. Thus over a length scale $\Delta x \sim H$ where this effect may be assumed to exist, the effect on h may be estimated as

$$\Delta h = a \frac{u_*}{u} H \tag{6}$$

where the constant a (which mainly accounts for the increase in u_* and decrease in $\bar{u}$ in the near wake) is perhaps somewhat larger than unity, but likely smaller than O(10). If the initial Ri-number is quite large so that it will remain large even immediately behind the fence the above estimate on Δh will be reduced by a factor of Ri^{-1}.

Regarding lateral enhancement of spread, it is probably of order Δh. Thus with y generally quite a lot larger than h due to gravity spread the absolute effect on the size of y may be neglected. Thus an estimate of the effect on the gas concentration may be gained from

$$\frac{\chi}{\chi_o} = \frac{Q}{yh} \Big/ \frac{Q}{(y+\Delta y)(h+\Delta h)} \simeq 1 - \frac{\Delta h}{h} \,. \tag{7}$$

which with eq. (6) translates into

$$\frac{\chi}{\chi_o} = 1 - \alpha \frac{u_*}{\bar{u}} \frac{H}{h} \quad .$$

Thus the effect on the concentration may or may not be significant according to what the magnitude of $au_*/\bar{u}$ and H/h is. If the Richardson number is significant the effect might be less as discussed above.

Experimental

The model gas used in the present experiments is CO_2. It is released through blow-down from ~ 65 bar through the shortest possible pipe (in the tank, the pipe is submerged in the liquid phase). The diameter of the pipe is 13.7 mm and the discharge is about 2.9 kg/s (this is being determined by an accurate weighing of the tank, which loaded contains 1 ton, in connection with a timing of open valve). According to standard formulae

$$v_o = c_F \sqrt{2\Delta p/\rho_L}, \tag{8}$$

where c_F is a loss coefficient, Δp is the tank pressure minus atmospheric pressure and ρ_L is the density of liquid CO_2. Since the release rate q is equal to $v_o \times \rho_L \times$ pipe cross section, the above-mentioned gauging results in $c_p \simeq 0.16$ and $v_o = 16$ m/s.

The initial vapour fraction is round about 50% for CO_2, but the estimation of this is complicated by the extra phase change and the question of surface energy bound in droplet formation. As discussed above, the droplets may or may not (as in fig. 7) have evaporated over the distance from the release point to the fence depending on the ambient conditions.

In the tests which have been conducted up until now, the fence has been positioned 50 m downwind of the release point, and the sampling has been done a further 5 m downwind. The lay out is shown in fig. 5. The surface of the test area consists of mowed grass. Sampling in the crosswind direction is done in 21 points separated 1.5 m, at a height of 0.8 m above the ground. Furthermore a number of samplers are positioned in a 7.5 m lattice mast, with a vertical separation of 0.8 m. The fence is 1.5 m high and 30 m long, and the porosity is about 50%. It consists of lattice work made of 1 $\frac{1}{2}$" wide lathing.

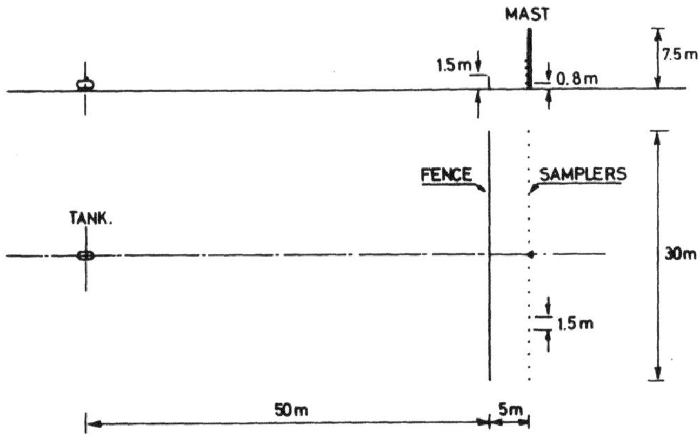

Fig. 5. Lay-out of experimental area.

The sampling units consist of a small centrifugal blower with the outlet choked, and a 50 liter plastic bag. The pump will fill the bag in about 2 min, which then determines the typical duration of a release. The pumps are started and stopped manually from a switch by the release valve on the CO_2 tank. Immediately after a test, the bags are collected and analyzed for CO_2 content on a gas chromatograph (COW-MAC 552, with porepack column, helium carrier gas, and hot-wire dectector). Typical concentrations (volume/volume) are of the order of 1%.

Some Results

We shall discuss three sets of concentration distributions: One obtained in a low windspeed situation (the one shown in fig. 6) with the fence present. The two others consecutively obtained in a higher windspeed, respectively, with (fig. 7) and without the fence present. Thus two evaluations can be made: the effect of windspeed and the effect of the fence. The lateral and vertical concentration distributions are shown in fig. 8. First the effect of windspeed is discussed.

Fig. 6. Figure showing a release from the pressurized CO_2 container (p ~ 65 bar), some time after stationary conditions have been obtained. What is seen is thus the full visible length of the plume under these conditions which are relatively calm (~ 2 m/s) and warm (~ 24 °C). The picture also illustrates the gravity slumping. The view is almost perpendicular to the fence at the right. Distance between tank and fence is 50 m.

Fig. 7. Conditions corresponding to higher wind velocity (~ 4 m/s) and lower ambient temperature (~ 16 °C) such that the visible length of the plume exceeds the distance to the fence. The crosswind and vertical line of samples are visible behind (to the left of) the fence.

According to (1) the value of x_j is 18 m and 9 m for $\bar{u} = 2$ and 4 m/s, respectively. Use of eq. (2) is marginally relevant in the $\bar{u} = 4$ m/s case but we will neglect it. On account of the difference in $\bar{u}$ and the slightly different distance to the fence (50 m - x_j) the estimate on plume width using (3) is 29 and 17 m for 2 and 4 m/s respectively. This is quite close indeed to the observed plume widths (fig. 8 (a) and (b)).

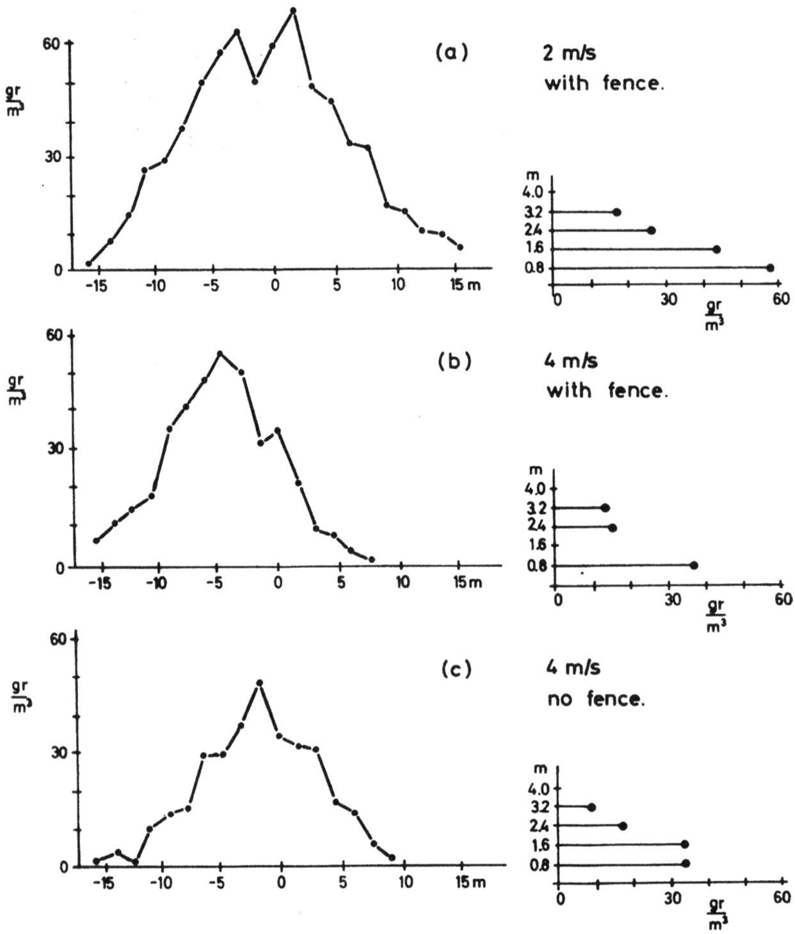

Fig. 8. Lateral and vertical concentration distributions in three different conditions. Zero on the lateral distribution marks the center of the sampling array, and the position where the vertical profile is obtained.

Regarding the plume height by the fence, we first realize that the distance x_s over which gravity slumping exceeds vertical entrainment in both cases are about 7 m. Therefore vertical entrainment has been dominating over most of the path towards the fence, with the result that plume height may not be very different in the two cases. Making the following computation (see fig. 9) using

$$\tfrac{1}{3}\chi_{peak}(\tfrac{1}{2}yh)\bar{u} = q, \qquad (9)$$

and relating the two cases by taking the ratio of maximum concentrations to 1.18 (which doesn't depend on absolute calibration) results in

$$\frac{h(2\ m/s)}{h(4\ m/s)} = \frac{17\times 4}{29\times 2}(1.18)^{-1} \cong 1, \qquad (10)$$

consistent with the above statement and the actual vertical concentration distributions measured in the two cases.

The vertical scale on fig. 8 is obtained from using eq. (9), by extrapolating the distributions in (a) to zero concentration (obtaining y = 30 m and h = 4 m). An absolute calibration is not very important in the present context as we are only dealing with relative magnitudes.

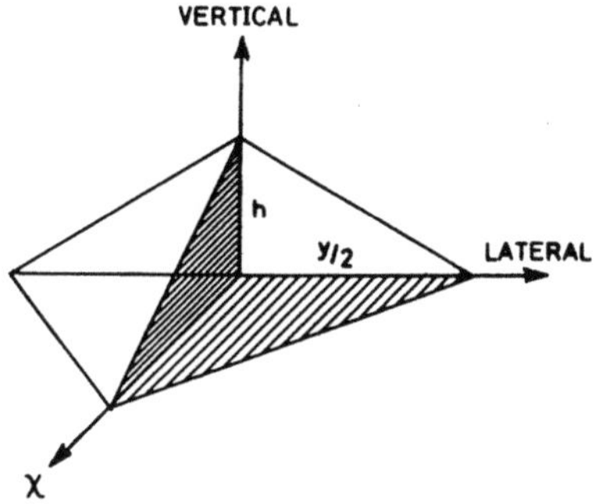

Fig. 9. Sketch of a simplified (triangular) two dimensional concentration distribution, which according to fig. 8(a) is a workable approximation.

Regarding the effect of the fence no firm conclusion can be made, especially because the wind speed in the case without the fence (c) possibly had a slightly larger value than in the case with the fence (b). Furthermore the vertical profiles are not obtained in the centerline, but at slightly different off axis positions (1.5 m and 5 m, respectively). In any case it may be safe to say that the vertical profile in the case with the fence has higher concentrations at the upper levels, despite these are taken at a more off axis position than the profile pertaining to the no-fence case. It thus appears that with the fence, the CO_2 is mixed to a higher level in accordance with expectation. However, the overall impression is that the ground level (0.8 m) concentration (the lateral profile) is highest in the case with the fence, which is contrary to expectation. This is probably due to the slight difference in wind speed already mentioned. The expectation using eqs. (6) and (7) with a = 2, a mean plume height of 2 m, and the drag coefficient $u_*/\bar{u} = 0.1$ (corresponding to a moved grass field) would have been $\chi/\chi_o = 0.85$.

Final remarks

The original concept behind these tests was to conduct them in pairs, with and without the fence in position, such that the effect of the fence could be found directly. However, the necessary time delay between the first and the second test in a pair in combination with the short-term variability in wind speed and direction, makes this approach less rational than it sounds at first. Future tests will be based on statistical comparison of a larger number of tests with and without fence, in which especially wind speed and ambient temperature is scaled out according to formulae developed above.

Additional observations which will be made include temperature measurements in the downwind plume direction, and over-head photography from an ajacent 123 m meteorological tower. Of related problems that warrant special study should be mentioned: entrainment dynamics of evaporating, aerosol-laden jets; droplet size in flash formation: and the special fluid mechanics in the outflow region.

Acknowledgement

The experimental programme is supported by Grindsted Products Ltd., Superfos Ltd., and BASF Vitaminfabrik Ltd.

References

1. Jensen, M.: Shelter effect. The Danish Technical Press. Copenhagen (1954) 264 pp.

2. Bradley, E.F. and Mulhearn, P.J.: Development of velocity and shear stress distributions in the wake of a porous shelter fence. Proc. 6th Int. Conf. Wind Eng., Gold Coast, Australia (1983).

3. Resplandy, A.: Etude expérimentale des propriétés de l'ammoniac. Chim. Ind. 102 (1969) 691-702.

4. Jensen, N.O.: On the calculus of heavy gas dispersion. Risø-R-439 (1981) 45 pp.

Large Scale Experiments on the Dispersion of Heavy Gas Clouds

J. McQUAID

Safety Engineering Laboratory
Health and Safety Executive
Red Hill, Sheffield S3 7HQ

Summary

A programme of experiments on the dispersion of fixed-volume releases of heavier-than-air gas at ground level in the atmosphere is described. The programme is in two phases, in both of which a volume of 2000 m^3 of gas is released instantaneously. The initial density of the released gas can be varied from neutrally buoyant up to 4.2 times the density of air. In Phase I of the programme, the dispersion of the gas cloud over uniform, unobstructed ground is studied, whilst Phase II studies the effect of some simple types of obstruction. This paper describes the purpose and design of the experiments and outlines the results achieved to date.

1. Introduction

In 1976, the Health and Safety Executive (HSE) instituted a programme of research on the dispersion of fixed-volume heavy gas clouds. The Heavy Gas Dispersion Trials (HGDT) project at Thorney Island is the large-scale constituent of the programme and is the subject of the present paper. Medium-scale field experiments conducted at the Chemical Defence Establishment (CDE), Porton Down have been described by Picknett [1]. Wind tunnel experiments conducted at the Warren Spring Laboratory have been described by Hall, Hollis and Ishaq [2].

The motivation for the programme was the absence of reliable experimental data with which to assess the validity of different predictive models. The uncertainty that prevailed is illustrated by the results of an invitation issued to mathematical modellers in advance of the trials. They were asked to provide predictions of the concentration distributions for an instantaneous release of 2000 m^3 of gas of initial relative density 2.0. The results are collated in Fig.1. This shows the maximum concentration

(C_{max}) at ground level on the path of the cloud centre as a function of downwind distance (x) from the source. Differences of up to two orders of magnitude in C_{max} at a given distance were predicted. Identification of the different models has not been included in Fig.1; to do so would be misleading since some of the models have since been modified.

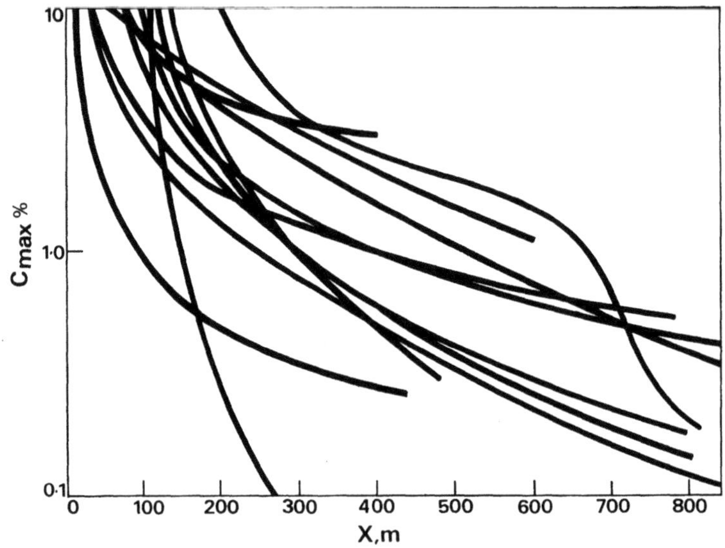

Fig.1 - Results of pre-trial predictions provided by dispersion models

The choice of a fixed-volume release of gas at ambient pressure and temperature was influenced by two considerations. Firstly, an examination of serious accidents suggested that instantaneous releases were more important than continuous releases (McQuaid [3]). Secondly, it was considered essential that the release conditions should be capable of close control to ensure that the effects of atmospheric conditions on dispersion could be isolated. This suggested a study of preformed clouds of gas, rather than, for example, clouds generated by the rupture of a container of pressurised liquefied gas.

The programme as originally planned was limited to experiments on clouds dispersing over uniform, unobstructed ground. After these experiments had commenced and the system design had been proved, a second programme of experiments was formulated to study the effects on cloud dispersion of several types of obstruction. The former experimental programme was thereafter designated as Phase I and the latter as Phase II. At the present time a further programme is envisaged in which the cloud will be released into a containing enclosure and its dispersion from this enclosure as a function of time will be studied.

2. Organisation of the Trials

The trials conducted by CDE were intended to provide information for the design of the large-scale trials. Nevertheless, the trials, which were funded by HSE, constitute a valuable contribution in their own right. CDE performed a design study for the large-scale trials which indicated that the cost would exceed the resources that HSE could make available. A proposal was therefore prepared by HSE for a jointly-funded project and invitations to participate were issued to other organisations. The outcome was financially successful, in that funding exceeded the contracted cost of the original Phase I programme, and technically successful, in that 16 trials were performed compared to the 5 trials contracted. Early in the trials programme, CDE had to withdraw and the contract for the trials was placed with the National Maritime Institute (now NMI Ltd). A notable feature of the trials, associated with the method of funding adopted, was the close, continual oversight provided by the sponsoring organisations. There were 36 sponsors and they are listed in the Annex. Most of them took an active part on the Steering Committee which met twice yearly over the 3 year duration of the project. A Technical sub-Committee met monthly to progress the planning of the trials and to resolve the many technical issues that arose.

3. The Trials Site

The site chosen was a former Royal Air Force station on Thorney Island in Chichester Harbour about 40 km east of Southampton on the south coast of England. The location of the site is shown in Fig.2. Hourly records of wind speed and direction at the site were available for 28 years and Pasquill stability categories for 10 years. The upwind fetch is clear for a distance of 1 km from the shoreline and the ground slope is nowhere greater than about 1°. The trials area is rough grassland except for two tarmacadam runways. Consideration was given to the coverage of the runways to eliminate the non-homogeneity of the ground surface. It was concluded that the change in surface roughness did not warrant such treatment of the runways, since it was intended that the grassed areas would be mown to maintain an aerodynamic roughness of approximately 10 mm. Measurements of the surface temperatures of the runways and grass indicated that differences up to $10^{\circ}C$ could be expected. It was decided that the temperature difference should be corrected and this was achieved by white-painting of the parts of the runways over which the cloud would travel.

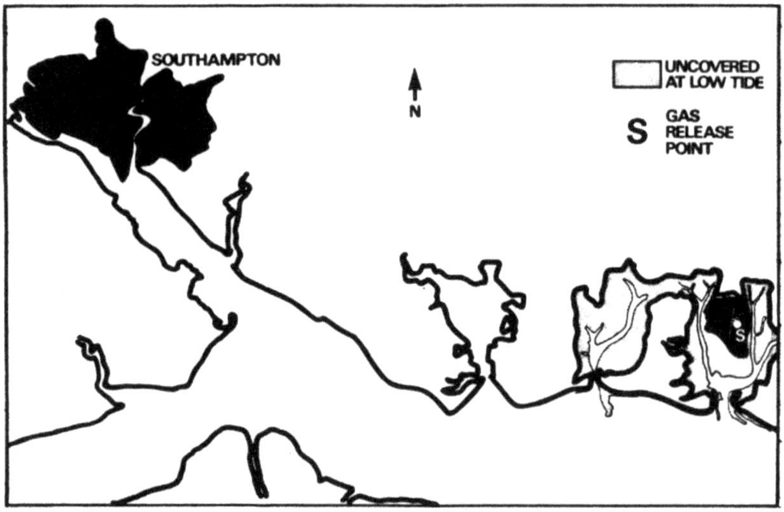

Fig.2 - Location of the trials site at Thorney Island

4. Design of the Experiments

Each experiment was intended to release a fixed volume of $2000m^3$ of gas at ambient pressure and temperature. The gas was contained in a twelve-sided bag 14 m across and 13 m high. The bag was fabricated from plastic sheeting supported on taut rigging from a central column with radial guy wires to ground anchors. The sides of the bag collapsed in concertina fashion to ground level in a time less than 2 secs, leaving an upright cylinder of gas momentarily stationary. The bag was fitted with a conical roof, also fabricated from plastic sheeting, which was withdrawn upwards immediately prior to release. The gas selected was a mixture of refrigerant-12 (dichlorodifluoromethane, CCl_2F_2) and nitrogen thus allowing initial relative densities to be selected in the range 0.97 to 4.2. The gases were stored in liquefied form. They were vaporised and mixed prior to being supplied to the gas bag. Filling of the bag took about 1 hour.

A ground plan of fixed masts had been designed by CDE based on the experience of the Porton trials and is shown in Fig.3. The results of the invitation to model developers, referred to in Section 1 above, did not produce any firm evidence that the CDE ground plan needed to be changed. The position of the ground plan on the site maximised the range of wind acceptance angles which would not be affected by fixed site features.

5. Instrumentation and Data Logging

5.1 Meteorological measurements - The characteristics of the approach flow were monitored at a 30 m high weather mast located 150 m upwind of the gas source. The position of the mast (designated 'A') is shown in Fig.3 and the instrumentation is described in Table 1. In addition, a cup anemometer and a wind vane were located on the mast at the last downwind position (designated 'D' in Fig.3). The sonic anemometers deployed in the field (see Section 5.3 below) also provided information on ambient wind structure during the periods before and after cloud passage.

Instrument Type	Mast Code (see Fig.3)							
	A	V	F	M1	M2 (a)	M2 (b)	M3	D
Cup Anemometer	5	-	-	-	-	-	-	1
Sonic Anemometer	2	-	-	1	2	2	3	-
Thermometer	5	-	-	-	-	-	-	1
Solarimeter	1	-	-	-	-	-	-	-
Relative Humidity Sensor	2	-	-	-	-	-	-	1
Wind Vane	1	1	-	-	-	-	-	1
Gas Sensor (1 Hz)	-	-	4	4	4	4	4	4
Gas Sensor (10 Hz)	-	-	-	2	2	1	3	-

Table 1. Instrumentation associated with the mast array in Fig. 3

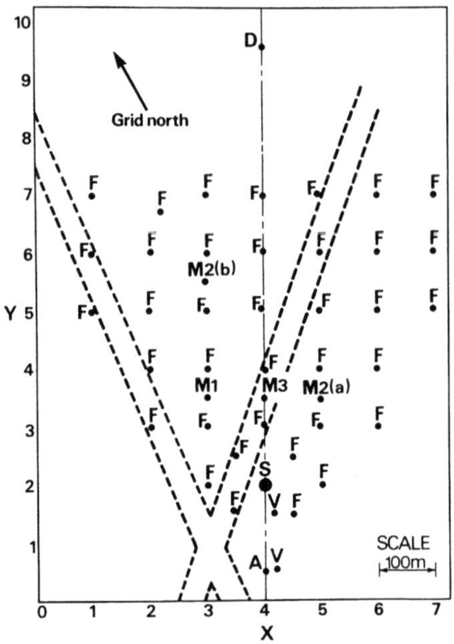

Fig.3 - Ground plan of mast array

5.2 Fixed masts - The main body of gas concentration sensors were deployed on the fixed masts, designated 'F' and 'D' in Fig. 3. The gas sensors employed a polarographic cell which measured the oxygen deficiency induced by the presence of the released gas. The use of this principle of operation dictated the choice of nitrogen for admixture with the heavy gas so that the released gas was totally deficient in oxygen. The cell is a commercial development of the design described by Bergman and Windle [4]. It has a lower limit of resolution of 0.1% of the released gas and a frequency response of about 1 Hz.

5.3 Mobile masts - Instrumentation for the measurement of turbulent fluctuating velocities and concentration was placed on trailer-mounted masts, designated 'M' in Fig. 3 and detailed in Table 1. Of the 8 sonic anemometers available, 3 were placed at heights intended to be above the cloud and the remainder within the cloud. For each of these latter instruments, a fast-response gas sensor was placed at the same height. These sensors were a modified form of those used on the fixed masts. Their response was enhanced by aspirating the diffusion head of the cell and by analogue processing of the out-put signal. Their frequency response was about 10 Hz.

5.4 Photography - The Porton trials had demonstrated the value of photographic records and extensive photographic coverage was included in the design of the Thorney Island trials. Cine, video- and still-film records were obtained from cameras at ground level and on a helicopter hovering at a height of 300 m. The gas in the container was marked with orange-coloured smoke prior to release.

5.5 Data capture - This was provided by a real-time central computer with a 60 M byte disc capacity. The data collection system comprised a total of 32 data terminals, each incorporating an analogue multi-plexer for 8 channels and an analogue-digital converter. The data terminals were disposed around the field convenient to the instrumented masts. The sampling rate was standardised at 20 times/sec/channel to suit the frequency response of the instruments with the fastest response. This arrangement had the over-riding advantage that any instrument could be connected to any channel.

6. Results of the Trials

The Phase 1 trials took place between July 1982 and June 1983. 15 heavy gas trials and 1 neutrally-buoyant trial were performed. A summary description of the trials is given in Table 2. The Phase II programme took place between July and October 1983. 10 trials were performed. The gas release arrangement was identical to that in Phase I. In 6 of the trials, a fence was placed on a 180° arc at 50 m radius on the downwind side of the gas source. Two fence types were used. The first was an impermeable screen 5 m high while the second was a series (either 2 or 4) of permeable screens 10 m high. 4 trials were performed with a 9 m cubical building, 3 with the building 50 m downwind of the source and 1 with the building 27 m upwind.

Processing of the results of the trials has concentrated so far on validation of the data and the organisation of the data records for analysis. This task is still continuing. A full evaluation of the success of the trials must await more detailed analysis, although early indications are encouraging. The number of gas sensors which responded to gas (up to a maximum of 73 in trial 008) will allow detailed mapping of the concentration in both horizontal and vertical planes. The atmospheric stability covered the range from B to F and wind speeds ranged from 1.7 to 7.5 m/s. The utilisation of the fast-response instrumentation was also very high, with data being obtained in most of the trials. The layout adopted for the instrumentation was vindicated to a remarkable degree. The gas sensor records show that the concentration data have high internal self-consistency, i.e. the record from any individual sensor is consistent with the records of its neighbours. This also applies to the fast-response gas sensors.

Trial Number	Wind Speed m/s	Pasquill Stability Category	Initial Relative Density	Number of Gas Sensors Which Responded to Gas
004	3.8	B	0.97	22
005	4.6	B	1.69	26
006	2.6	D/E	1.60	46
007	3.2	E	1.75	57
008	2.4	D	1.63	73
009	1.7	F	1.60	62
010	2.4	C	1.80	11
011	5.1	D	1.96	26
012	2.6	E	2.37	65
013	7.5	D	2.00	47
014	6.8	C/D	1.76	50
015	5.4	C/D	1.41	38
016	4.8	D	1.68	45
017	5.0	D/E	4.20	62
018	7.4	D	1.87	60
019	6.4	D/E	2.12	67

Table 2. Details of Phase I Heavy Gas Dispersion Trials

A feature of the project has been the inclusion of a number of subsidiary investigations. The statistical aspects of heavy gas cloud dispersion were studied by Chatwin [5], who also investigated the possible effects of wind shear on a heavy gas cloud (Chatwin [6]). The initial motion of the cloud has been studied by Rottman and Simpson [7]. Each of these investigations has been oriented towards an examination of features peculiar to the form of experiment chosen for study, i.e. release of an initially stationary, fixed-volume cloud of gas. The purpose of the investigations was to assist with the interpretation of the trials results. Pre-trials wind tunnel simulations were performed by Hall, Hollis and Ishaq [2] and their results are currently being compared to those from the field trials.

7. References

1. Picknett, R.G.: Dispersion of dense-gas puffs released in the atmosphere at ground level. Atm. Env. 15 (1981) 509-525.
2. Hall, D.J.; Hollis, E.J.; Ishaq, H.: A wind tunnel model of the Porton dense gas spill field trials. Rep. no. LR 394 (AP), Warren Spring Laboratory, Stevenage 1982.
3. McQuaid, J.: Dispersion of heavier-than-air gases in the atmosphere: review of research and progress report on HSE activities. HSL Tech.Paper 8, Health and Safety Executive, Sheffield 1980.
4. Bergman, I.; Windle, D.A.: Instruments based on polarographic sensors for the detection, recording and warning of atmospheric oxygen deficiency and the presence of pollutants such as carbon monoxide. Ann. Occ. Hyg. 15 (1972) 329-337.
5. Chatwin, P.C.: The use of statistics in describing and predicting the effects of dispersing gas clouds. J. Haz. Mat. 6 (1982) 213-230.
6. Chatwin, P.C.: The incorporation of wind shear effects into box models of heavy gas dispersion. Proc. IUTAM Symp. on Atmospheric Dispersion of Heavy Gases and Small Particles. Berlin: Springer-Verlag 1984.
7. Rottman, J.W.; Simpson, J.E.: The initial development of gravity currents from fixed-volume releases of heavy fluids. Proc. IUTAM Symp. on Atmospheric Dispersion of Heavy Gases and Small Particles. Berlin: Springer-Verlag 1984.

ANNEX

List of Sponsors of the HSE Heavy Gas Dispersion Trials

HSE
British Gas Corporation
Department of Energy, UK
Battelle Institut, Federal Republic of Germany
American Petroleum Institute
ICI Plc
Insurance Technical Bureau
British Petroleum International Ltd
Comitato Nazionale per l'Energie Nucleare, Italy
United States Coast Guard
Commissariat a l'Energie Atomique, France
Institute for Air Research, Norway
Shell UK
Gaz de France
Electricité de France
Central Electricity Generating Board
Gas Research Institute, USA

Commission of the European Communities
Safety and Reliability Directorate, UKAEA
Amoco
Mobil
Texaco
Department of Trade, UK
TNO, Netherlands
Ministry of Public Health and Environmental Hygiene, Netherlands
Union des Industries Chimiques, France
E.I. Dupont de Nemours & Co., USA
National Maritime Institute (now NMI Ltd)
Government of Sweden (Principal Dept's National Defence Research Institute, Fire Research Board, Civil Defence Administration)
Esso, UK
DSM, Netherlands
TransCanada Pipelines
Arabian Americal Oil Co., Saudi Arabia
Atmospheric Environment Service, Canada
Britoil
Cie Francaises des Petroles

Improved Understanding of Heavy Gas Dispersion and its Modelling

S. Hartwig, Universität Wuppertal; G. Schnatz, Battelle Institut; W. Heudorfer, Universität Wuppertal

Summary

The first results of the evaluation of experimental data show that a heavy gas cloud is decoupled partly from the dynamics of the atmospheric boundary layer, even after the short gravity spreading phase.
As expected, vertical diffusion coefficients are distinctively smaller as in the atmospheric boundary layer.
Turbulence induced by the vapour cloud has a noticeable effect on diffusion.
More data are needed and must be evaluated to support these findings.

Introduction

As you all will know, there are two main pathways in the development of heavy gas dispersion models, the more integral or so-called slab models and the differential type of models or K- models. Each of these consists of a variety of different types. A few examples are shown in figure 1. Because of growing interest of industry and the scientific community in heavy gas dispersion models within the last two or three years, some detailed review papers have been published. (Blackmore (1), Hartwig (2), Havens (3), Jagger (4)).
Both types of models have their merits and disadvantages.For slab models the requirements of input data are less stringent than for numerical models. As a direct consequence,the predictions and results of these models are less detailed and sometimes insufficient for the problem in question. Nevertheless slab models are often very useful, not so demanding in computertime,and fairly often used for consequence analysis.
~~Recently~~ considerable improvements have been achieved in these types of models. For instance, a stricter application of the conservation law of energy compared with previous approaches resulted in better time prediction of heavy gas concentration after an instantaneous spill (Schnatz (5)).

In investigating experimental data,we found a distinctly better agreement between the improved type of slab model and measured concentration ratios in the first phase of a heavy gas spill. The basis and principles of the differential type of heavy gas dispersion models are fairly well understood (Jagger (4)). But similar to the situation in the dispersion of tracers the "closure problem" exists in the dispersion of heavy gases,or in a more parametrized formulation, the data set of diffusion coefficients to be used within and outside the cloud is poorly or hardly known. In the following section I will discuss what our group has learned by data evaluation in this field.

Discussion of Experimental Data; General Features

In talking about instantaneous releases of heavy gas spills, let us focus on the slumping of a vapour cloud.
If we look at a sensor for the heavy gas at a given point, we see a distribution of concentration over time at that fixed point as shown in fig. 2. We have chosen a moderately long averaging time of 80 sec. If we look for the density fluctuations of the same sensor,taking the average values of fig. 2 as a baseline, we so to speak arrive at fig. 3.
This picture indicates that we still have strong eddies inside the cloud caused either by the dynamics of the cloud or by eddies transmitted from the atmospheric boundary layer.
More detailed investigation discloses that it is not an "either- or", but "as well as"which means that turbulence is caused by the dynamics of the vapour cloud and by boundary layer eddies.
As an explanation of fig. 4 you see at the same location as the gas sensor the vertical wind fluctuation over the time. Now, if we add fig. 2 to that picture showing the mean gas concentration at the same height and location, a quenching of amplitude is obvious (fig. 2 and 4 => 5).
This supports the supposition that a smaller eddy diffusion coefficient is valid inside the cloud than outside.
Before I try to estimate the value of the diffusion coefficient, I will discuss some other general features of an instantaneously released vapour cloud.
The correlation of vertical wind and density fluctuation gives the turbulent mass flux as shown in fig. 6. As an overall feature we can differentiate between three phases of the flux if we distinguish after its sign:

I. upward motion of mass during the initial vortex phase;
II. downward motion due to gravity;
III. upward motion due to dispersion and dilution.

An interesting result occurs if we try the same correlation with sensors 0.4m apart as shown in fig. 7. The feature is entirely different which suggests that such correlations are meaningless. For me that is an indication that at least part of the vapour cloud turbulence is due to its dynamics and not to the outside turbulence specturm.

Eddy Diffusion Coefficient

As it travels along, the vapour cloud has a concentration profile over the height. At its beginning and near its spill source, it will have a pronounced profile whereas at later times and at greater distances the profile will be less distinctive. This process of smoothing is caused by turbulent diffusion. In this way the change of profile gives us a strong indication of the rate of the diffusion coefficient.
In fig. 8 and 9 two examples of the changing concentration profiles are given. The abscissa shows concentration in arbitrary units, the ordinate the height. The parameter for different profiles is the duration of time after the release.
If we assume that the change of concentration is governed mainly by the equation of continuity and the gradient approach, it is possible to calculate the vertical eddy diffusion coefficient effective in the vapour cloud (fig. 10) in a box model type notation.
Coefficients calculated in this manner give an upper limit of their values because some simplifications are used. These are:
- the process is assumed to be linear which is not entirely true;
- the mixing is assumed to be stochastic which is a simplification;
- it is assumed that we have a closed system which also is a simplification;
- horizontal eddy diffusion is neglected.

But despite these simplifications these coefficients give us valuable indications of the order of magnitude of eddy diffusion.
In fig. 11 and 12 you can see the results of the calculation, giving average values $K_z = 3 \cdot 10^2 cm^2/sec$ and K_z $6 \cdot 10^2$ for the two sets of profiles shown in fig. 11 and 12.
We calculated the K_z figures from the undisturbed atmospheric boundary layer by applying the following relationship (WU 1965):

$$K_z = \left[\left(\frac{\partial u}{\partial z} \right)^2 - \frac{g}{T} \frac{\partial \theta}{\partial z} \right]^{1/2} \ell^2$$

$\frac{\partial u}{\partial z}$ = velocity gradient

g = gravity constant

T = mean temperature of the boundary layer

θ = potential temperature

ℓ = mixing length

The average value is $K_z = 10^4$ cm^2 / sec and $3 \cdot 10^3$ cm^2 / sec respectively, roughly one order of magnitude higher than the eddy diffusity inside the cloud.

References

1. Blackmore, D.R., M.N. Herman and J.L. Woodward: Heavy Gas Dispersion Models, Journal of Hazardous Materials, 6 No. 1/2, 1982.

2. Hartwig, S.: Identfication of Problem Areas Related to the Dispersion of Heavy Gases; von Karman Institute Lectures Series 1982-03, Brüssel 1982.

3. Havens, J.A., T.U. Spicer: Further Analysis of Catastrophic LNG Spill Vapour Dispersion in Heavy Gas and Risk Assessment II, Reidel Publishing Company Dordrecht, 1983.

4. Jagger, S.F.: Formulation of the Dense Gas Dispersion Problem in Heavy Gas And Risk Assessment II, Reidel Publishing Company Dordrecht, 1982.

5. Schnatz, G., J. Kirsch and W. Heudorfer: Investigation of energy fluxes in heavy gas dispersion, in Heavy Gas and Risk Assessment II, Reidel Publishing Company Dordrecht, 1982.

6. Wu, S.S.: A Study of Heat Transfer Coefficients, J. Geophys, Research 70, p 1801, 1965.

Input Variables for different Types of Models

	integral (Slab models)	**differential** (K-theory models)
Models	HEGADAS II Cox and Carpenter Eidsvik Fay	ZEPHYR HEGAS SIGMET-N MARIAH DISCO MONA
INPUT DATA	integral values of: — wind speed — stability — atmospheric temperature — atmospheric pressure	vertical profiles of time dependent: — wind vector — temperature — turbulence and its modification by heavy gas

Fig. 1 Different types of models and the necessary input data

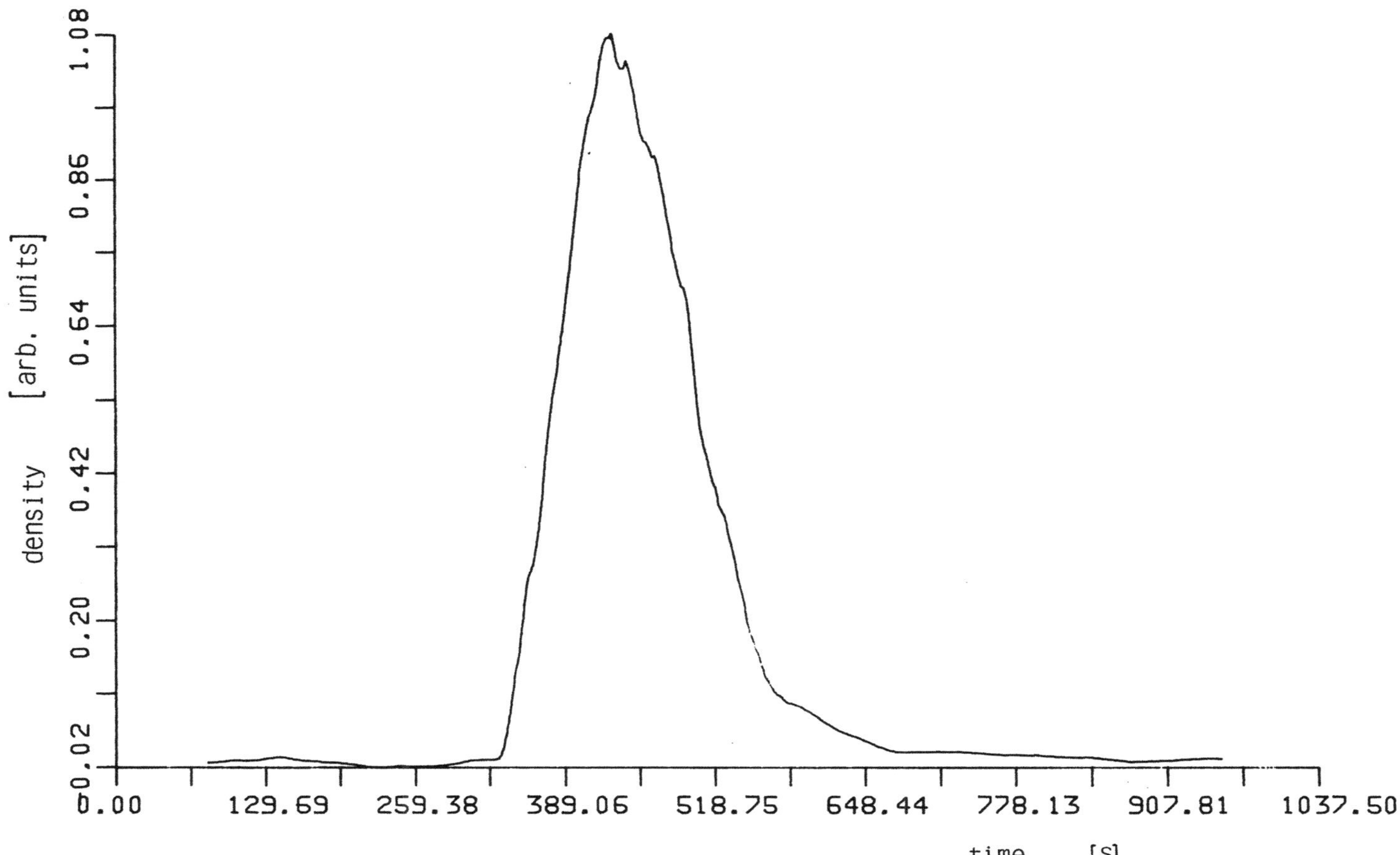

Fig. 2 Concentration values of heavy gas cloud at fixed sensor location over time

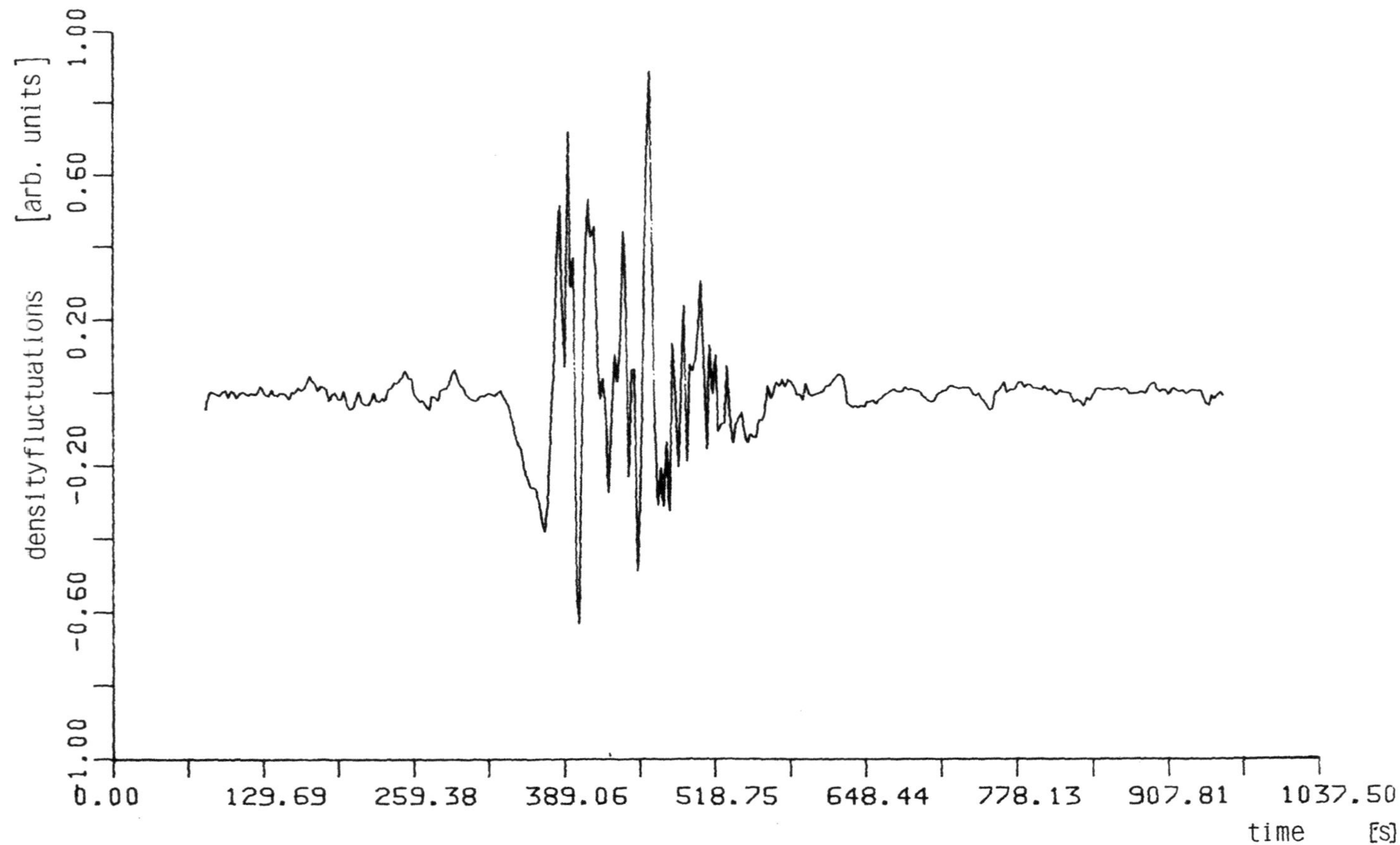

Fig. 3 Fluctuation of concentration values at the same location as fig. 2

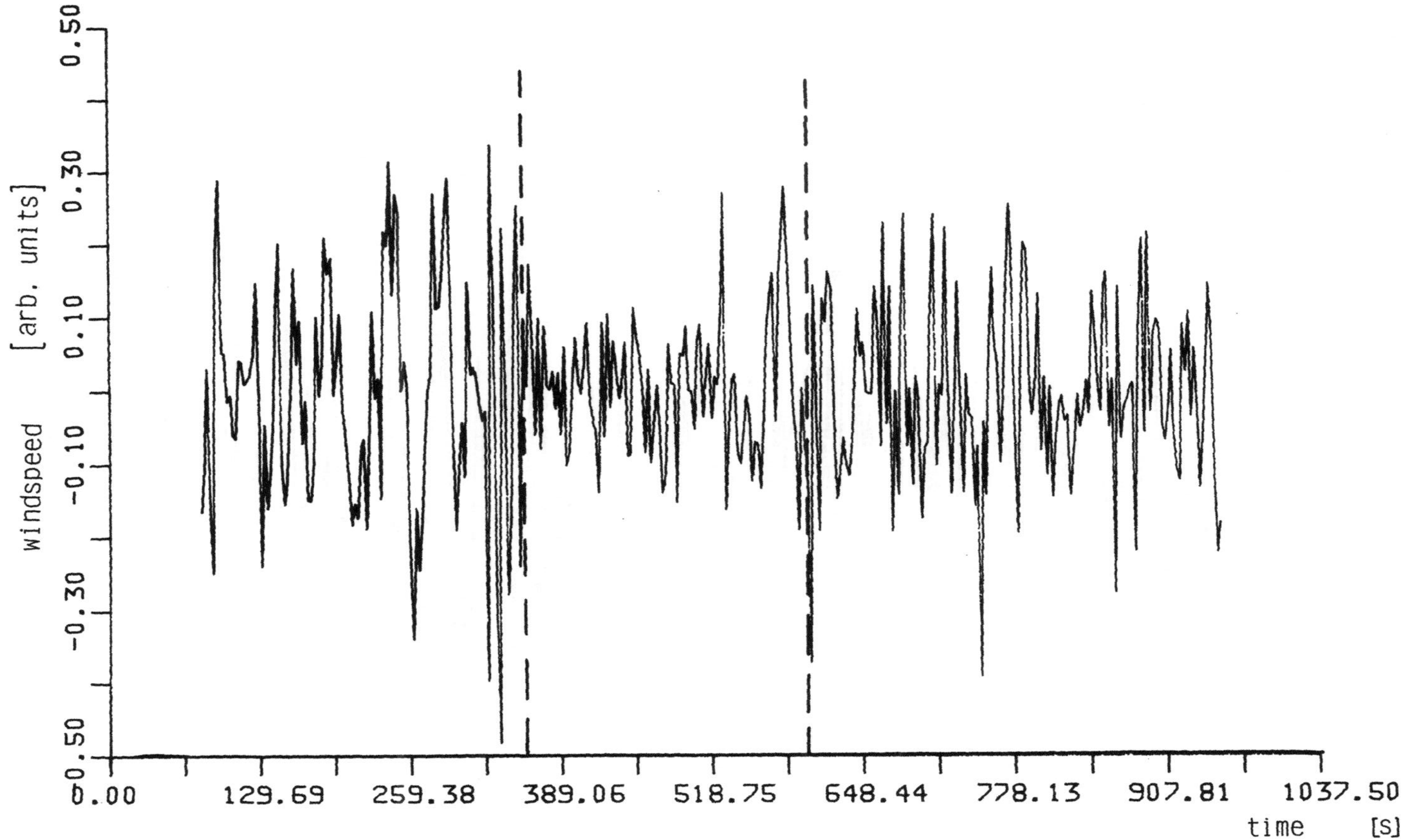

Fig. 4 Fluctuation of vertical wind speed over time

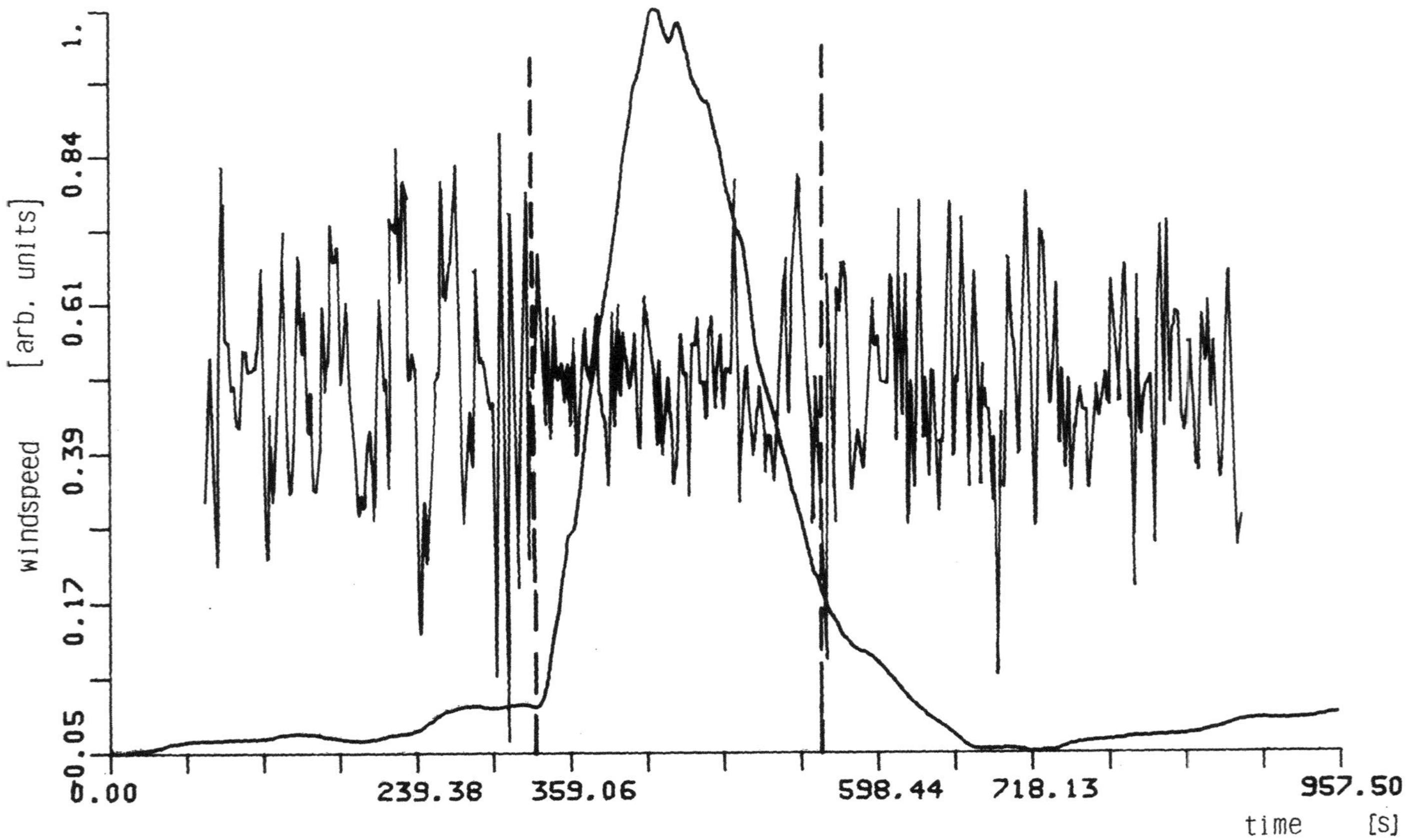

Fig. 5 Quenching of windfluctuations inside the vapour cloud

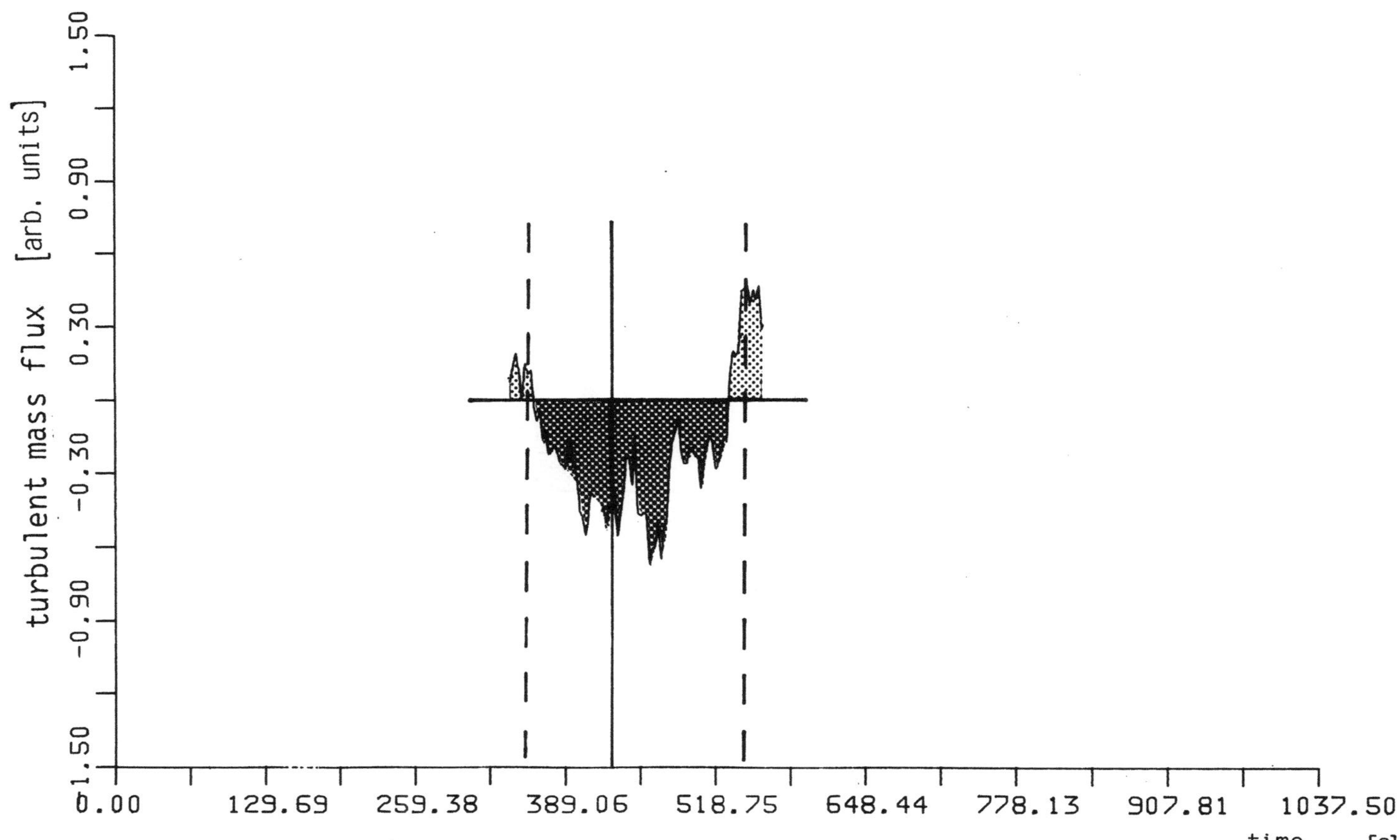

Fig. 6 Sign of turbulent mass flux during passing of the vapour cloud at fixed location

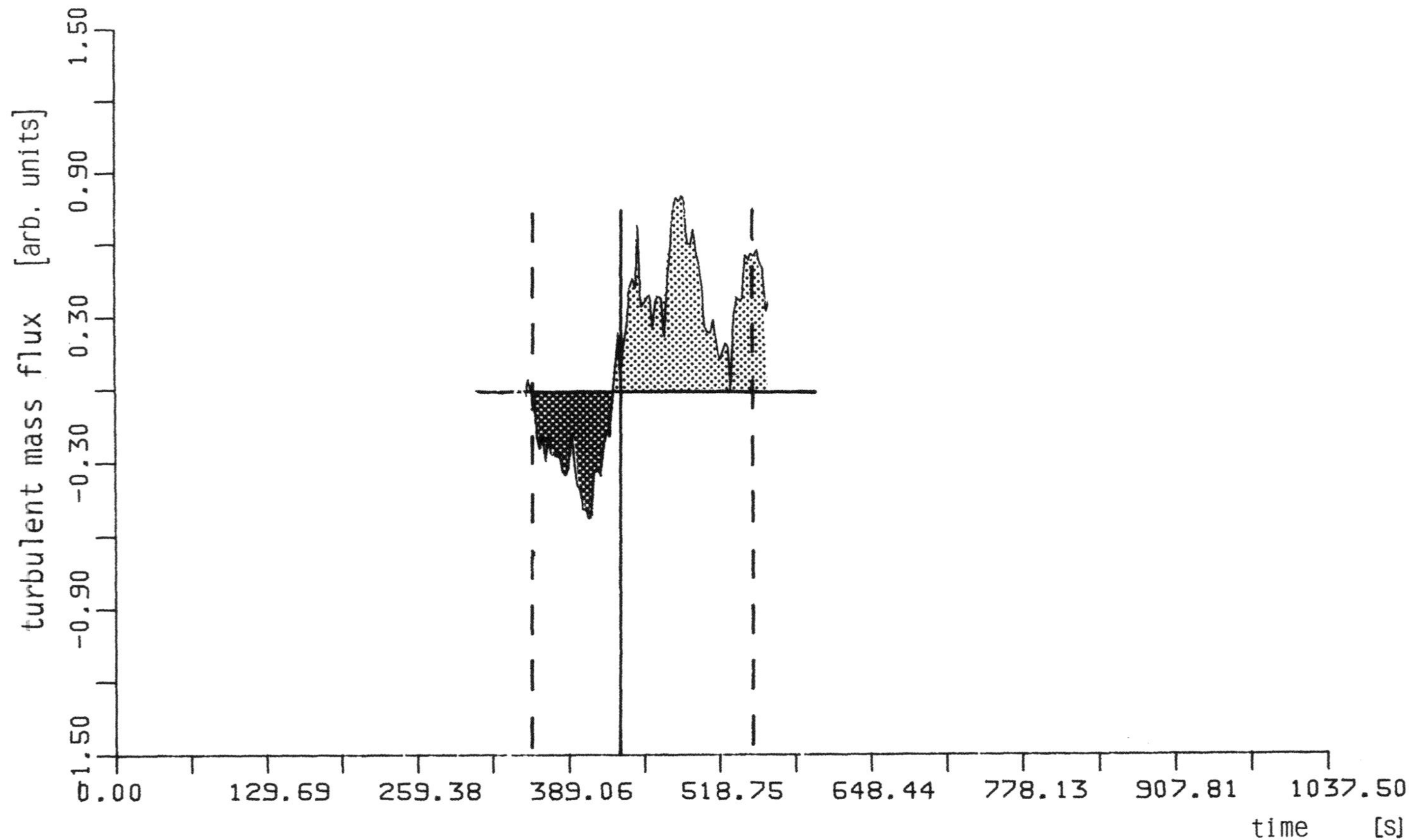

Fig. 7 Same as fig. 6, but sensors wider apart

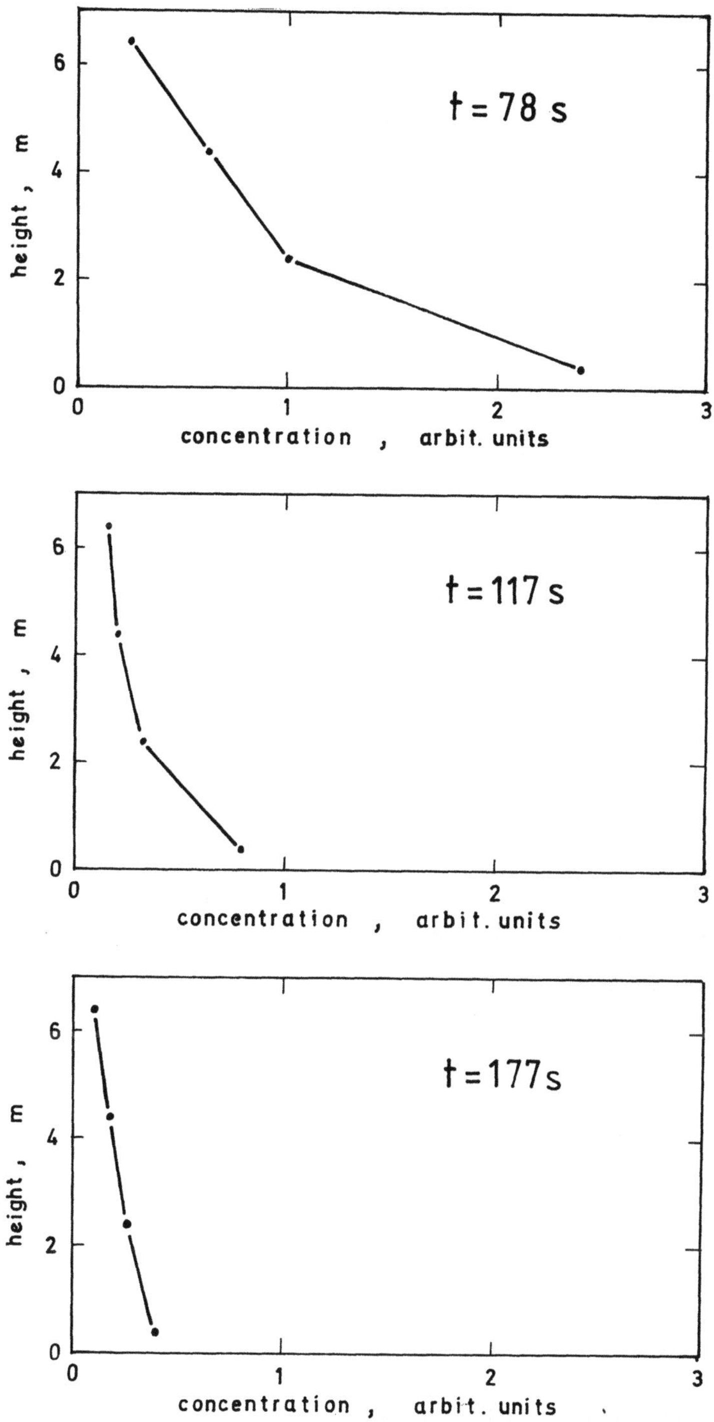

Fig. 8 Concentration profiles in dependency of time, trial I

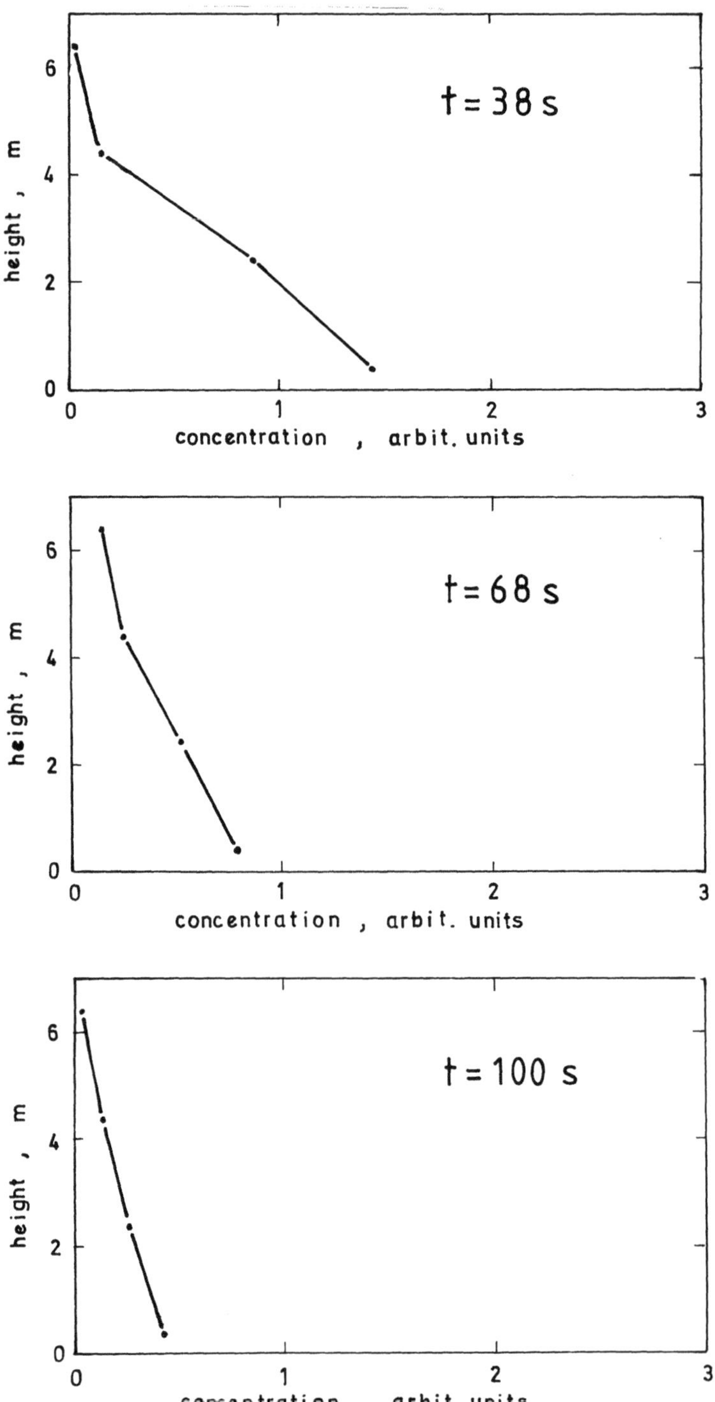

Fig. 9 Concentration profiles in dependency of time,trial II

Combination

$$\frac{\Delta z^2}{\Delta t} \cdot \left[\frac{(c_2^1 - c_3^1) - (c_2^0 - c_3^0)}{(c_4^0 - c_3^0) - (c_2^0 - c_1^0)} \right] = K$$

c_s^t — t: time increment; s: space increment

Fig. 10 Calculation of eddy diffusion coefficient written in box notation

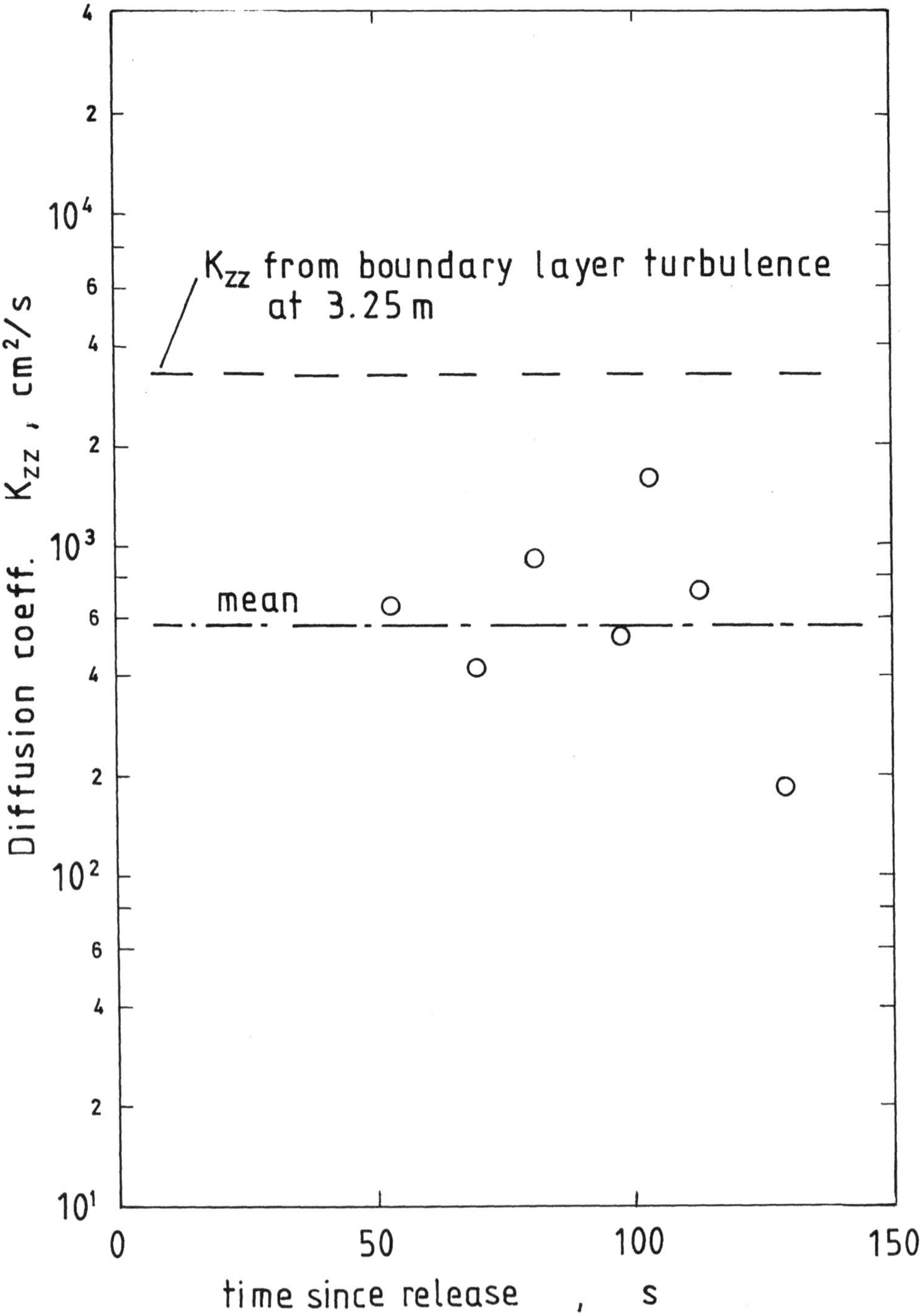

Fig. 11 Eddy diffusion coefficient inside and outside the cloud, trial I

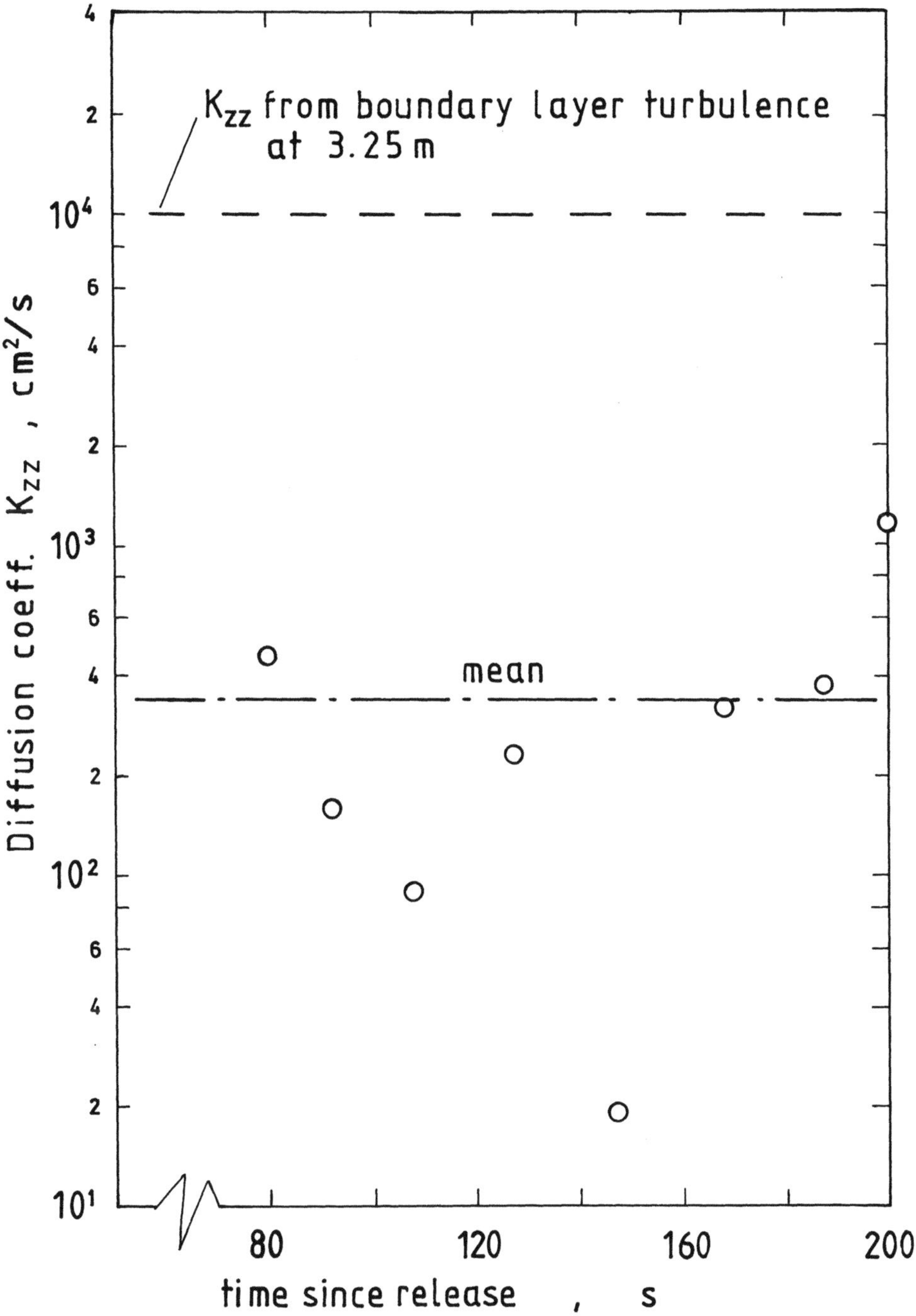

Fig. 12 Eddy diffusion coefficient inside and outside the cloud, trial II

Effects of a Spill of LNG on Mean Flow and Turbulence under Low Wind Speed, Slightly Stable Atmospheric Conditions*

HOWARD C. RODEAN

Lawrence Livermore National Laboratory,
Livermore, California

Summary

Of the many liquefied natural gas (LNG) spill experiments in the 1980 Burro and 1981 Coyote series at the Naval Weapons Center (NWC), China Lake, California, only one was observed to affect the mean flow and turbulence in the near-surface atmospheric boundary layer. This experiment, Burro 8, was conducted under atmospheric conditions that permitted the gravity flow of the cold, dense gas to be almost independent of the atmospheric boundary layer. The mean flow kinetic energy was damped proportionately more than the turbulent kinetic energy. These effects of the Burro 8 LNG spill were observed at only one instrument station.

Three large populations of measured Reynolds stresses are used to demonstrate two data analysis procedures that can lead to an understanding of the physical processes that resulted in the above observations. One involves the use of correlation coefficients to identify active causal links among the Reynolds stresses. The other consists of determining empirical relations among the three invariants of the Reynolds stress tensor.

Introduction

The Burro and Coyote series of LNG spill experiments were conducted in 1980 and 1981, respectively, at the NWC. These experiments were executed by the Lawrence Livermore National Laboratory (LLNL) and the NWC under the sponsorship of the U.S. Dept. of Energy. Koopman et al. [1] reported that one spill, Burro 8, had a significant effect on the mean flow field around the spill pond. Rodean and Cederwall [2] found that the turbulence at one instrument station was strongly affected by the Burro 8 spill. These observations are described in the first part of this paper.

The latter part of this paper concerns two procedures for turbulence data analysis that are demonstrated using three large

* Work performed under the auspices of the U.S. Dept of Energy by the LLNL under contract number W-7405-ENG-48.

populations of measured Reynolds stresses. It is proposed that the use of these procedures can lead to an improved understanding of turbulence phenomena, including the above spill effect.

Sources of Data

The descriptions of the effects of the Burro 8 spill on the mean flow and turbulence are based on Koopman et al. [1] and Rodean and Cederwall [2], respectively.

The remainder of this paper concerning relations among the six Reynolds stresses and the three Reynolds stress tensor invariants is based on three sets of turbulence data. The first two sets were obtained during the Burro [1] and Coyote [3] LNG spill experiments. The first set consists of turbulence measurements made on three Burro experiments at four stations and three elevations for a total of 27 samples of wind velocity data. The second set is from one station for four Burro and six Coyote experiments for a total of 42 samples of data. Each sample from the Burro and Coyote experiments consists of 10 min of wind velocity data from bivane anemometers. The third set of turbulence data used in this paper consists of 36 sets of Reynolds stresses published by Müller [4]. These turbulence measurements were made by hot-wire probes in a three-dimensional boundary layer in a low speed wind tunnel. In this paper, Müller's y and z coordinates have been reversed to conform to the meteorological practice used for the Burro and Coyote data.

Effects of Density Differences on Mean Flow and Turbulence

The Burro 8 experiment was the only one of the Burro and Coyote series in which the LNG spill was observed to affect the mean flow and turbulence. Burro 8 was conducted under the lowest wind speed (1.8 m/s) and most stable conditions (Ri = +0.121) of all these experiments, permitting the gravity flow of the cold, dense gas to be almost independent of the atmospheric boundary layer. The array of wind field anemometers showed that the wind was temporarily diverted around the Burro 8 spill. A momentum displacement analysis at stations T1 (upwind) and T2 (downwind) for Burros 8 and 9 showed that the atmospheric flow at T2 was temporarily displaced upward during the Burro 8 experiment. The mean wind speeds at T1 and T2 during Burro 8 are plotted in Fig. 1. Note that the wind speed at 1 m (actually 1.36 m) at T2 temporarily dropped to almost zero.

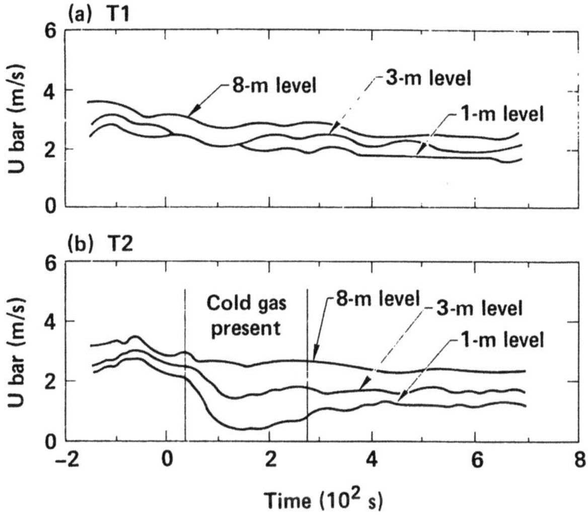

1. Mean wind speed during Burro 8: (a) at station T1 upwind of the spill pond and (b) at station T2 downwind of the spill pond [1].

Rodean and Cederwall [2] processed the 10-min data samples for T1 and T2 for Burros 7-9 to obtain the sequences of nine 1-min averages for I_1 shown in Fig. 2. The variable I_1 is the normalized first invariant of the Reynolds stress tensor where

$$I_1 = q^2/U^2 \tag{1}$$

and

$$q^2 = \langle u_i u_i \rangle . \tag{2}$$

Here q^2 is twice the turbulent kinetic energy in units of velocity squared and I_1 is the ratio of the turbulent kinetic energy to the mean flow kinetic energy. It is shown in Fig. 2 that the variation of I_1 at z = 1 m was very different for Burro 8 at T2: there was a very large temporary increase in I_1 that was coincident with the presence of natural gas. The same effect was present, but to a lesser degree, at z = 3 m. These temporary increases of I_1 were associated with corresponding decreases of q^2 and proportionately greater decreases of U^2.

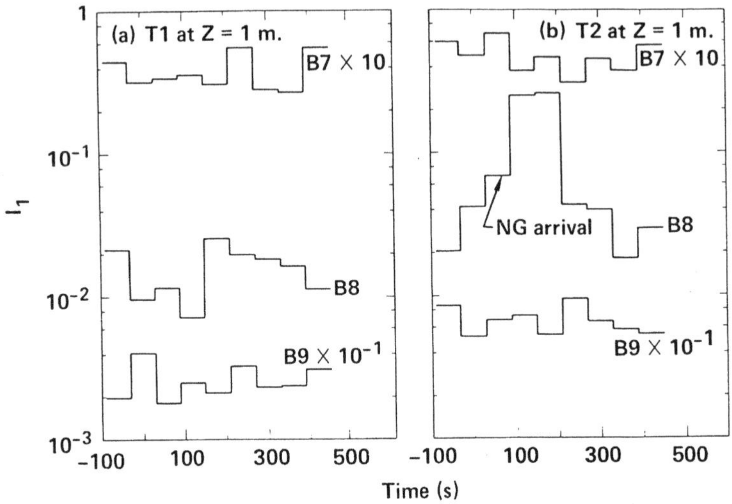

2. Reynolds stress tensor invariant I_1 vs time for turbulence stations T1 and T2 at z = 1.36 m, based on 1-min averages of the wind-velocity data for Burros 7-9 [2].

Statistical and Causal Links Among Reynolds Stresses

The turbulent transport equations in three dimensions involve 10 quantities: six Reynolds stresses, three scalar fluxes, and the scalar variance. The transport equations define a total of 27 causal links that connect these ten quantities. These links involve three types of terms in the equations: production, redistribution, and buoyancy. The causal processes can be in either or both directions along these links. According to Launder [5], "In a sense it is up to Nature to work out what particular kinds of mechanism it needs to produce just the right amount of transport." Stewart [6] examined the cause-and-effect relations among some of the Reynolds stresses and proposed the following cycle (in the nomenclature used in this paper): "..the presence of <ww> in a velocity gradient [$\partial U/\partial z$] causes production of <uw>, which in turn interacts with the mean gradient [$\partial U/\partial z$] to produce <uu>. Redistribution by pressure fluctuations then transfers energy from uu to ww (and to vv)." In the above, u, v, and w are velocity fluctuations in the longitudinal (x), lateral (y) and vertical (z) directions, respectively, and U is the mean (longitudinal) velocity.

In their analysis of the Burro turbulence data, Rodean and Cederwall [2] recalled Stewart's proposal when they observed that their data formed the scatter-diagram patterns shown in Fig. 3. These patterns suggest linear relations, on the average, between <uw> and <uu> and between <uw> and <ww>. The correlation coefficient is connected with the linear dependence between two random variables [7], so correlation coefficients were calculated for pairs* of the six Reynolds stresses. The highest correlation coefficient values for the Burro data are for the pairs (<ww><uw>), (<uw><uu>), and (<uu><vv>), consistent with Stewart's causal links from <ww> to <uw> to <uu> to <vv>.

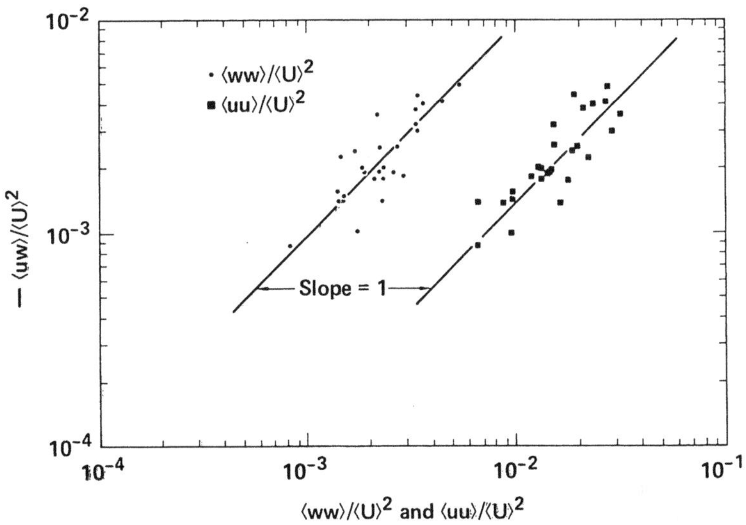

3. The Reynolds stresses <uu> and <ww> vs <uw> based on time averages of the Burro 7-9 data from stations T1-T4 [2].

Rodean [8] evaluated the turbulence data from the three sources described above, selected three populations of Reynolds stresses

*The six Reynolds stresses, taken two at a time, form a total of 15 pairs; in a graph, the six stresses can therefore be connected by 15 links. The 15 links consist of nine production links and three redistribution links defined by the Reynolds stress transport equation plus three noncausal links. The production links involve all six Reynolds stresses, but the redistribution links involve only the three components of the turbulent kinetic energy. The noncausal links are between a turbulent energy component and a shear stress involving the other two velocity fluctuation components (<uu> and <vw>).

for analysis, and calculated complete sets of correlation coefficients for all 15 links connecting the six Reynolds stresses. In no case did he find statistical evidence for any of the three noncausal links. His results are summarized and compared with Stewart's proposal in Fig. 4.

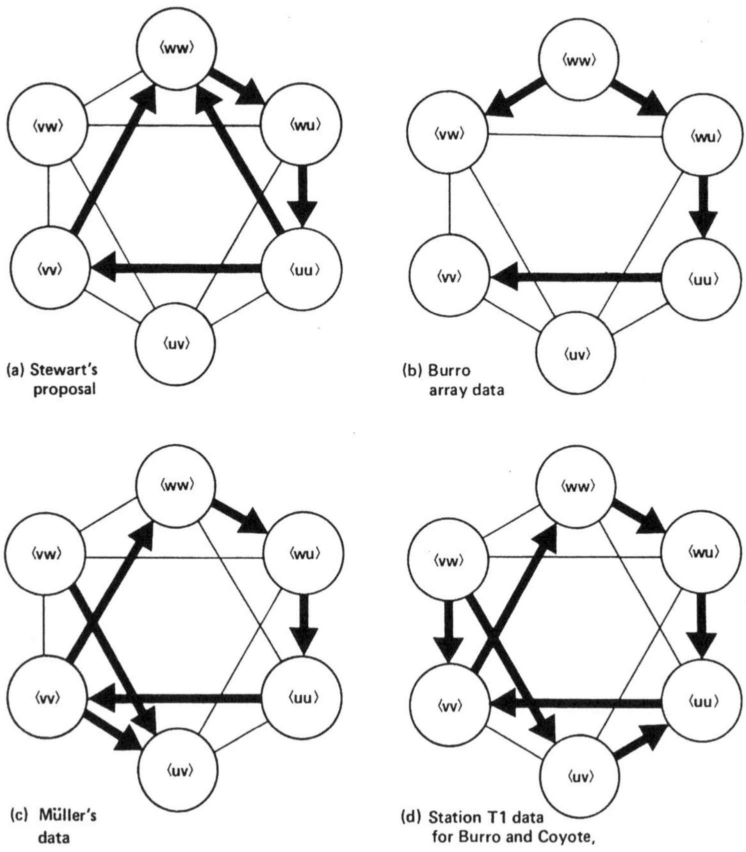

4. Coincident causal and statistical links together with the direction of the dominant causal process [8].

The directed graphs in Figs. 4b-4d have defects not present in Fig. 4a: all or some of the production and redistribution cannot be sustained. There is no feedback to <ww> in Fig. 4b, and there is no input to <vw> in Figs. 4c-4d. As noted by Rodean, buoyancy forces were present in the Burro and Coyote experiments and could have supplied the necessary input to <ww> for Fig. 4b and to <vw> for Fig. 4d. There were no buoyancy effects in Müller's experiments, but Rodean found some evidence for the <ww> → <vw> production term for Fig. 4c. With these additions, the directed graphs in Figs. 4b-4d are consistent with sustained production and redistribution.

Relations Among the Reynolds Stress Tensor Invariants

The Reynolds stress tensor has three invariants which are independent of the orientation of the coordinate axes [9]:

$$I_1 = [<uu> + <vv> + <ww>]/U^2, \qquad (3)$$

$$\begin{aligned} I_2 = & [(<uu><vv> - <uv>^2) \\ & + (<vv><ww> - <vw>^2) \\ & + (<ww><uu> - <uw>^2)]/U^4, \end{aligned} \qquad (4)$$

$$\begin{aligned} I_3 = & [<uu><vv><ww> - <uu><vw>^2 \\ & - <vv><uw>^2 - <ww><uv>^2 \\ & + 2 <uv><vw><uw>^2]/U^6. \end{aligned} \qquad (5)$$

Schumann [10] examined the necessary conditions for realizability of Reynolds stress turbulence models, and noted that the three invariants must be equal to or greater than zero.

Rodean and Cederwall [2] plotted I_2 and I_3 vs I_1 for the Burro data (Fig. 5), and found strong trends for I_1 to be proportional to $\sqrt{I_2}$ and to $\sqrt[3]{I_3}$. Rodean [8] extended this analysis to the T1 data for the Burro and Coyote experiments and to Müller's data. The ratios $<\sqrt{I_2}>/<I_1>$ and $<\sqrt[3]{I_3}>/<I_1>$ for the three data sets, together with tneoretical upper and lower limits, are given in Table 1. The upper limits are for ideal isotropic turbulence, and the lower limits are those given by Schumann. Note that the ratios for Müller's data are quite close to those for isotropic turbulence, and that the ratios for the Burro and Coyote data are characteristic of more anistropic turbulence. It is clear from the correlation coefficients (almost unity) that there are strong linear relations between I_1 and $\sqrt{I_2}$ and between I_1 and $\sqrt[3]{I_3}$ for these data sets.

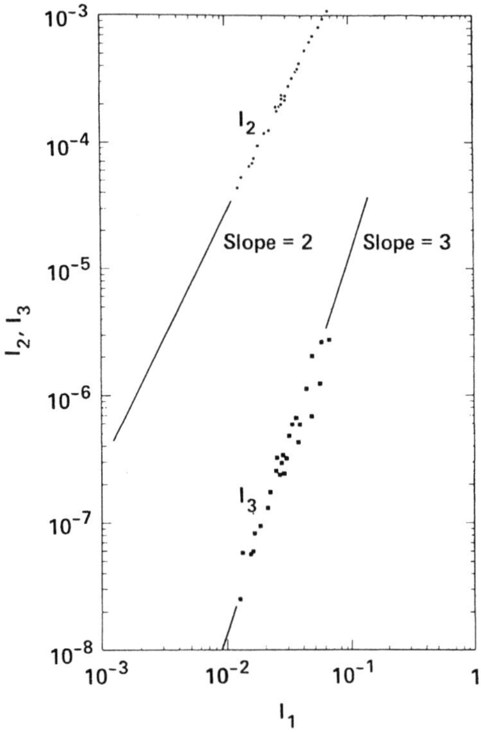

5. The Reynolds stress tensor invariants I_2 and I_3 vs I_1 based on averages of the Burros 7-9 data from stations T1-T4 [2].

Table 1.
Reynolds Stress Tensor Invariant Statistics.

Data or Theoretical Limit	$\langle\sqrt{I_2}\rangle/\langle I_1\rangle$	$\langle\sqrt[3]{I_3}\rangle/\langle I_1\rangle$	$r(I_1\sqrt{I_2})$	$r(I_1\sqrt[3]{I_3})$
Upper Limit	$1/\sqrt{3}$ = 0.5774	1/3 = 0.3333	--	--
Müller's Data	0.5763	0.3289	0.9999	0.9996
Burro Array Data	0.5131	0.2303	0.9986	0.9705
Station T1 Data	0.4761	0.1921	0.9969	0.9854
Lower Limit	0	0	--	--

Conclusions

One LNG spill, Burro 8, had significant effects on both the mean flow and the turbulence at and downwind of the spill pond. The atmospheric conditions permitted the gravity flow of the cold, dense gas to be almost independent of the atmospheric boundary layer. The Burro 8 data are not sufficient for a quantitative description of the effects of the LNG spill on turbulence. More comprehensive measurements are needed to obtain data for the range of density differences associated with heavy gas spills. Data for turbulent scalar fluxes as well as Reynolds stresses are required. As demonstrated with three large populations of measured Reynolds stresses, a statistical procedure can be used to identify active causal links among the Reynolds stresses. This procedure can be extended to include the buoyancy terms associated with heavy gas spills. For given conditions of turbulent flow (as demonstrated with three sets of data), there are distinct relations among the three invariants of the Reynolds stress tensor: I_1 is linearly proportional to $\sqrt{I_2}$ and to $\sqrt[3]{I_3}$. The statistical procedure for identification of active causal links among the Reynolds stresses and the relations among the Reynolds stress tensor invariants can be used in the development of turbulence closure models.

References

1. Koopman, R.P.; Cederwall, R.T.; Ermak, D.L.; Goldwire. H.C., Jr.; Hogan, W.J.; McClure, J.W.; McRae, T.G.; Morgan, D.L.; Rodean, H.C.; Shinn, J.H. Analysis of Burro series 40-m³ LNG spill experiments. J. Haz. Mat. 6 (1982), 43-83.

2. Rodean, H.C.; Cederwall, R.T. Analysis of turbulent wind-velocity and gas-concentration fluctuations during the Burro series 40-m³ LNG spill experiments. Lawrence Livermore National Laboratory, Livermore, California, UCRL-53353, 1982.

3. Goldwire, H.C., Jr.; Rodean, H.C.; Cederwall, R.T.; Kansa, E.J.; Koopman, R.P.; McClure, J.W.; McRae, T.G.; Morris, L.K.; Kamppinen, L.; Kiefer, R.D.; Lind, C.D.; Urtiew, P.A. Coyote series data report, LLNL/NWC 1981 LNG spill tests: dispersion, vapor burn, and rapid phase transition. Lawrence Livermore National Laboratory, Livermore, California (in press), 1983.

4. Müller, U.R. Measurement of the Reynolds stresses and the mean-flow field in a three-dimensional pressure-driven boundary layer. J. Fluid Mech., 119 (1982), 121-153.

5. Launder, B.E. Reynolds stress closures-status and prospects. AGARD Conference Proceedings, No. 271, 1979.

6. Stewart, R.W. The problem of diffusion in a stratified fluid. Adv. in Geophysics, 6 (1959), 303-311.

7. Feller, W. An introduction to probability theory and its applications. Vol. I, 3rd ed., New York: John Wiley and Sons, Inc., pp. 236-237, 1968.

8. Rodean, H.C. Statistical and causal relations among the six Reynolds stresses in boundary layers. Submitted to J. Fluid Mech. Lawrence Livermore National Laboratory, Livermore, California, UCRL-89402, 1983.

9. Jaeger, J.C.; Cook, N.G.W. Fundamentals of rock mechanics, London: Chapman and Hall, Ltd., pp. 9-24, 1971.

10. Schumann, U. Realizability of Reynolds-stress turbulence models. Phys. Fluids, 20 (1977), 721-725.

139r/22r

Temperature Measurements in a Negatively Buoyant Round Vertical Jet Issued in a Horizontal Crossflow

A. BADR

Commissariat à l'Energie Atomique

IPSN/DAS/SAER
B.P. 6 . 92260 Fontenay-aux-Roses - France

Summary

The problem considered here is the vertical discharge of round negatively buoyant jets through a horizontal crossflow. Laboratory experiments are performed in a flume : fresh water is emitted vertically through warm water. Temperature measurements are undertaken along verticals in the jet axis. It is found that the mean temperature values can be plotted as similarity diagrams. Some experimental correlations are deduced for the maximum height of jet rise. Turbulent temperature quantities do not appear obeying to similarity in the ascending part of the jet. Asymmetry is observed for the various profiles. The intermittency zone was also investigated in some cases.

1. INTRODUCTION

In this work we are concerned with the continuous vertical discharge of an effluent whose density is greater than that of the receiving body and where a horizontal uniform current is assumed to be present. This study is restricted to the region preceding the point where the heavy jet would ultimately fall to the ground. For this purpose, experiments have been performed in a flume : fresh water was discharged through a chimney into flowing warm water. This arrangement resulted in a negative buoyant force and an upward initial momentum. Temperature measurements were performed and the following quantities were determined from the measurements :

- the thermal axial trajectory
- the jet width
- the vertical mean temperature profiles
- the vertical temperature mean standard-deviation profiles.

Defining the non-dimensional numbers K and Fr_o as :

$$K = U_o/U_\infty \quad \text{(ratio of initial to ambient velocity)}$$

and $$Fr_o = U_o^2 \,/\, \left(g \frac{\rho_o - \rho_\infty}{\rho_\infty} R_o \right) \quad \text{(initial densimetric Froude number)}$$

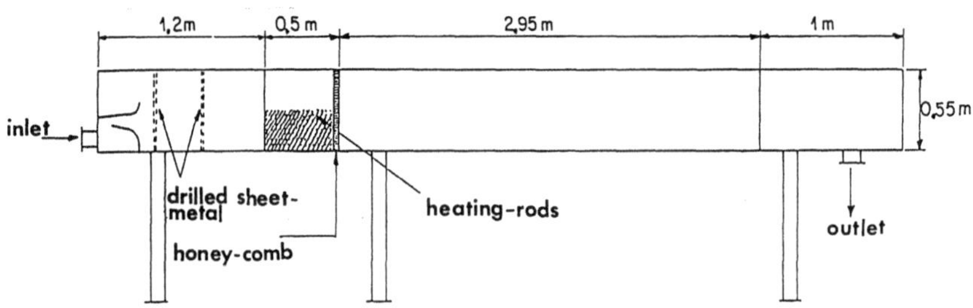

Figure 1 : Schematic view of the flume

14 experimental runs were undertaken with :

$$2.5 \leqslant K \leqslant 14 \quad \text{and} \quad 156 \leqslant Fr_o \leqslant 2020$$

In the following sections, we present the experimental apparatus and the results obtained (for more details, see BADR / 1 /).

2. EXPERIMENTAL SET-UP AND APPARATUS

2.1 Flume : the experiments were conducted in a 5.65 meter long flume shown schematically in Figure 1. The flume has a cross-section of 50cm wide by 55 cm deep. The flow speed U_∞ within the flume could be controlled between 4cm/s and 10cm/s. Some grids, a honey-comb and other devices ensured the tranquillization and uniformity of the flow in the flume. Velocity measurements have shown that in the central part of the flume cross-section the relative velocity variation $\delta U_\infty / U_\infty$ is generally less than 10 %, and the velocity fluctuations were found to be less than $0.03U_\infty$. 128 cylindrical heating rods were placed just before the honey-comb. They provided a maximum power of 40kw which proved sufficient to increase the temperature of the water in the flume from 12°C to 40°C in less than two hours. Once the desired temperature T_∞ was reached, the heating apparatus was put off and T_∞ was kept constant by means of a heating wire placed at the flume inlet. The temperature distribution during the steady state was found to be practically uniform with a mean standard-deviation $\sqrt{\overline{T'^2}}$ of about 0.02°C.

2.2 Jet discharge : the jet is emitted vertically through the crossflow by means of a 9cm chimney, coated with PVC and having and internal diameter of 1 cm. This chimney is fixed at the bottom of the flume on its longitudinal axis at 317.5 cm from the flume entrance. The temperature T_o

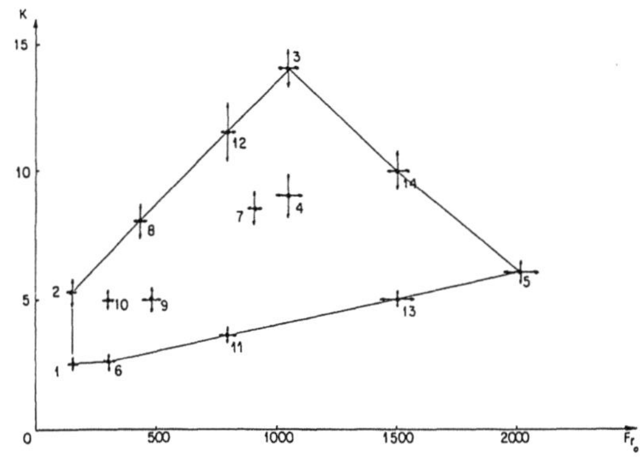

Figure 2 : Experimental domain of measurements

of the jet water varied between 11°C and 15°C, the PVC coating allowed an almost uniform temperature distribution at the jet exit. The jet initial Reynolds number was varied between 2000 and 5000. The greatest value was chosen in order to get jet large structures smaller than the transverse dimension of the flume. The smallest value of the jet initial Reynolds number was chosen in accordance with TOWNSEND's / 2 / principle for having trajectories not depending on the initial Reynolds number.

2.3 Instrumentation and measurement procedure : the temperature measurements were undertaken by means of 0.5 mm diameter chromel-alumel inox coated thermocouples. One thermocouple located at the flume entrance allowed the checking of T_∞ . The other one located in the jet discharge circuit, allowed to check T_o . Finally, one thermocouple was used to measure the temperature along the jet axis. When the steady state was reached, the measurements in the jet were made point by point along verticals in the jet axis which supposed to coincide with the longitudinal flume axis. The thermocouples had a time response of 30ms and the results could be read directly on a digital voltmeter. The domain of measurements is shown in Figure 2 with the uncertainties associated for each run. Along with measurements of T and $\sqrt{\overline{T'^2}}$, a brief investigation of the intermittency zone was conducted for runs n°4 and 10.

2.4 Errors associated with the measurements : errors during the measurements could be due to : - non-uniformities of U_∞ .- Variations of T_∞ (~ 0.1°C) .- Variations of T_o ($\lesssim 0.3$°C) .- Location of the point of measurement (2 to 3 mm) .- Displayed values : the absolute error due to the thermocouples was 0.1°C . - Averaging time : T and $\sqrt{\overline{T'^2}}$ were averaged during 1 mn. However many measurements were repeated to be sure of the results convergence. Also some runs were repeated completely to check the reproductibility of the measured values.

3. EXPERIMENTAL RESULTS

3.1 Axial trajectory : the jet trajectory is defined as being the locus of minimum temperature on the jet longitudinal plane of symmetry. One example is shown in Figure 3 for run n°9 : the trajectory is in non-dimensional coordinates and the points represent the experimental values (z_{cl} is determined with a relative error of about 10-15%).

3.2 Axial temperature decrease : one typical result is shown in Figure 4 for run n°9. The axial temperature difference $\Delta T_{cl} = |T_\infty - T_{cl}|$ is non-dimensionalized with respect to the initial temperature difference $\Delta T_o = |T_\infty - T_o|$. It is found that there is a fast decrease of ΔT_{cl}. (Note that the relative error on ΔT_{cl} is about 15-20%).

3.3 Half-width : the half-width R shown in Figure 5 for run n°7 is defined as the distance to the jet axis where $(T_\infty - T) = (T_\infty - T_{cl}) / e$, e being the Neper number. The relative error in the determination of R is high (about 30 %).

3.4 Vertical mean temperature profile : an example of vertical temperature profile is shown in Figure 6 for run n°10. $\Delta T = |T - T_\infty|$ is non-dimensionalized with respect to ΔT_{cl} , and radial distances r are non-dimensionalized with respect to the lower and upper half-width R^*_- and R^*_+. The star * denoting that this time the half-width is defined as being the distance to the jet axis where $\Delta T = 0.5 \Delta T_{cl}$. Two values for R^* have been chosen because it was observed that the length scales characterizing the big vortices differ between the upper and the lower region of the jet. The main features on the vertical mean temperature profile have been :

- similarity,
- asymmetry of the temperature distribution. For the whole of the experiments, it was observed that $R^*_- \simeq 1.5 R^*_+$.

3.5 Vertical temperature mean standard-deviation profile : a typical result is shown in Figure 7 for run n°10. The standard-deviation $\sqrt{\overline{T'^2}}$ is non-dimensionalized with respect to ΔT_{cl}. The characteristics of this profile are the following :

- no definite similarity : the best regrouped points correspond to the near horizontal portion of the jet trajectory.

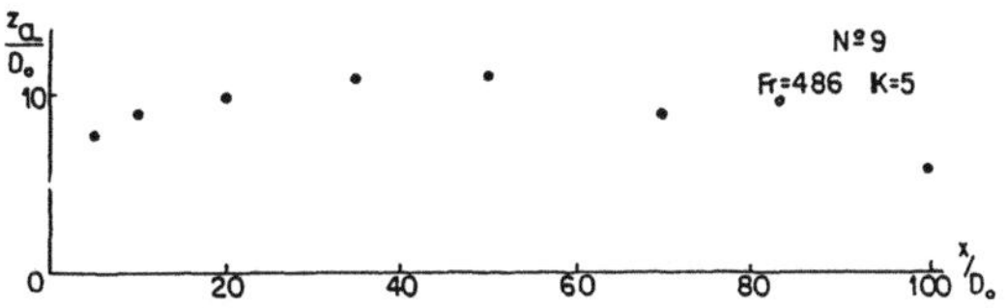

Figure 3: Axial Trajectory

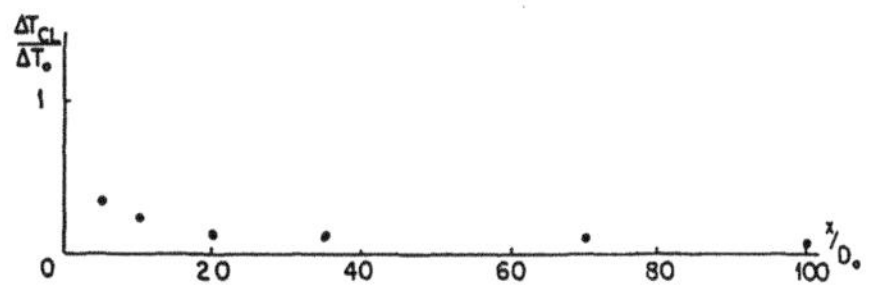

Figure 4: Jet Axial Temperature (Run N°9)

- Very pronounced asymmetry, especially near the source.

- Two extrema for $\sqrt{T'^2}$ are observed in the upper and lower region of the jet. The greatest values for $\sqrt{T'^2}$ being found in the ascending phase of the jet.

3.6 Intermittency region : the jet flow is of intermittent nature, especially toward the boundary region. There, also ΔT can be very small, the possibility of getting instantaneous could puffs may be important from a practical point of view. This is why some measurements for the intermittency factor I were undertaken. Theoretically, $I(x,z) = \lim \frac{1}{t'} \int_0^{t'} I(x,z,t)dt$ (t and t' are times)

where $I(x,z,t) = 1$ if $|T(x,z,t) - T_\infty| > T_k$

and $\quad I(x,z,t) = 0$ if $|T(x,z,t) - T_\infty| \leq T_k$

Here T_k is an estimate of the lowest temperature difference that can be accurately measured. The results for I are shown in Figure 8 for run n°10. A graphical recording of the delivered signal was also made for the same run at two positions in the upper and lower region (see Figure 9).

4. DISCUSSION OF EXPERIMENTAL RESULTS

4.1 Axial trajectory and temperature measurements : in order to get a greater universality in the results obtained, the following length scales are used (in accordance with an approach developed by WRIGHT / 3 / and LIST and IMBERGER / 4 / :

$$l_m = \frac{\sqrt{\pi R_o^2 \; U_o^2}}{U_\infty}$$ scale relating to the interaction of a momentum-dominated jet with the crossflow

$$l_b = g \, \frac{\rho_o - \rho_\infty}{\rho_\infty} \; \frac{\pi R_o^2 \; U_o}{U_\infty}$$ scale similar to l_m for buoyancy-dominated flow

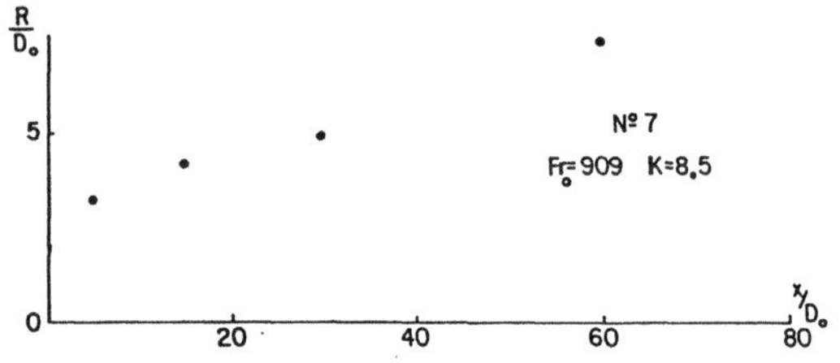

Figure 5: Jet Half-Width

Figure 6: Vertical Mean Temperature Profile

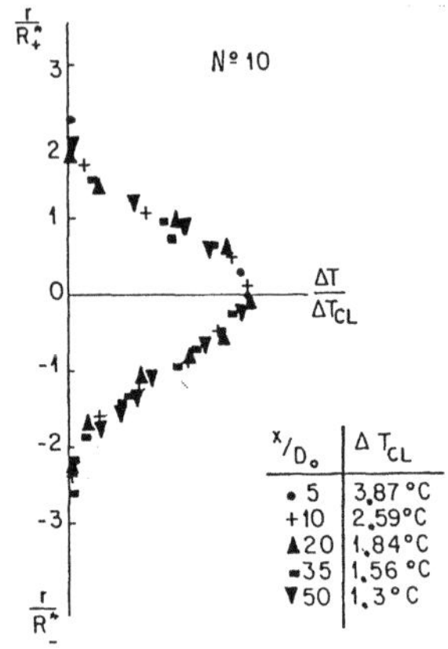

The behaviour of plumes and jets are generally momentum-dominated near the source and buoyancy-dominated farther. The intermediate region momentum or buoyancy-dominated depending on whether l_m/l_b is greater or smaller than 1. Here $l_m/l_b > 1$ at all runs.

Referring to an integral model developed by CHU / 5 / which neglects dynamic drag and is not applicable when K is smaller than 4, it has been possible to assemble most of the results concerning the trajectory into one curve (see Figure 10) using the following non-dimensional coordinates :

$$(x / l_m Fr_\infty , z_{cl} / l_m Fr_\infty^{1/3}) \text{ with } Fr_\infty = \frac{U_\infty}{g \frac{\rho_o - \rho_\infty}{\rho_\infty} R_o} \propto l_m / l_b$$

At the exception of runs n°1, 6 and 11 for which K<4 the degree of unification obtained for the results is satisfactory. One practical consequence is that $z_{max} / l_m Fr_\infty^{1/3}$ is constant when $K < 4$. CHU found this constant equal to 1, and in the experiments presented here it is approximately equal to 1.1. Other experiments performed by STOLZENBACH and ADAMS / 6 / led to a constant equal to 1.6. However their velocities ratio K varied from 10 to 100.

Concerning the axial temperature decrease, when it was drawn as in Figure 11, it was found to obey to a relation of the form
$\Delta T_{cl} / \Delta T_o \propto (x / D_o)^{-2/3}$.

And at z_{max} , it was observed that (when $K \geq 4$): $\Delta T_o / \Delta T_{cl} \; K Fr_\infty^{2/3} \simeq 0.5$ which is comparable to the value obtained by CHU (0.4).

4.2 Vertical profiles of T and $\sqrt{\overline{T'^2}}$:the peculiarities of these profiles lie in the asymmetry observed, the extrema for $\sqrt{\overline{T'^2}}$ and the similarity (or lack of similarity) of the profiles.

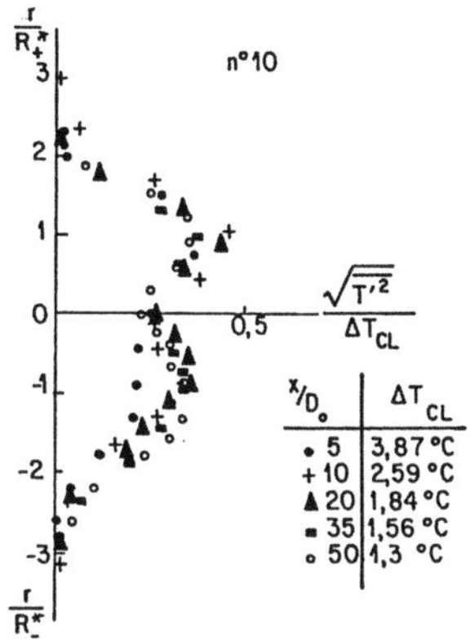

Figure 7 : **Vertical Temperature Standard-Deviation Profile**

The question of asymmetry and unequal extrema for $\sqrt{\overline{T'^2}}$ seem to be associated each other. Concerning the presence of extrema for $\sqrt{\overline{T'^2}}$, it is clear that they are due to the crossflow influence. If one examines a budget of $\overline{T'^2}$ established by ANTONIA et al. / 7 / in the case of an almost neutrally buoyant jet issued in a coaxial crossflow (Fig. 12), it is noticed that there is a maximum for the exchanges at $r/R^* \simeq 0.8$ especially for the production term.

The result corresponds well to the observations made here:the extrema for $\sqrt{\overline{T'^2}}$ are found for $0.7 \leqslant r/R^* \leqslant 1$. It is also noticed that only one extremum for $\sqrt{\overline{T'^2}}$ is found at distances very close to the source. It confirms the fact that the crossflow interacts differently between the zone of attack to the jet and the lee region behind it.

Also noticeable is the fact that the highest values for $\sqrt{\overline{T'^2}}$ appear in the upper region of the jet where $\partial T/\partial z > 0$ which implies a stable thermal configuration. Moreover, when one examines the balance equation for the turbulent kinetic energy, it appears that the buoyancy term is a sink for turbulence in the upper zone - where heat is transferred below - and a source in the lower region. On the other hand, and due to the unequal length scales between the upper and lower region of the jet, the production of turbulent kinetic energy by interaction with the mean motion - a term proportional to the velocity gradient - is greater in the upper region of the jet. A more precise study concerning the turbulent budgets is needed to clarify this point.

Finally, and concerning the profiles of T and $\sqrt{\overline{T'^2}}$, although the self-preservation is obvious for T, it doesn't appear clearly for $\sqrt{\overline{T'^2}}$. If we refer to a study of plane buoyant jets made by KOTSOVINOS / 8 / , it seems that the evolving contributions, within the jet flow, of the mechanical and

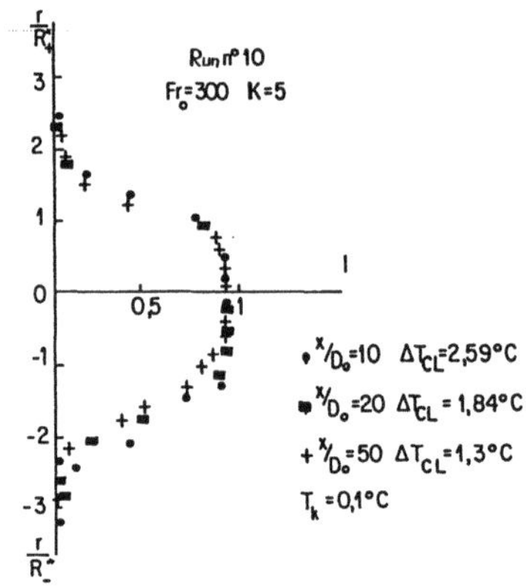

Figure 8 : Intermittency Factor Profile

thermal energy to the production of $\overline{T'^2}$ is behind this lack of similarity. However, since an equilibrium state must be ultimately reached, one can expect similarity at some stage.

4.3 Intermittency region : the intermittency factor measured allows a statistical information while the graphical recording of the instantaneous signal delivered by the thermocouples gives a direct insight. Concerning the intermittency factor, the following remarks can be made (see Figure 8):

- at fixed r/R^* , I decreases with increasing x/D_o.
- In the flat central zone $I \simeq 1$. However some ambient fluid is seen from time to time.
- The central zone ends sharply at $r/R^* \simeq 0.9$ to 1 , that is approximately where $\sqrt{\overline{T'^2}}$ is maximum.

The graphical recording shown in Figure 9 corresponds to a lapse of time of 550s. The puffs of jet fluid are more frequent in the lower zone than in upper one, but they correspond to lower $\Delta T = |T - T_\infty|$. The puffs in the lower zone are such that $\Delta T \simeq 0.5 \Delta T_{cl}$ while they can be greater than $2 \Delta T_{cl}$ in the upper zone. In both cases, these cold puffs are associated with the jet energetic structures of low frequencies. These results show in an obvious manner the possibility of getting relatively cold puffs in the outer jet region.

5. CONCLUSION

The use of length scales characterizing buoyant jets behaviour has provided a mean for unifying most of the results concerning the trajectory. Some practical correlations have then been deduced, but more work is needed to better precise the range of these correlations. It has been shown that the mean temperature in the jet can be described by radially similar pro-

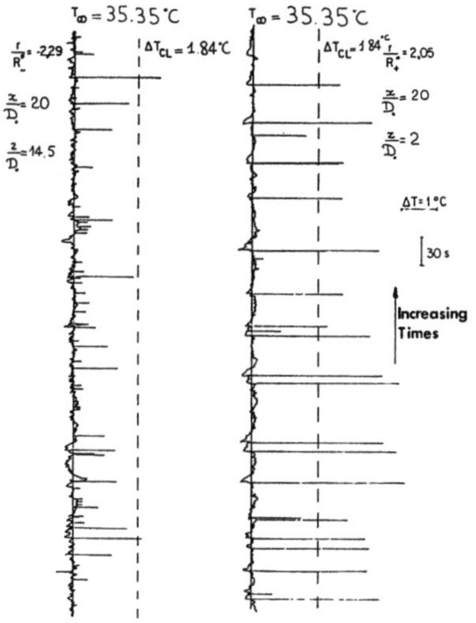

Figure 9 : Part of the Graphically Recorded Signal For Run No 10

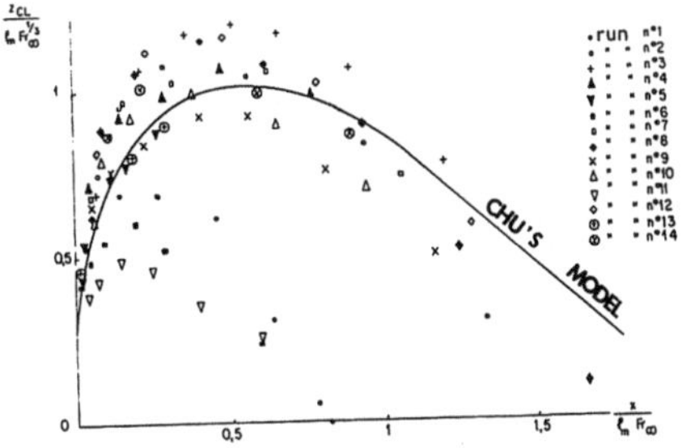

Figure 10 : Experimental Results for the Axial Trajectory Compared to CHU's (1975) Model

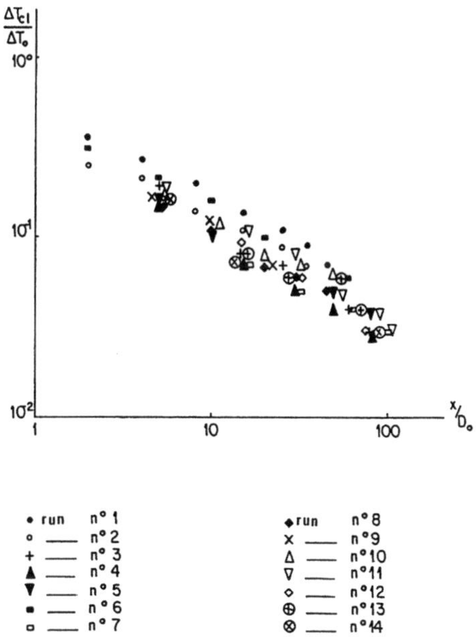

Figure 11: Axial Decrease of $\Delta T_{cl}/\Delta T_o$

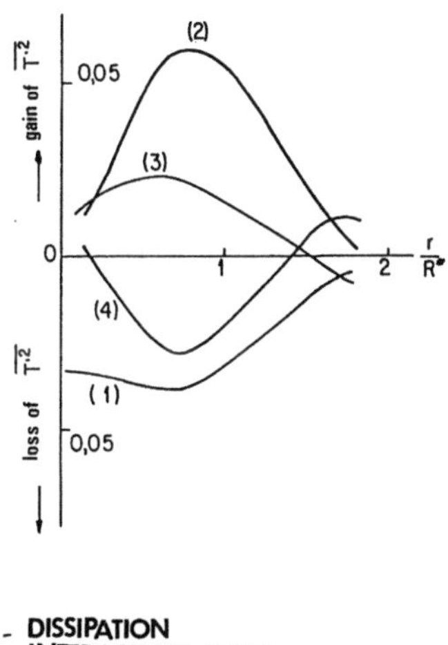

1 - DISSIPATION
2 - INTERACTION WITH THE MEAN TEMPERATURE GRADIENT
3 - ADVECTION
4 - DIFFUSION

Figure 12 : $\overline{T'^2}$ Budget (Antonia et al. 1975). $F_{r_o} = 48500$

files. That did not seem to be the case for the temperature mean standard-deviation, at least in the ascending part of the jet.

A pronounced asymmetry has also been observed as well as unequal extrema for $\sqrt{\overline{T'^2}}$ at $r/R^* \simeq 0.8$ to 1, that is where the interactions between the jet and the crossflow are the most intense. A detailed turbulence budget is needed to clarify some peculiarities encountered with these profiles, especially concerning the fact of having the greatest extremum for $\sqrt{\overline{T'^2}}$ in the upper jet region. These differences of behaviour between the upper and lower jet region were confirmed once more when examining the intermittency region, and cold puffs with relatively large $|T - T_\infty|$ have been observed inside this region.

ACKNOWLEDGMENTS

The writer would like to thank Dr. D.GRAND from the CEA/CEN.G/STT and Dr. E.J. HOPFINGER from the University of Grenoble whose guidance and comments were essential.

References

1. Badr, A. : Contribution à l'étude des jets radioactifs et des jets lourds émis en présence d'un courant traversier. Thèse de Docteur-Ingénieur . Institut National Polytechnique de Grenoble (1981).

2. Townsend, A.A. : The structure of Turbulent Shear Flow . Cambridge University Press (1956).

3. Wright, S.J. : Effects of Ambient Crossflow and Density Stratification on the Characteristic Behavior of Round Turbulent Buoyant Jets . W.M. Keck Laboratory of Hydraulics and Water Resources. California Institue of Technology. Pasadena, Report KH-R-36 (1977).

4. List, E.J. and Imberger, J. : Turbulent Entrainment in Buoyant Jets and Plumes". Journal of the Hydraulics Division, proc. ASCE, HY9,pp.1461-1474 (1973).

5. Chu, V.H. : Turbulent Dense Plumes in Laminar Crossflow . Journal of the Hydraulic Research, 13, n°3, pp.263-279 (1975).

6. Stolzenbach, K.D, Adams, E.E. : Submerged Discharges of Dense Effluent. Proceedings of the 2nd International IAHR Symposium on Stratified Flows. Trondheim. Norway, vol.2, pp.832-844 (1980).

7. Antonia, R.A., Prabhu, A; Stephenson, S.E. : Conditionally Sampled Measurements in a Heated Turbulent Jet. Journal of Fluid Mechanics, Vol. 72, part 3, pp.455-480 (1975).

8. Kotsovinos, N.E. : A study of the Entrainment and Turbulence in a Plane Buoyant Jet". Ph.D.Thesis. California Institute of Technology, Pasadena. 1975.

Gravity Spreading and Air Entrainment by Heavy Gases Instantaneously Released in a Calm Atmosphere

J. A. Havens and T. O. Spicer

Chemical Engineering Department
University of Arkansas, Fayetteville

This work was supported by U.S. Coast Guard Contract DTCG23-80-C-20029 with the University of Arkansas. The opinions or assertions contained herein are the private ones of the writers and are not to be construed as official or reflecting the views of the Commandant or the Coast Guard at large.

Summary

Measurements of gravity spreading and dilution of right circular cylindrical volumes of dense gas released instantaneously in still air are described. Effects of volume released (34, 54, and 135 liters), initial density (2.19, 2.95, and 4.27 relative to air), and initial volume height-to-diameter ratio (0.4, 1.0, and 1.6) are reported. Ground level peak gas concentrations measured are compared with previously reported smaller scale experimental measurements, and scaling relationships are demonstrated.

Introduction

Extensive experimental data useful for testing mathematical models of atmospheric dispersion of denser-than-air gases have recently become available from field test (1,2,3) and wind tunnel test (4,5) programs. The data generally confirm expectations that for large releases at low wind speeds an important part of the gas dilution process results from turbulence generated by gravity-driven flows. The relative importance of turbulent mixing due to the gravity-driven flow and the turbulent mixing properties of the atmospheric flow is determined by the (heavy) gas source flux into the atmosphere, its thermodynamic properties (primarily temperature and molecular weight), and the atmospheric flow conditions. If the gas flow into the atmosphere alters the preexisting boundary layer flow, the resulting dispersion may not be accurately described by models developed for passive atmospheric dispersion processes. Limiting cases for such behavior result for large release rates at low wind speed and in calm air releases. Field test data for low-wind, large scale releases are still scant, and wind tunnel simulations of large releases at low wind speed are difficult due to the requirement for impractically low wind tunnel velocities.

Since there are strong indications that an important part of the initial dilution (to ~1% gas) of rapid large releases of heavy gases in low wind speeds is the result of turbulence generated by the resulting gravity-driven flows, we believe that an understanding of the dilution processes extant in calm air releases is prerequisite to the accurate prediction of dispersion of large scale releases in low winds.

We report here initial measurements of the gravity spreading and dilution of right circular cylindrical volumes of isothermal gas released instantaneously into calm (laboratory) air.

Description of Experimental Techniques

Our laboratory has been designed to allow the time-dependent and instantaneous release of up to about 1 m^3 dense gas in the center of a flat 7.2 m diameter circular area. Concentration measurements are provided at eight or fewer positions anywhere in the spill area (including vertically to a height of about 1 meter). Photographic documentation of smoke-marked releases utilizes motor-driven 35 mm cameras and a high speed 16 mm movie camera. Our initial studies have focused on instantaneous releases of cylindrical gas volumes with different size, initial density, and initial height-to-diameter ratio. In the following we describe the methods used to effect an instantaneous release and the measurement of gas concentrations at fixed positions in the developing cloud.

Release Method

Figure 1 shows a sector of the release area surrounding a 54 liter gas container. The gas container is a 1/8 inch thick polycarbonate sheet rolled to form a cylinder with vertical exterior support ribs which extend above the cylinder to a fitting attached to the end of a rod in a pneumatic cylinder. The pneumatic cylinder is rigidly mounted in a framework hung from roof support beams. A solenoid valve operated by the computer control and data acquisition system admits air under the pneumatic cylinder piston for a designated time period, moving the gas container vertically past the gas volume. The cylinder travel time is controlled by the operating pressure of the air supply line and by the length of time the solenoid valve is maintained open. Cylinder travel time is measured by timing the passage of a reflective tape marker on the cylinder between light beams projected from optical fibers mounted to the side of the container.

Container removal rates were studied using smoke-marked gas volumes to determine operating conditions required to leave a freestanding, minimally perturbed, cylindrical gas volume after the container had risen above the gas. Figure 2 shows high speed 16 mm movie frames of the vertical travel of a 135 liter container (D / H = 1.0) initially containing smoke-marked argon. The second frame indicates that the bottom of the container is past the top of the gas volume 0.18 seconds after its vertical movement began, and the gas is shown to be freestanding with essentially no movement or perturbation.

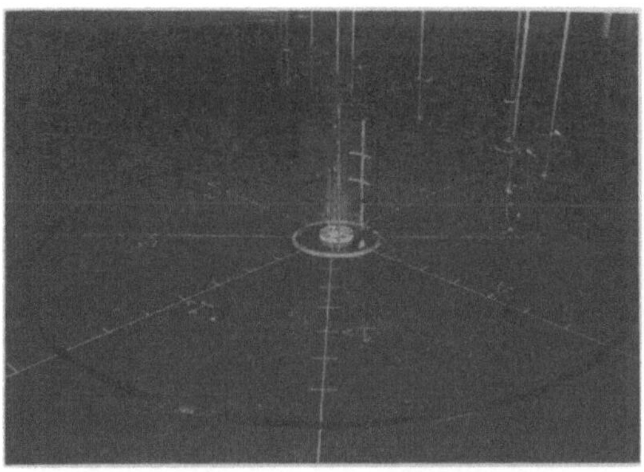

Figure 1. Gas container and concentration sensor placement.

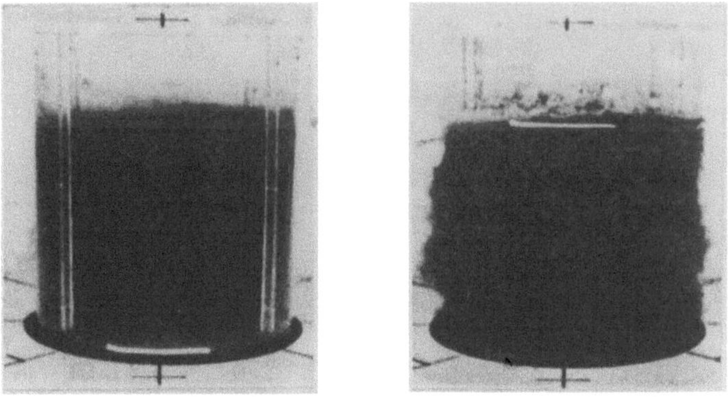

Figure 2. Freestanding gas cylinder initial condition.

Figures 3 and 4 show a filled 135 liter container and the released spreading cloud (taken 1.2 seconds after release) respectively. Figures 5 and 6 show overhead views of a 135 liter Freon-12 release. The gas container is just hidden under the square plate which is part of the release mechanism framework. The edge of the spreading gas cloud has advanced to a radial distance of 1.5 m in Figure 5 and 2.0 m in Figure 6. The radial symmetry of the cloud is clearly indicated. Observations of the cloud's movement beyond the edges of the area photographed confirm radial cloud advance to distances at which the peak gas concentration at floor level has decreased to at least 1% of the initial value.

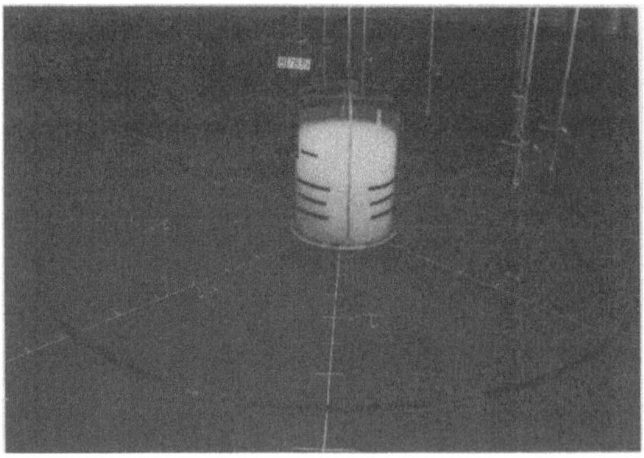

Figure 3. 135 liter Freon-12, H/D = 1.0.

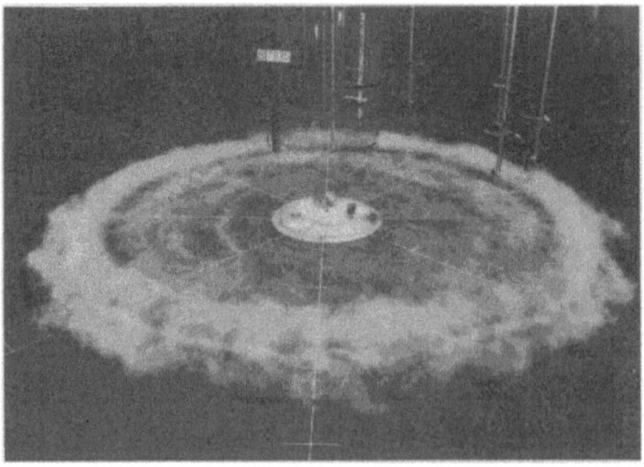

Figure 4. 135 liter Freon-12 release, t = 1.2 s.

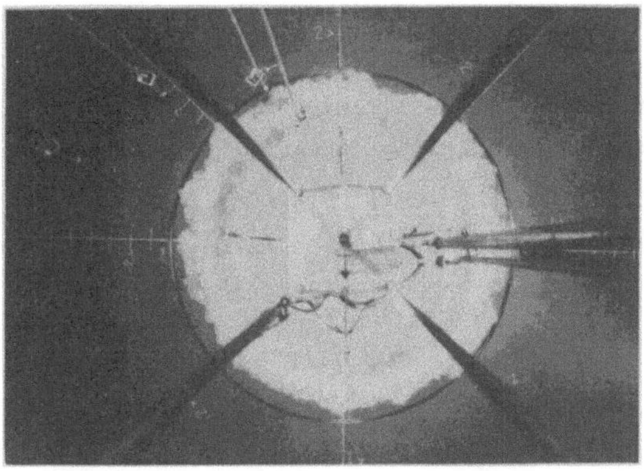

Figure 5. 135 liter Freon-12 release, t = 1.2 s.

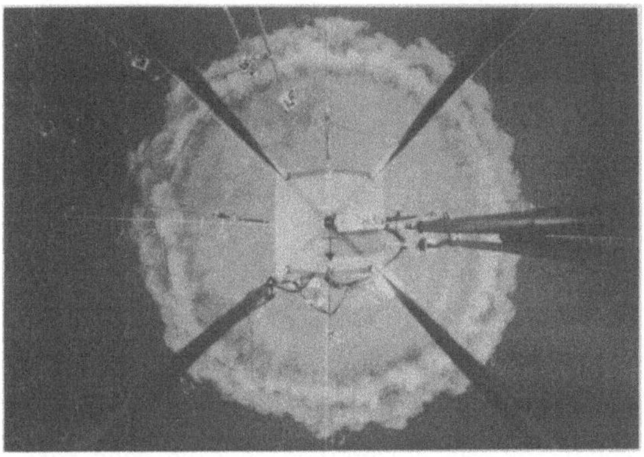

Figure 6. 135 liter Freon-12 release, t = 1.7 s.

For the releases discussed here, the spreading gas rapidly formed a torus or doughnut shape as observed in previous field (3) and wind tunnel (4,5) still-air releases.

Gas Concentration Measurement

Figure 1 shows a sector of the spill area indicating the placement of support rods on which the gas sensors are mounted; sensors are positioned to avoid interference in the flow caused by other sensors. Figure 7 shows a sensor mounted on a support rod. A vacuum pump aspirates gas through a

4 mm diameter sample port fitted with a fibrous filter; the sample flows over a 4 μ wire or 25 μ film mounted on a TSI 1260 probe, and then through a 400 μ diameter choke. The aspiration rate with the 400 μ choke, used in most of the measurements described here, is approximately 1.5 liters/minute, although some measurements have been made with aspiration rates as low as 300 ml/min. The high aspiration rates have been used to maximize the resolution of the peak concentrations in the cloud.

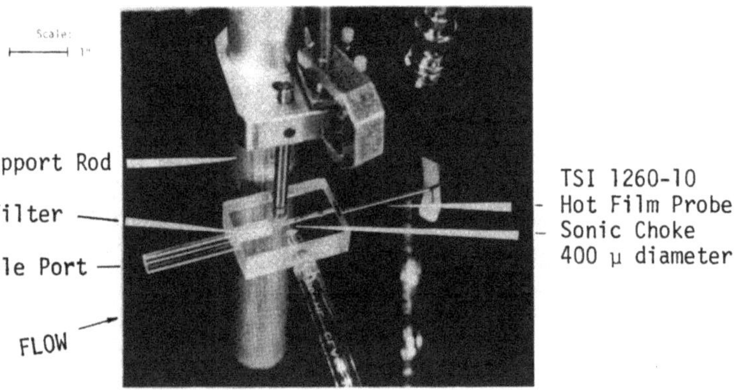

Figure 7. Aspirated hot film gas concentration sensor.

The hot wires or films were operated at overheat ratios of 1.32 and 1.16 respectively, corresponding to an operating temperature of about 85°C. This operating condition was determined to give good resolution of the concentration of Freon-12/air mixtures without appreciable deterioration of the sensors experienced in high Freon-12 concentration, high overheat ratio usage.

The output of the TSI 1053B anemometers is fed to a reference voltage shifting circuit, through a low pass filter (100 Hz) and amplifier, and input to a DEC MINC/11-23 computer data acquisition system. Gas concentration measurements were made at 240 Hz for 50 seconds after release for the experiments reported here.

The gas container is filled by introducing the test gas at the bottom of the open-topped cylinder through a distribution plate with eight radial outlets to minimize mixing effects due to gas jetting. Horizontal slots cut in the container wall determine the gas height when filled.

Experiments indicate that gas addition at 10 liters/minute for a period sufficient to add twice the gas container volume results in a heavy gas interface at the location of the horizontal slot which is very sharp, with the gas concentration below the slot estimated to be greater than 98% pure heavy gas. During cylinder filling, the excess gas overflows through these horizontal slots down the exterior wall of the container and is dispersed by four small (100 CFM) axial flow instrument cooling fans placed at floor level about one cylinder radius away, at 90° angles.

After filling, all personnel move to an adjacent room housing the data acquisition and experimental control computer system, and the room containing the spill area is sealed off. A calming period of ten minutes is observed, after which the computer sequences the actuation of photographic lights and cameras (when used), the raising of the gas container, and the data acquisition from the gas sensors (when used).

Gas calibration mixtures are fed to the gas sensors just before cylinder filling and again at the end of a series of three repeat releases. Clear air readings are also made between releases to correct for sensor drift which may result primarily from change in air temperature or pressure and secondarily from other factors such as sensor aging and electronic circuitry drift. We estimate the gas concentration measurements reported to be accurate to within about 2% of reading in the range 50-100%, 4% of reading in the range 25-50%, 10% of reading in the range 5-25%, and 20% of reading in the area of 1% concentration, based on analysis of the sensors' drift characteristics. Primary measurement of the mixture concentration prepared by the rotameters was done by gas chromatography.

Test Results

Experiments to date have been designed to study the scaling characteristics of spills of different size, the effect of height-to-diameter ratio, and the effect of initial gas density, for instantaneous isothermal releases. Table 1 shows the experiments reported here.

Each of the experiments was repeated at least twice. (One repeat of Experiments 3 and 9 was discarded due to a failure to seal the gas container bottom during filling.) The experiments appear very reproducible. Figures 8 through 10 show typical ground level gas concentration measurements at three successive nondimensionalized radial distances ($R/V^{1/3}$)

from the release center. Three concentration time series, taken in three successive experiments, are shown for each radial position. In most cases, the variation of maximum concentrations is close to or within the expected accuracy of the concentration technique. Gas concentration measurements near the release indicate a complex, but remarkably repeatable, structure which appears to correlate well with complex flow patterns exhibited in the cloud leading edge near the release. Distinct peaks and valleys in the concentration records, as in Figure 8, appear to correlate with observed complex roll and wave structures in the cloud's leading edge near the release (see Figures 4, 6). As the cloud spreads radially, the complex frontal movement is seen to diminish, both in the photographs and in the gas concentration measurements.

TABLE 1
GAS RELEASE EXPERIMENT CONFIGURATIONS

Exp. No.	Gas and Density (ρ/ρ_{air})	Volume (m^3)	H/D	No. Runs
1	Freon-12, 4.27	.034	1.0	3
2	Freon-12, 4.27	.054	1.0	3
3	Freon-12, 4.27	.135	1.0	2
4	Freon-12, 4.27	.054	0.4	3
5	Freon-12, 4.27	.054	1.6	3
6	Freon-12/air, 2.95	.034	1.0	3
7	Freon-12/air, 2.95	.054	1.0	3
8	Freon-12/air, 2.95	.135	1.0	3
9	Freon-12/air, 2.19	.135	1.0	2

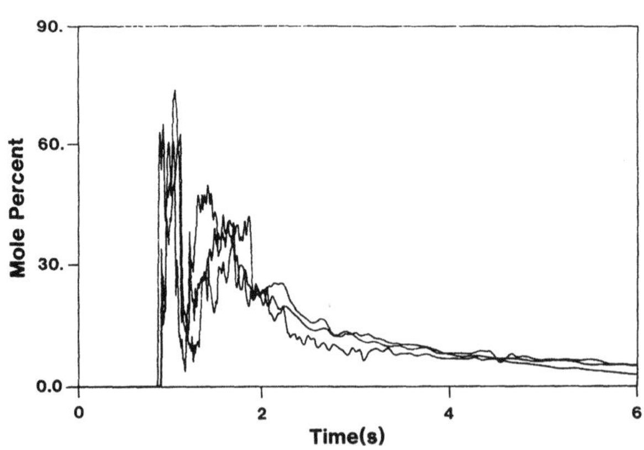

Figure 8. Three repeat gas concentration measurements at R* = 2.6 - Freon-12, V = 54 liters.

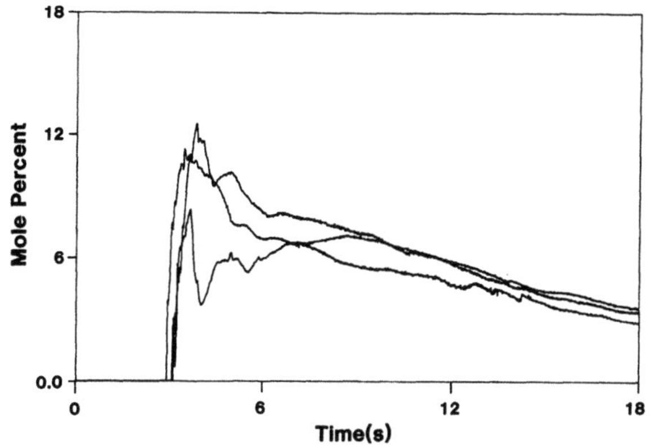

Figure 9. Three repeat gas concentration measurements at R* = 5.8 - Freon-12, V = 54 liters.

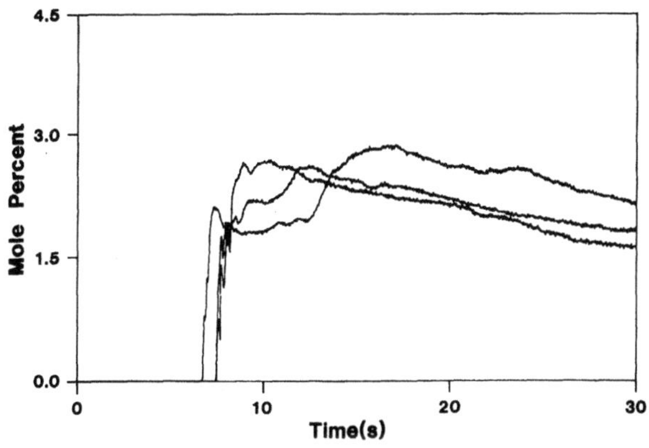

Figure 10. Three repeat gas concentration measurements at R* = 9.2 - Freon-12, V = 54 liters.

Hall (4) has suggested that calm air releases of cubic volumes of gas should have a characteristic length of $L = V^{1/3}$ and a characteristic time scale $T = V^{1/6} / \sqrt{g\Delta}$ where g = gravitational acceleration and $\Delta = (\rho_g - \rho_a) / \rho_a$. Figure 11 shows a plot of $t^{*\dagger}$, the non-dimensional cloud arrival time, vs. $(R^{*2} - R_0^{*2}) / 2^{\dagger\dagger}$ for the three release volumes tested

---

$^\dagger t^* = t/T$
$^{\dagger\dagger} R^* = R/L$

thus far. Integration of the "gravity intrusion" formula

$$\frac{dR}{dt} = K\sqrt{g\Delta H} \tag{1}$$

indicates that cloud radius should be proportional to $\sqrt{t}$. The line of slope 1 dashed on Figure 11 corresponds to that dependence of R on t for times greater than $t^* \cong 20$. Extrapolation of the dashed line indicates a value of $K \simeq 1.2$ in Equation (1).

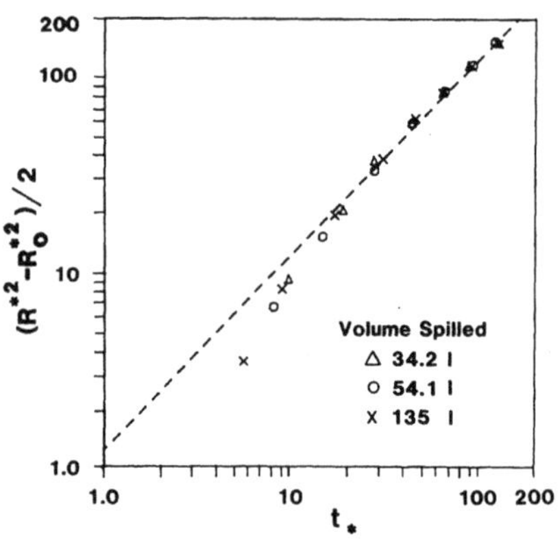

Figure 11. Cloud travel distance vs. time of arrival.

Figure 12 shows the peak measured concentration (nondimensionalized) plotted against R*, the nondimensionalized distance from the spill center. The data for all three spill volumes collapse well using the scaling rules proposed by Hall. The individual points on Figure 12 are averages of three repeat experimental runs, and the dashed lines define the outer boundaries on all peak concentrations measured for the H/D = 1 releases.

Figure 13 compares the results shown in Figure 12 with data from calm air/ Freon-12 (H/D = 1) releases reported by Hall (4) and Loymeyer and Meroney (5). Hall's releases were 2.7 liters, and Lohmeyer and Meroney's releases were .035, .165, and .450 liters. The results of our experiments indicate greater concentrations than Hall at the same dimensionless distances and higher concentrations than Lohmeyer and Meroney's results except for the 0.035 liter releases. We do not yet know whether the differences between our results and those reported by Hall and Lohmeyer and Meroney are due to scaling effects or to other factors associated with the experimental

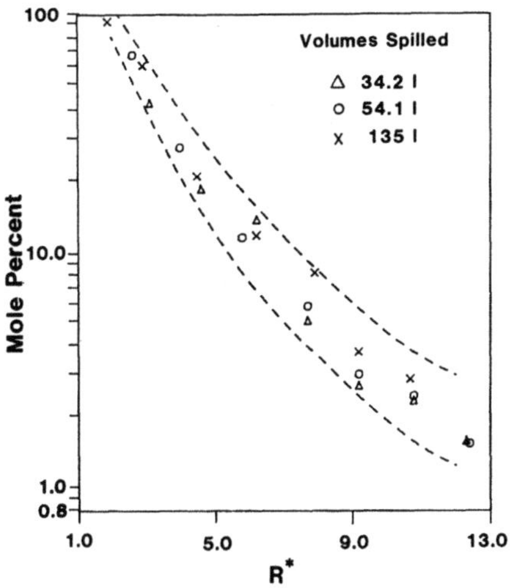

Figure 12. Concentration vs. R*--Freon-12, V = 34, 54, 135 liters.

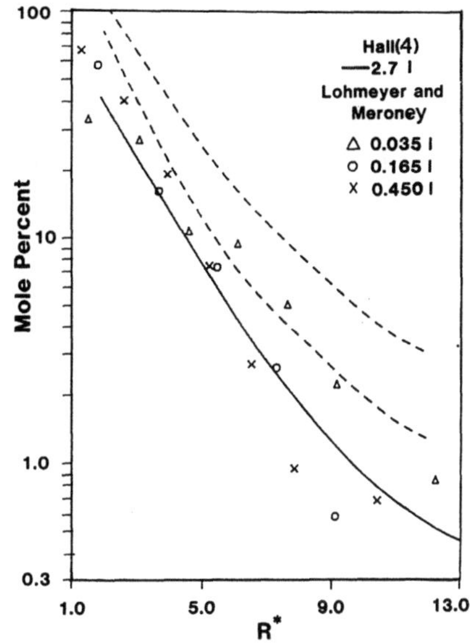

Figure 13. Comparison with data of Hall (4) and Lohmeyer and Meroney (5).

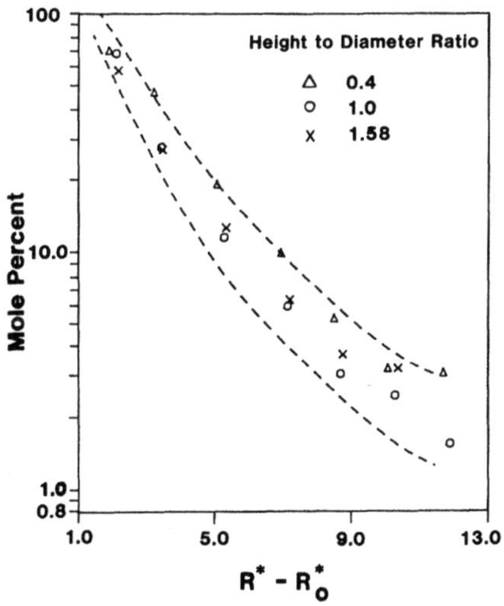

Figure 14. Concentration vs. R* - R^*_0--Freon-12, H/D = 0.4, 1.0, 1.6.

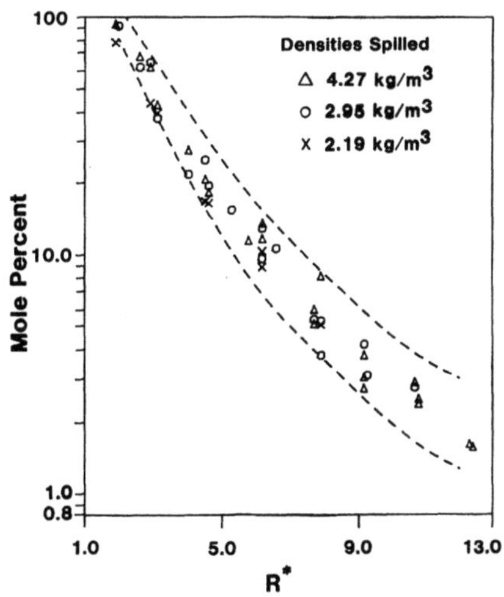

Figure 15. Concentration vs. R* - Three gas volumes and three initial densities.

technique. However, we believe that the differences may be partially associated with difficulties in making the aspirated hot wire measurements in the very thin cloud layers experienced and with different frequency response characteristics of the sensors used.

Figure 14 shows a plot of concentration vs. $R^* - R_0^*$ for three 54.1 liter Freon-12 releases with initial height-to-diameter ratios of 0.4, 1.0, and 1.6. Within the experimental data scatter, the different H/D ratio results are very similar, although there appears to be a slight bias toward higher concentrations with lower H/D ratio.

Figure 15 shows a plot of concentration vs. R^* for the three volumes tested, including releases with initial densities 4.27, 2.95, and 2.19 kg/m^3. All of the ground level peak concentration vs. dimensionless distance data taken for different volumes and initial densities fall on the same plot within the limits of experimental error and experiment reproducibility.

We are analyzing the results of these experiments to determine entrainment specifications for incorporation into the gravity-spreading phase of a dispersion model being developed for the U.S. Coast Guard and plan additional experiments on smaller and larger gas volumes. We also plan to make measurements of the vertical concentration distribution in these releases as well as local velocity measurements using pulsed wire anemometry.

References

1. Koopman, R. P. et al., BURRO Series Data Report, LLNL/NWC 1980 LNG Spill Tests, University of California ID-19075, Vols. 1,2, December 1982.
2. Puttock, J. S. et al., Field Experiments on Dense Gas Dispersion, Journal of Hazardous Materials, Vol. 6, Nos. 1 and 2, July 1982.
3. Picknett, R. G., Dispersion of Dense Gas Puffs Released in the Atmosphere at Ground Level, Atmospheric Environment, Vol. 15 (1981), pp. 509-525.
4. Hall, D. J. et al., A Wind Tunnel Model of the Porton Dense Gas Spill Field Trials, LR 394(AP), Warren Spring Laboratory, Department of Industry, Stevenage, UK.
5. Meroney, R. N. and A. Lohmeyer, Gravity Spreading and Dispersion of Dense Gas Clouds Released Suddenly into a Turbulent Boundary Layer, Report to Gas Research Institute under Contract 5014-352-0203, August 1982.

A Wind Tunnel Model of the Porton Dense Gas Spill Field Trials

D. J. HALL [1], E. J. HOLLIS [2], H. ISHAQ [3]

1. Dept of Industry, Warren Spring Laboratory
2. Health and Safety Executive, Sheffield, UK
3. Vacation Student, Hatfield Polytechnic, UK

Summary

This paper is a condensed description of a study of wind tunnel model simulations of dense gas clouds released instantaneously in the atmosphere, undertaken to assess the validity of the wind tunnel modelling technique as a dispersion prediction method [1]. The comparative data used was that obtained from the field trials carried out at Porton Down in 1976/7. In addition a number of scaling experiments were carried out in order to further investigate the modelling technique.

Introduction

The spreading and dispersion of heavier-than-air gas clouds released into the atmosphere is a complicated phenomenon, and it is self-evident that neither an understanding of the dispersion process nor effective dispersion prediction in practical situations will be possible without recourse to both physical (i.e. wind and water tunnel) and numerical modelling. It is also desirable to validate models by testing against full scale data wherever possible. This paper describes a wind tunnel validation experiment using the heavy gas dispersion field trials carried out for the UK Health and Safety Executive at Porton Down during 1976/7 [2,3]. At that time it was the only well-documented field data available on heavy gas releases and we were asked by the Health and Safety Executive to carry out a wind tunnel model study as a validation exercise. Six selected trials were modelled in the experiments and, in addition, some experiments to investigate model scaling methods were carried out.

There is insufficient space to fully describe the whole experimental programme so the present paper is restricted to presentation of some representative results and a general discussion of the results of the work, a full description can be found in ref. [1].

The Porton Down Field Trials

The field trials used a heavy gas source in the form of a cubical tent of about 3.5 m side containing 40 m^3 of gas, the sides of which were collapsed rapidly under gravity to expose a cube of heavy gas to the prevailing wind conditions. A total of forty two individual trials were run covering a wide range of source gas densities, windspeeds and other parameters, including surface roughness. Figure 1 shows the trials plotted against their two main operating parameters, gas density ratio (here expressed as $\Delta\rho/\rho_{air} = (\rho_{gas} - \rho_{air})/\rho_{air}$) and windspeed. The six trials chosen for modelling are shown, numbered, on the figure, they were chosen for having produced adequate data and for covering a good spread of the range of measurements. Trials 3 and 37, which lie close together on the plot, were chosen because trial 37 was on a flat surface and trial 3 on a 1 in 13 upward slope, thus the effects of slope on the gas cloud development could be observed. In this particular case the model showed no obvious difference between the two sets of measurements, the effect of the slope appeared to be small. The gas clouds were marked with smoke and filmed, so that their appearance and rate of spread could be determined. Also a limited number of gas concentration measurements were made within the cloud, using both continuous monitors and integrating samplers.

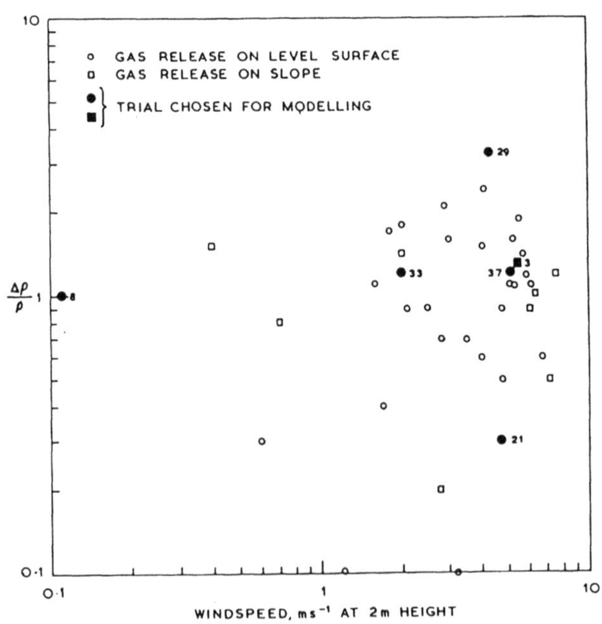

Fig. 1. Operating Conditions for Porton Trials

The Wind Tunnel Model

The experiments were carried out to a scale of 1/25, so that the model of the source tent had a side of 140 mm. The source model was a carefully constructed replica of the original tent, with the exception that the collapsing walls were driven by a pneumatic cylinder whose piston rod passed up through the centre of the tent, the collapse time being scaled appropriately.

The model scaling requirements are, briefly, that for a representative model both the initial gas/air density ratio, $\Delta\rho/\rho$, and the Froude Number, $U/(gL)^{1/2}$, (where U = reference windspeed, L the scale and g acceleration due to gravity) should match the full scale values. Thus windspeeds on a $^1/_2$ scale model must be $^1/_5$ of the full scale values. It was readily possible to achieve this in the present experiments and all the experiments were scaled in this fashion. However, if experiments at much smaller scales are required, and this will be an inevitable requirement of modelling large scale releases, the model windspeed requirement becomes impracticable and severe Reynolds Number effects can occur. An alternative form of scaling is to use the Richardson Number, $Ri = g(\Delta\rho/\rho)L/U^2$, as the sole scaling parameter, which allows the substitution of increased source gas density in place of a greatly reduced model windspeed. Strictly, however, this form of modelling is only valid for small values of $\Delta\rho/\rho$ in the released gas, a condition which certainly does not apply near the source of a typical heavy gas release. In practice errors from this cause may not be very large, and one of the purposes of the experiments was to examine this source of error by looking at releases with various source gas densities and windspeeds, but the same Richardson Number.

Results of the Full Scale Comparison Experiments

Model/full scale comparisons were made with plan and elevation photographs of the gas cloud, measured travel times and cloud spread rates and with concentration measurements made within the cloud. Figures 2-6 show results from one of these comparisons, trial No. 37. All times and distances on the figures are full scale values, the model data being scaled appropriately. The operating conditions for trial 37 were:

	Full Scale	Model (scale 1/25)
Windspeed at 2 m height	5.1 m s^{-1}	1.02 m s^{-1}
Gas Density	1.89 x air	1.89 x air
Surface Roughness, z_0	10 mm	0.4 mm
Stability Class	C-D	Neutral

Results of all the trial comparisons were generally similar to those shown here for trial 37. The model consistently produced gas clouds of very realistic appearance compared with the full scale and with very similar rates of spread and downwind travel, as can be seen from Figs 2, 3 and 4.

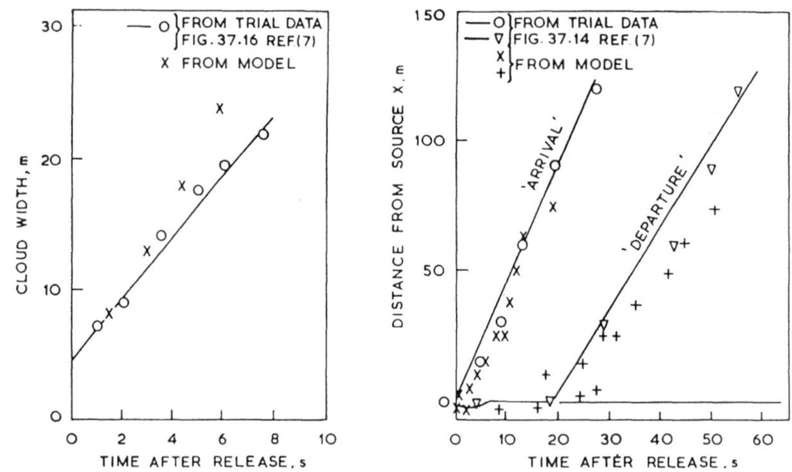

Fig. 2. Porton Trial No. 37. Comparison of model/full scale cloud size and travel times

Comparisons with the full scale concentration measurements proved more troublesome for a number of reasons and the results of the comparison were more uncertain. Firstly, there can be a high degree of naturally occurring repeat variability in the measurements. Though only one-shot data was (inevitably) available for the full scale trials, some repeat measurements could be made in the model. The degree of repeat-run variability altered with the Richardson Number of the release, trials with low release Richardson Numbers (such as trial 37) showed the highest levels of variability. Figure 7 shows results of twenty repeat runs for a particular condition of relatively low Richardson Number, the measurements being made at the ground about two source heights directly downwind of the source. The difference between the largest and smallest peak concentrations in the traces is almost an order of magnitude. Because of this all the model comparisons used triplicated model measurements to get some idea of the degree of variability. Secondly the full scale measurements were limited in number (typically to about six per trial) and sometimes of uncertain accuracy. Also it appeared that the detector time constant was relatively long compared with many features of the gas cloud, which would appear in the model measurements but not in the full scale. The layout of samplers in trial 37 is shown in Fig. 5, and results

Approximate Times

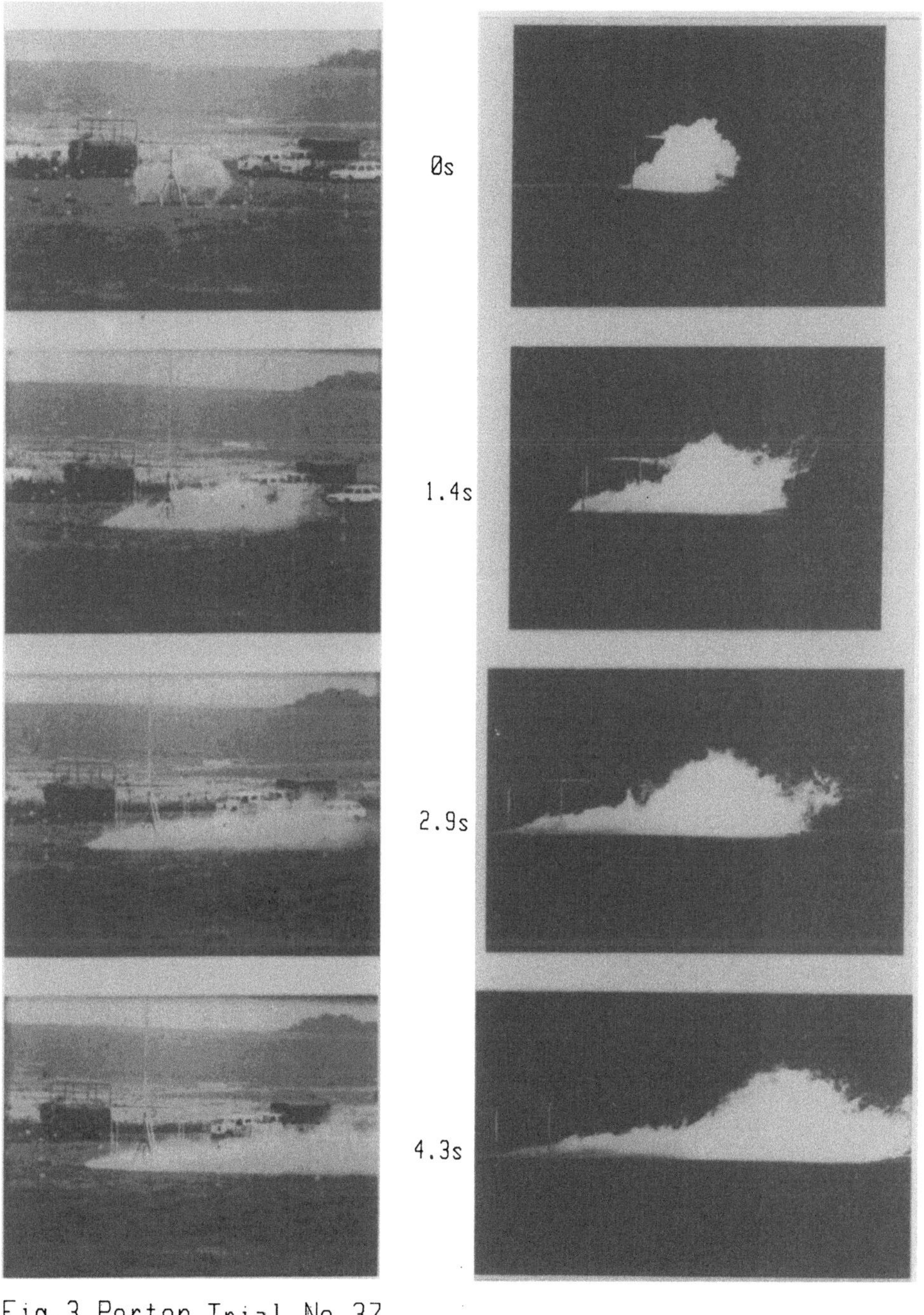

Fig.3 Porton Trial No.37

Model/Full Scale Visualisation Comparison. Plan.

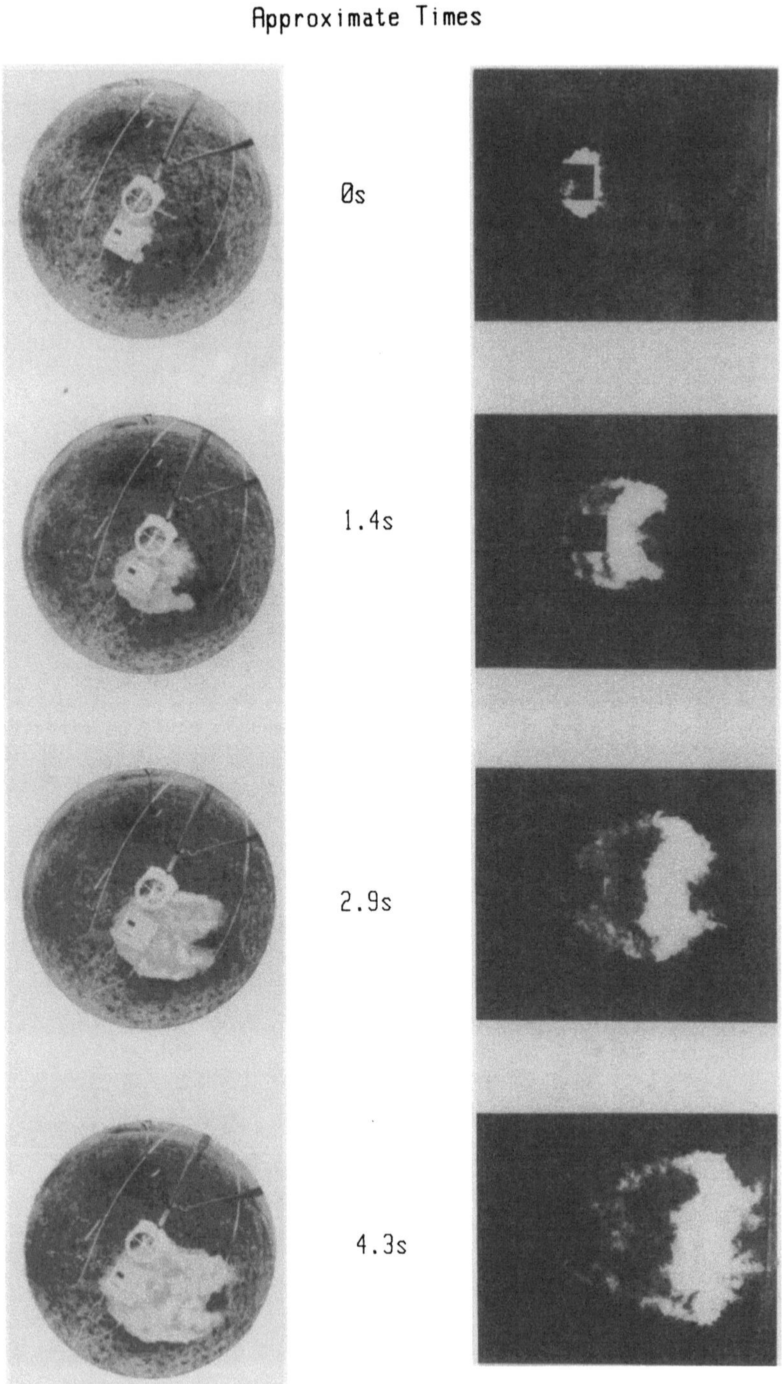

Fig.4 Porton Trial No.37

Model/Full Scale Visualisation Comparison. Plan.

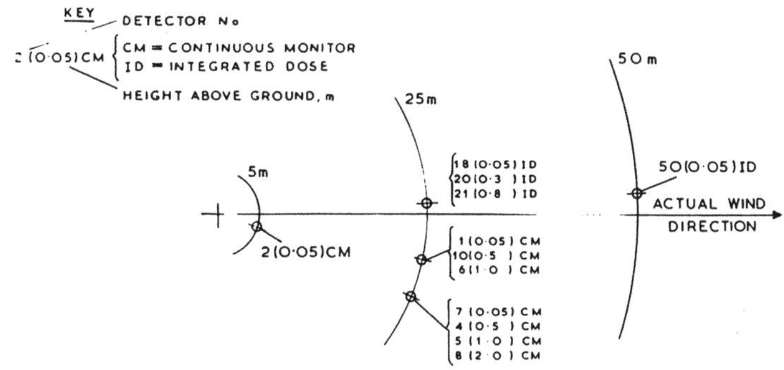

Fig. 5. Porton Trial No. 37. Layout of Continuous Concentration Monitors and Integrated Dose Samplers

of the model/full scale concentration sampler comparisons in Fig. 6. The quality of the comparisons was variable, some were very good, others were poor. Of the total, half had maximum levels of concentration in agreement within a factor of two. Considering the various difficulties in the comparison, mentioned above, and, in addition, the fact that no model/full scale comparison of an atmospheric event can expect to be a very precise affair, it was felt that the results were as good as could be expected.

Results of the Scaling Experiments

A number of experiments were carried out in which releases with the same Richardson Numbers but different source gas densities and reference wind-speeds were compared to see if the use of Richardson Number alone was a sufficient modelling criterion. In addition the effects of changes in surface roughness on the cloud behaviour was examined. The general conclusions were, firstly, that Richardson Number modelling alone was adequate for gas cloud models and, secondly, that in most of the circumstances examined the effects of surface roughness were relatively small.

Figure 8 shows an example from the experiments. The measurements are of the peak concentration measured at the ground at increasing distances from the source for releases into still air. There are measurements for two different source gas densities and two different surface roughnesses. Within the level of accuracy of the experiment there is no observable effect due to either source gas density or surface roughness.

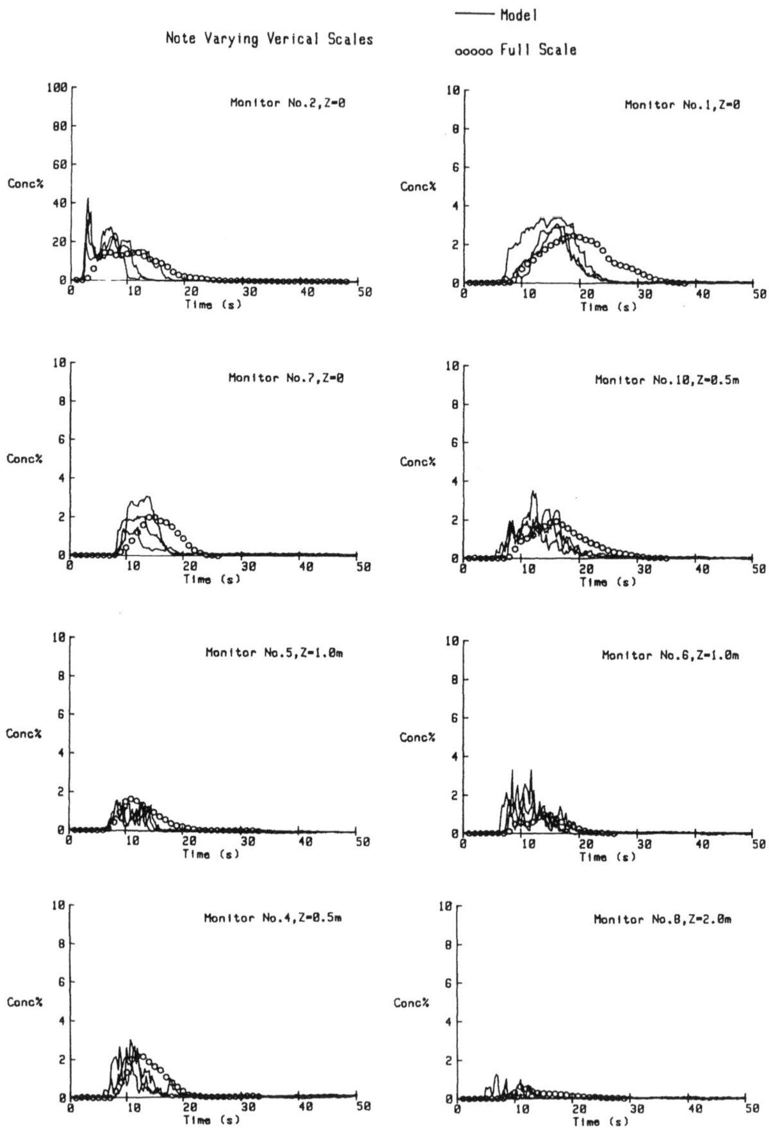

Fig. 6. Porton Trial No. 37. Comparison of Continuous Monitor Measurements with Model Results

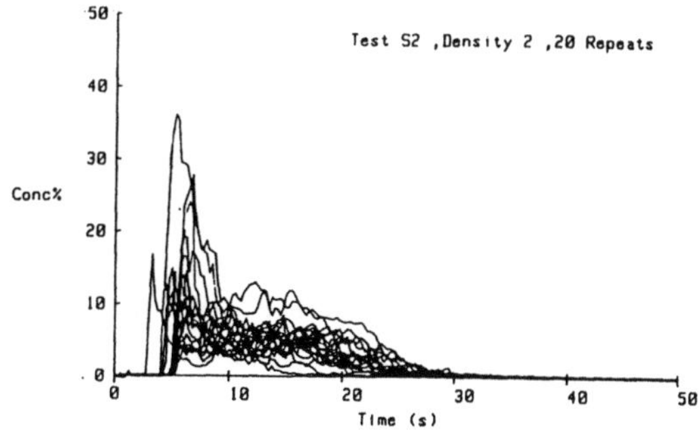

Fig. 7. Variability in Repeat Runs on Model

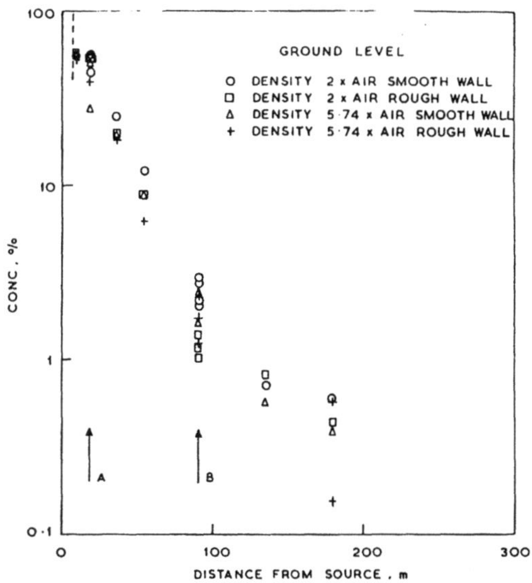

Fig. 8. Scaling Experiment. Peak Ground Level Concentrations for Different Gas Density and Surface Roughness

In conclusion, one final observation is of interest. Though the experiments covered a wide range of release Richardson number, it was noticed that the distributions of peak concentration with distance from the source did not vary very widely. Figure 9 shows a plot of the distance to the point at which peak concentrations in the clouds had reduced to 2% of the source value (expressed as a proportion of source test heights) against the bulk Richardson number of the release (plotted as 1/Ri). The measurements all fall between about 7-20 source heights and show a distinct peak at a Richardson Number of about 4. The reason seems to be due to the relative effects of the clouds self induced gravity-driven dispersion and the combination of cloud drift and additional dispersion that comes from whatever wind is blowing. In still air only the cloud's self induced dispersion affects the gas concentrations within it. Light winds (thus high Richardson Numbers) cause the cloud to drift downwind without greatly increasing its dispersion, so that downwind distances to the point of 2% maximum concentration are increased. Stronger winds not only carry the cloud downwind but also increase the rate at which it disperses and the distance to the 2% concentration boundary then decreases again.

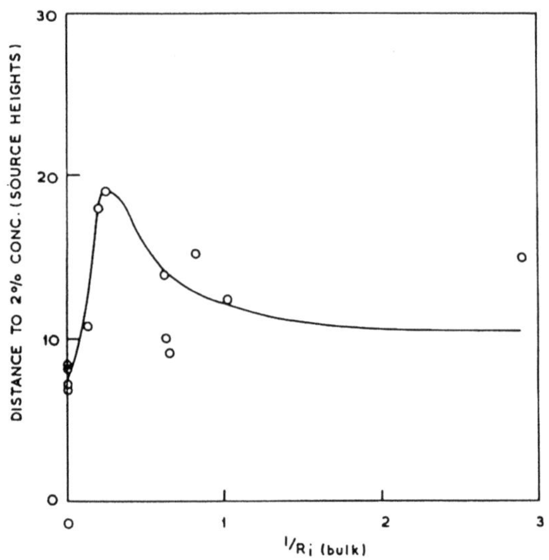

Fig. 9. Effect of Bulk Richardson Number on Downwind Distance to 2% Peak Concentration for Release of a Cube of Material

References

1. Hall, D.J.; Hollis, E.J.; Ishaq, H.: A Wind Tunnel Model of the Porton Dense Gas Spill Field Trials. Warren Spring Laboratory, Stevenage, UK. Report No. LR 394 (AP) 1982.

2. Picknett, R.G.: Field Experiments on the Behaviour of Dense Clouds. Porton Down, UK. CDE Report No. IL 1154/78/1, 2 and 3. Contract Report to the HSE, Sheffield, Sept 1978.

3. Picknett, R.G.: Dispersion of Dense Gas Puffs Released in the Atmosphere at Ground Level. Atmos. Environ. 1981, 15, pp 509-525.

Modelling of Heavy Gas Plumes in a Water Channel

S. C. CHEAH, S. K. CHUA, J. W. CLEAVER, A. MILLWARD

Department of Mechanical Engineering
The University of Liverpool
P.O. Box 147, LIVERPOOL L69 3BX

Summary

An experimental study has been made of the turbulent mixing of a negatively buoyant salt solution injected into a neutral unobstructed turbulent boundary layer in a water channel flow. Particular emphasis is placed on plumes developed from low momentum sources fitted flush to the surface.

Flow visualisation and conductivity probes, to measure the mean concentration have allowed plume boundaries to be determined for a wide range of flow velocities, source flow rates, density ratios, source sizes and differing surface roughnesses. Ground level concentration profiles indicate a two stage decay which becomes more pronounced as the relative heaviness of the plume is increased.

A comparison of plume boundary data and mean concentration data with available wind and water tunnel data show reasonable agreement and indicate that water channel modelling of heavy gas spills may be an attractive alternative to wind tunnel modelling.

Introduction

Physical modelling of heavy gas releases has been made in meteorological wind tunnels [1,2,3] and water tunnels [4,5]. In both cases suitable modelling of the turbulent structure of the atmospheric flow in the vicinity of the release is required. Although it is usually impossible to obtain the same Reynolds number on the model as the full scale there is evidence to suggest that the loss of high frequency turbulence in model tests does not seriously change the plume development and that the portion of the spectrum which has the greatest effect on dispersion remains largely invariant over a wide range of Reynolds numbers. In wind tunnel studies Hall [1], Neff and Meroney [3] have shown that partial simulation can be obtained if there is equality of the volume flux ratio Q/UL^2, the Froude number $U/(gL)^{\frac{1}{2}}$ and the density ratio $(\rho_H-\rho_L)/\rho_L$, where ρ_H,ρ_L are the density of the heavy and ambient fluids, Q the source volume flow rate, U is a reference velocity, g the gravitational constant and L some characteristic length. A more convenient grouping can be obtained by combining the Froude number with the density ratio to give a form of Richardson number $(\rho_H-\rho_L)gL/\rho_L U^2$. Equality of the volume flux ratio and the Richardson number can usually be obtained but at the expense of reducing the Reynolds number to such an extent that the turbulence intensities in the model shear layer may be reduced. The added effect of the stabilising influence of the density gradients places serious, but at this stage unknown, limitations on the range of conditions which can be modelled.

The modelling parameters apply to both wind and water tunnels but the latter, when used in conjunction with salt solutions to simulate the heavy gas, offers several potentially useful advantages. Point concentration measurements and flow visualisation are readily made, the technique is free of hazards and in principal higher model Reynolds numbers can be obtained. In practice the gain in model Reynolds number is lost due to the smaller scale of most tunnels and the fact that the relative source density is restricted to about 1.4. For Richardson number equality this means operating at quite low velocities. With these limitations in mind the present work describes experiments which are designed to assess the usefulness of water tunnel modelling. The specific case of an isothermal plume arising from a low momentum ground level source is investigated and the effect of varying surface roughness and source size on the resulting plume boundary and mean concentration is considered.

Experimental Facilities

All the tests were carried out in a recirculating water channel. The working section is 1.4 m wide, 0.84 m deep and 4 m long, with a velocity distribution uniform to better than ± 0.5% throughout the working section. A detailed description is given in references [6] and [7].

A false floor was fitted 31 cm below the free surface and because of the relatively short length of the working section a Counihan [8] type boundary simulation system was installed in order to establish a thickened boundary layer over a short distance. Tests were carried out on smooth and rough surfaces with the distributed roughness consisting of $\frac{1}{16}$" and $\frac{1}{8}$" expanded P.V.C. mesh or 'Lego' boards. For the latter two rough surfaces the roughness length was found to be approximately 0.01 cm. Measurements of the mean and fluctuating velocity components indicated that a boundary layer height of 25 cm was achieved after about seven boundary layer heights downstream of the leading edge. Beyond this point the boundary layer characteristics remain reasonably invariant and the mean velocity distribution could be represented by a power law profile: further information can be found in references [9, 10]. The results of the measurements indicated that the flow closely models a naturally developed boundary layer over a rough terrain. In all cases a flush circular source, of 5 or 10 cm diameter, was mounted in the floor. Uniformity of the source flow was ensured by filtering the salt solution through a source box consisting of spherical beads and a fine gauze fitted flush with the surface.

Concentration measurements were made with a single electrode conductivity probe [10] and the output from the probe was processed digitally to obtain the required statistics. Sampling intervals were typically of order 0.005s and the data was averaged over a period of about 1 min. The probe tip diameter was visually estimated to be 12.5 μm which, based on the work of Gibson Schwarz [11] gives a sampling volume of about 0.1 mm^3.

Flow Visualisation

The ground level extent of the plume was obtained by introducing dye into the source flow. Visual estimates of the upstream and lateral displacement of the plume boundary in the vicinity of the source are given in Figures 1 and 2. Following Britter [4] the length scales were made non-dimensional with the buoyancy length scale $L_B = g'Q/U^3$, where $g' = (\rho_H - \rho_L)g/\rho_L$. The overall trends are in broad agreement with available data [2,3,4,12,13] obtained from both wind and water tunnels. There is some evidence to show that the roughened surfaces tend to increase the upstream and lateral spread of the plume but the influence of differing roughnesses and boundary layer simulation methods is small. A change in source diameter from 5 cm to 10 cm is also insignificant within the accuracy to which the plume profile can be measured.

For the heavier plumes a central cigar shaped region was observed which indicated little plume dilution. The extent of this region could be linked to the distinct transition zone observed in the ground level concentration profiles. It is surmised that the vigorous mixing towards the edges of the plume arose from the horseshoe vortex system generated by the

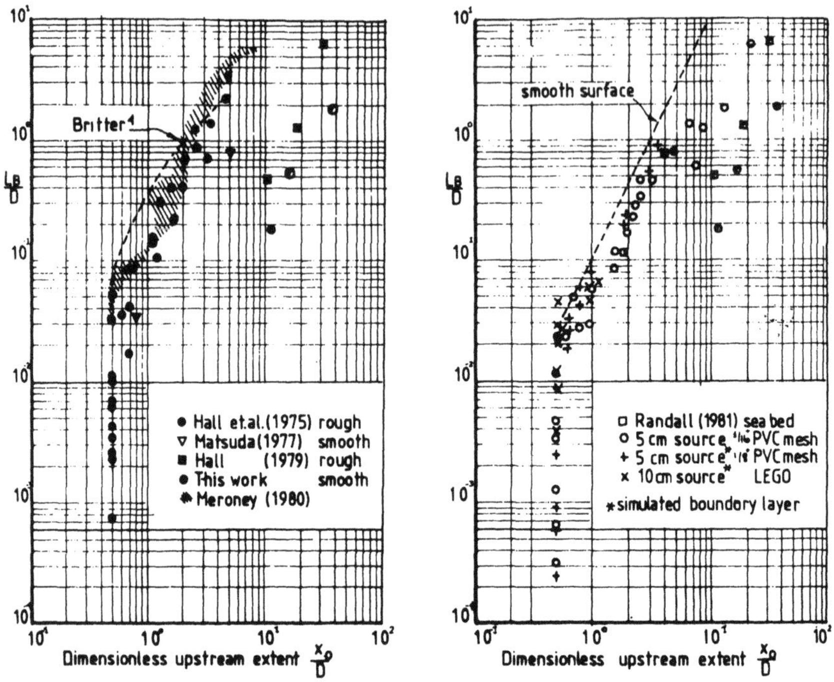

Fig. 1. Effect of buoyancy length on the upstream extent of plume front of the plume.

A comparison of the ground level plume development downstream of the source is shown in Figure 3 for a range of density $\left(0 < \frac{\rho_H - \rho_L}{\rho_L} < 0.25\right)$, source flow rate ($2 < Q < 100$ cm^3/s) and mean flow velocities ($8 < U < 40$ cm/s). The axes are chosen on the basis of Britter's [4] analysis and a relatively good collapse of data is observed. For the heaviest of the plumes both sets of data are in good agreement with the predictions of Britter's analysis. As might be expected, however, as the plume becomes diluted the assumptions in the analysis become invalid and both the wind and water tunnel data display similar trends away from the expected variation. Each set of data remains well correlated even for relatively passive plumes.

Mean Concentration Distributions

Time mean ground level concentration distributions along the centreline, downstream of the source, are shown in Figure 4 for differing surface

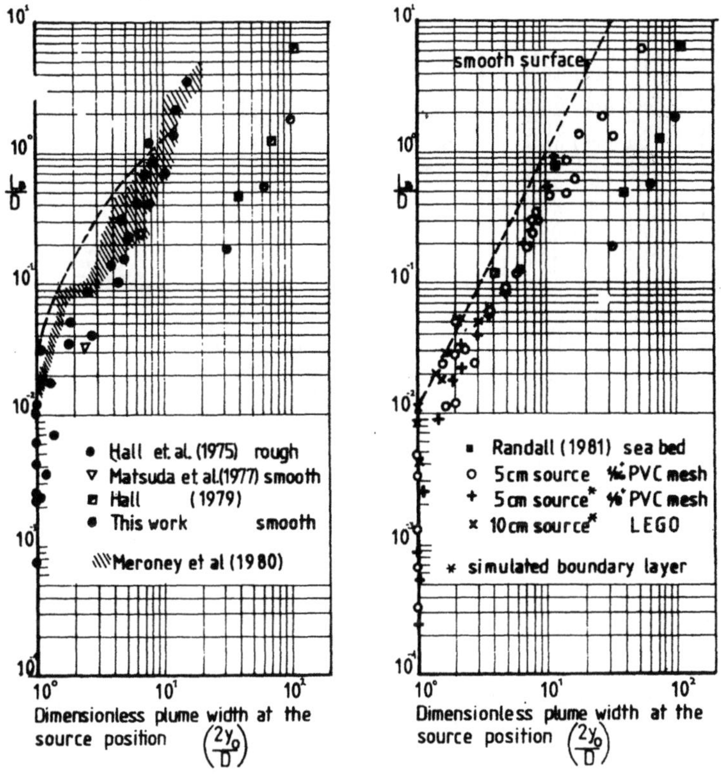

Fig. 2. Effect of buoyancy length on plume width

roughnesses. For all the relatively heavy plumes two distinct regions of dilution are observed. For the larger values of the Richardson number, or the buoyancy length L_B, the source fluid is highly stratified and mixing in the near vicinity of the source is severely restricted. With decreasing L_B the rate of dilution steadily increases and the transition zone approaches the source. For the smallest values of L_B the rate of dilution closely resembles that of a passive plume and two separate zones are barely noticeable.

A comparison of the data for the two expanded P.V.C. rough surfaces shows that the overall concentration levels are lower for the rougher surface. It is tempting to suggest that this is due to the increased turbulence levels giving rise to greater mixing. However, it must be noted that the expanded mesh is porous and thus allows plume development within the surface roughness. In addition the concentration probe cannot penetrate to the base of the rough elements and in consequence the rougher the

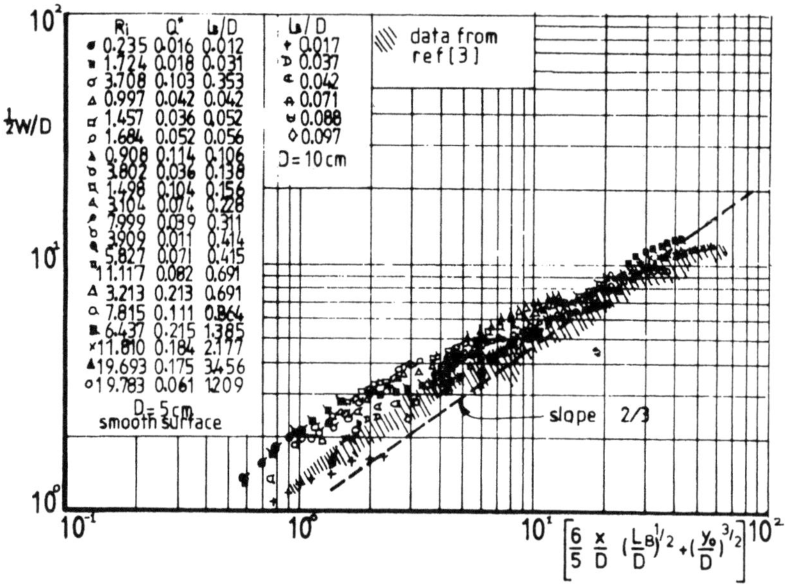

Fig. 3. Lateral extent of plume boundary

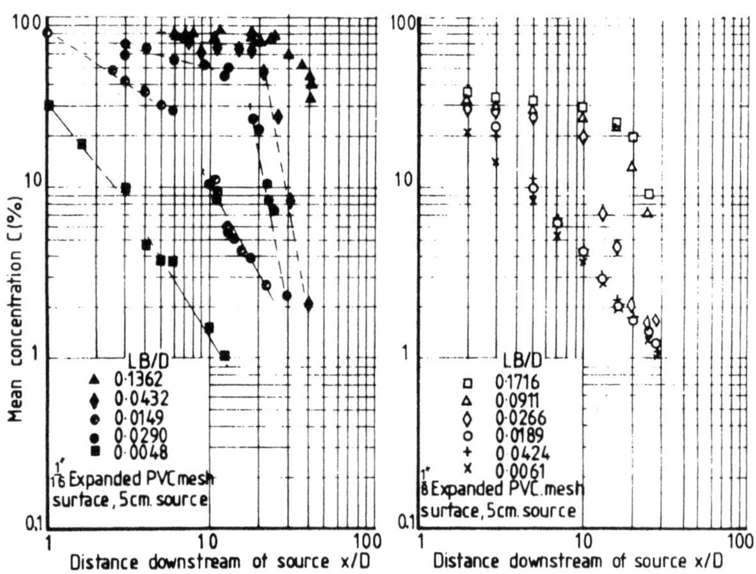

Fig. 4. Variation of centreline ground level mean concentration with distance downstream of source

surface the higher the elevation of the concentration measurement relative to the base of the surface. Rapid changes in concentration are observed close to the surface so that the reduction seen in Figure 4 may well arise from the position of the probe rather than through the effect of increased mixing.

In an attempt to resolve this query the roughness elements were replaced by 'Lego' boards which allowed the concentration to be measured at the base of the elements. Figure 5 shows that for the lightest of the plumes the mixing is sufficiently energetic for the difference between the top and bottom of the elements to be insignificant. For heavier plumes this difference becomes more apparent and suggests that dispersion within the roughness elements is reduced. Measurements [9] of the mean and fluctuating velocity field indicate only small differences between the two types of surface and therefore it must be concluded that much of the differences in the concentration arises through the difficulty of locating a true ground level reading.

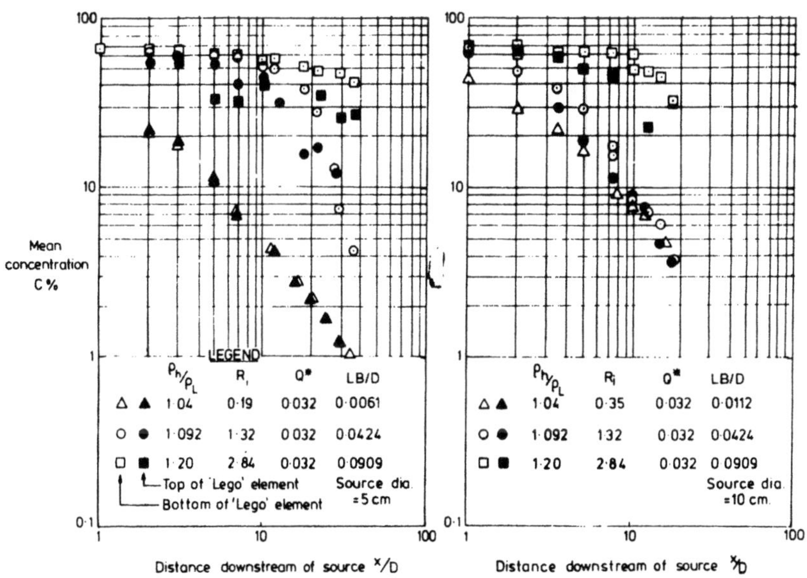

Fig. 5. Comparison of mean ground level concentration distributions taken at the top and bottom of roughness elements for a 5 cm and 10 cm source

To assess the value of using a water channel to model a heavy plume development requires full-scale measurement. These are lacking at the moment and therefore the best that can be done is to compare the present data with equivalent wind tunnel experiments [2,3]. One immediate difference between wind and water tunnels results appear to be the lack of a two stage decay of concentration in wind tunnels, excepting the data of Neff, et al. [14] who noted a similar effect when observing the plume developed from a dike.

In order to make a more quantitative comparison consider the following elementary model. Assume that the flow may be considered one-dimensional and that it is sufficiently wide, compared with its height, for mixing with the external flow to occur only along the upper surface. The mean value of the concentration at any downstream position x is then given by

$$C = \frac{Q}{Q + \int_{-x_0}^{x} V_E W dx} \tag{1}$$

where V_E is the entrainment velocity at the top of the plume and W the plume width.

The flow visualisation data shows that the dimensions of the plume scale well with L_B, even for the lightest of the plumes. Thus equation (1) can be written

$$\frac{C}{1-C}\left(\frac{L_B}{D}\right)^2 \frac{1}{Q^+} = \frac{1}{C_1 \int_{-x_0/L_B}^{x/L_B} \frac{V_E(s)}{U_\infty} W^+(s) ds)} \tag{2}$$

where D is the source diameter, U_∞ is the free stream velocity, $Q^+ = Q/U^2D$, $W^+ = W/L_B$ and $C_1 = U_\infty/U$ which is approximately constant.

Assume [15] that the mixing across the top of the layer can be characterised by an entrainment velocity of the form

$$\frac{V_E}{U^*} \sim \frac{1}{R_i^{*\,3/2}} , \tag{3}$$

where U^* is the friction velocity and $R_i^* = \frac{g(\rho_H - \rho_L)h}{\rho_L U^{*2}}$. There is little evidence available concerning the height of the plume, h; however, in order to relate it in terms of the overall variables it is assumed to be of order of the height to which a vertically directed plume would reach in a quiescent surrounding. The work of Turner [16] suggests it is of the form

$$\frac{h}{D} \sim \left(\frac{\rho_H}{\rho_L}\right)^{3/4} \left(\frac{D}{L_B}\right)^{1/2} Q^{+3/2} \qquad (4)$$

For a fully developed flow it is anticipated that $U^* \sim U_\infty$ and in conse- equation (2) may be recast in the form

$$\frac{C}{1-C} \left(\frac{\rho_L}{\rho_H}\right)^{9/8} \left(\frac{L_B}{D}\right)^{5/4} \frac{1}{Q^{*7/4}} = F\left(\frac{x}{L_B} ; \frac{D}{L_B}\right) \qquad (5)$$

The function F arises from the integral in equation (2) and its dependence on $\frac{x}{L_B}$ and $\frac{D}{L_B}$ is suggested from the flow visualisation data. It will also include any variation of h with $\frac{x}{L_B}$ that may occur. Considering the simpilistic nature of the analysis it is encouraging to note in Figure 6 the degree to which the wind and water channel data is correlated. The data of Hall [2] is in itself well correlated but lies significantly below the bulk of the other data. Hall operates with a much higher source momentum than the other data. This is likely to provide a greater barrier to the approach flow and the stronger vortex system undoubtedly give rise to better mixing of the flow and hence lower relate concentrations. In making a comparison of the present work with the wind tunnel experiments the data taken with the smallest roughness elements and that taken at the base of the Lego elements have been omitted in Figure 6. Best agreement with the data of Neff-Meroney [3] is obtained for those conditions which appear to be well mixed downstream of the source. Where significant stabilisation of the brine solution appear to exist (i.e., largest L_B's) the data shows departures from the overall correlation, indicating significantly reduced mixing in the source region.

If, in equation (1) allowance is made for the possible effect of molecular diffusion an additional entrainment term must be included. It is readily shown that this will be a function of the non-dimensional group $\mathcal{D}/L_B U$ where $\mathcal{D}$ is the diffusivity. This is usually a small quantity but it is interesting to note that water channel modelling gives similar values to full-scale spills whereas wind tunnel modelling results in a value some three orders of magnitude greater. With the thin layer present in the dispersion of heavy gases it is highly probable that in wind tunnel

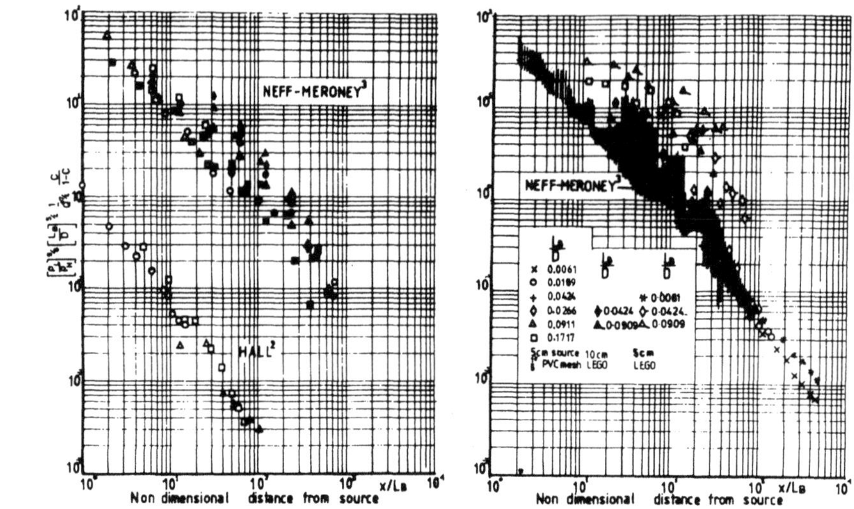

Fig. 6. Non-dimensional mean concentration distribution for wind tunnel and water tunnel data

modelling some initial mixing due to molecular diffusion is already present immediately downstream of the source and may be one of the reasons why wind tunnel data gives generally smaller values of the concentration and does not display the two stage decay experienced in the water channel. In addition, account must also be taken of the fact that in wind tunnel modelling the viscosity of the heavy gas is invariably less than the ambient air flow, whereas in water channel modelling the reverse is true. This may give rise to a more stable layer in the water channel modelling.

Concluding Comments

Comparison of ground level profiles of the edge of the plume indicate that both full-scale wind and water tunnel data show reasonable agreement when the characteristic length scale is taken as L_B, and supports the overall conclusions of Britter [4].

Lack of suitable full-scale concentration data in continuous plumes prevents similar conclusions being made but comparison with equivalent wind tunnel data shows modest overall agreement but it must be noted that only the water channel measurements give rise to a two stage decay in the concentration. Where differences occur there is evidence to suggest that considerable stabilisation occurs in the source region where the mixing is likely to be significantly depressed.

Acknowledgment

The authors wish to acknowledge that the work described in this paper is part of a project funded by the Science and Engineering Research Council (Marine Technology Directorate) through Marinetech North West and also through the award of a British Gas Scholarship.

References

1. Hall, D. J; Barrett, C. F.; Ralph, M.O.: Experiments on a model of an escape of a heavy gas, Warren Springs Laboratory, Report No. LR217/AP (1974).

2. Hall, D. J.: Further experiments on a model of an escape of heavy gas. Warren Springs Laboratory, Report LR312/AP (1979).

3. Neff, D. E.; Meroney, R. N.: The behaviour of L.N.G. vapour clouds: wind tunnel tests on the modelling of heavy plume dispersion. Chicago: Gas Research Institute Report No. 80/145 (1982).

4. Britter, R. E.: The ground level extent of a negatively buoyant plume under a turbulent boundary layer. Atm. Environ. 14 (1980) 779-785.

5. Emblem, K.; Brovoll, T.; Skauvik, I.; Steinskog, T.: Gas dispersion physical models. River and Harbour Laboratory, Norwegian Institute of Technology (1979).

6. Preston, J. H.: The design of high speed, free surface water channels. Proceedings NATO Advanced Study Institute on Surface Hydrodynamics, Bressanone, Italy (1966) 1-82.

7. Millward, A.; Nicholson, K.; Preston, J. H.: The use of jet injection to produce uniform velocity in a high speed water channel. J. Ship Research 24 2 (1980) 128-132.

8. Counihan, J.: An improved method of simulating a neutral atmospheric boundary layer in a wind tunnel. Atm. Environ. 3 (1969) 197-214.

9. Cheah, S. C.; Cleaver, J. W.; Millward, A.: Water channel simulation of the atmospheric boundary layer. Atm. Environ. 8 (1983) 1439-1448.

10. Chua, S. K.: The dispersion of a negatively buoyant plume in a turbulent shear flow. Ph.D. Thesis (1982) University of Liverpool.

11. Gibson, C. G.; Schwarz, W. H.: Detection of conductivity fluctuations in a turbulent flow field. J. Fluid Mechanics 16 (1963) 357-364.

12. Matsuda, T. et al: Fundamental studies on the diffusion of flammable gas from an area source. Mining and Safety (Japan) 23 10 (1977) 1-15.

13. Randall, R. E.: Measurement of a negatively buoyant plume in the coastal water off freeport. Ocean Engineering 8 4 (1981) 407-419.

14. Neff, D. E.; Meroney, R. N.; Cermak, J. E.: Wind tunnel study of negatively buoyant plume due to an LNG spill. Colorado State University (1975) Report CER76-TIDEN-RNM-HEC22.

15. Turner, J. S.: Buoyancy effects in fluids. Cambridge University Press (1979).

16. Turner, J. S.: Jets and plumes with negative or reversing buoyancy. J. Fluid Mechanics 26 (1966) 779-792.

The Entrainment of Small Particles by a Turbulent Spot

F.G.J. ABSIL / G.L.H. BEUGELING

Lab. for Aero- and Hydrodynamics
Delft University of Technology
The Netherlands

Summary

In a windtunnel trajectories of particles ($d \simeq 1 - 100$ μm) entrained from a flat plate by a turbulent spot in a laminar boundary layer ($U_\infty \simeq 7$ m/s) are recorded using high-speed photography (at 2000 frames/s). Scanning the films using a motion-analyser shows several types of particle movement and the measured entrainment velocities and heights indicate the particles stay within the spot structure.

Introduction

Particle entrainment by an airflow may occur in many situations for instance during conveyance or storage of ores. In order to prevent dust nuisance or to quantify the particle flux into the airflow knowledge of the entrainment mechanism is required. Suppose above a flat particle bed ($\rho_p/\rho = O(10^3)$) the free-stream velocity is increased. Then there is a critical stage where first particle movement is observed. This stage has been determined experimentally for many bed and flow conditions. Results are given in a Shields-diagram where dimensionless shear stress $\rho u_*^2/(\rho_p - \rho)gd$ is plotted versus particle Reynolds number $u_* d/\nu$. This diagram shows several regimes that may be predicted by a force-balance set up for a single spherical particle on a flat surface (Figure 1). In vertical direction there are two holding forces, particle weight ($F_g \sim d^3$) and adhesion ($F_{ad} \sim d$). For relatively large particles ($d \gtrsim 200$ μm) weight, which then dominates over adhesion, is counteracted by an aerodynamic lift force ($F_L \sim u_*^2 d^2$). This leads in the critical stage to $(\tau_o)_{crit} \sim d$. Very small particles ($d \lesssim 30$ μm) are embedded in the viscous sublayer ($y^+ \lesssim 7$) where the flow may be considered turbulent smooth. For these particle sizes weight is negligible to adhesion and theoretically in a stationary sheared Stokes flow no lift-force is predicted. But as entrainment of particles from the sublayer has been observed during experiment other effects have to be taken into account. These effects may be of either inertial, leading to a Saffman [1] lift force, or instationary kind, when the sublayer is disturbed by the presence of wall structures, resulting in the "updraft under a burst"-lift-force (Cleaver/Yates [2]). Both lift forces show a proportionality $F_L \sim (u_* d)^3$ which in the critical stage leads to $(\tau_o)_{crit} \sim d^{-4/3}$. From observation of Figure 1 one may see there is an intermediate region where the critical wall shear stress shows a minimum (d = 75 - 100 μm, u_{*min} = 15 - 20 cm/s).

Research in the fully developed boundary layer using visualisation and conditional sampling techniques has revealed the occurrence of coherent structures both in the wall ($y^+ \leq 100$) and outer region (see the review article by Cantwell [3]). Near the wall there is the burst cyclus, consisting of energetic motions (ejections and sweeps) that generate the major part of turbulent energy. Especially the ejection phase creates a flow movement (approximating an inverse axisymmetric stagnation point flow) that may lead to the "updraft under a burst" lift force acting on the particles. Several research workers have put forward the idea that the existence of near-wall phenomena and the initiation of particle entrainment (both occurring streakwise) may be linked. Cleaver/Yates [2] present a model predicting entrainment from a flat monolayer of polydisperse particles using mean size and spreading of bursts and a lift-up criterium according to the afore mentioned force-balance .

When determining experimentally the influence of a coherent structure on particle entrainment there is the problem of isolating a single structure. In the fully developed turbulent boundary layer they are generated randomly and in continuous interaction. This requires detection and tracing of a single structure. The problem is avoided by considering a turbulent spot in a laminar boundary layer. The spot is a structure naturally occurring when the flow is in the transitional stage from laminar to turbulent but can also be generated artificially by disturbing the laminar flow in the unstable Re_x-region. Experiments visualizing the spot and determining its ensemble-averaged structure (Refs. [4], [5], [6], [7]) show that several spot features indicate a resemblance to boundary layer structures. Arguments confirming this analogy are: sizes (of order δ_t) and convection velocities ; both show longitudinal streak-vortices near the wall with equal mean lateral spreading ; the spot shows an array of vortices similar to the ones observed by Head/Bandyopadhyay [8] in the turbulent boundary layer. Working from this analogy experiments have been performed at the Delft Lab. for Aero- and Hydrodynamics where in a windtunnel a turbulent spot in a laminar boundary layer passes over a flat particle bed. The trajectories of entrained particles are recorded using high-speed photography.

Experimental Set-up

In the rectangular test section ($30 * 40$ cm^2) of an open windtunnel a 1.5 m long flat plate was placed as sketched in Figure 2. Over the flat plate having a sharp leading-edge a laminar flow ($U_\infty \simeq 7$ m/s) with negligible pressure gradient was established. At x = 30 cm ($Re_x = 1.5 * 10^5$) behind the plate leading edge a 0.15 mm thick cupper tripping wire was mounted just above the plate, perpendicular to the flow, embedded in a profile except for the central 15 mm. The wire, being in the field of permanent magnets underneath the plate was pulled towards the wall by switching on an electric current. This generated a point-like flow-disturbance that grew downstream to a turbulent spot. The spot passed over a sheet of emery paper ($28 * 23$ cm^2) glued to the plate with its leading edge at x = 60.6 cm. At x = 61.6 cm in the plane of symmetry (z = 0) its grains were scratched away from a $15 * 2$ mm^2 strip. Here a powder bed settled after being injected through the top of a 1 m long, 35 cm wide cylinder placed over the paper. In the cylinder bottom there was a disk with an opening $15 * 2$ mm^2 corresponding to the strip on the emery paper. This device produced reasonably homogeneous flat beds with their toplayer nearly flush with the emery paper. The bed-material consisted of either cement-powder ($\rho_p = 3.2$ gr/cm^3; $d_{50} = 65$ μm) or a fraction of Durcal-powder ($CaCO_3$, $\rho_p = 2.75$ gr/cm^3, $d_{50} = 5$, 10, 15, 30 and 60 μm).

Films were recorded with a Strobodrum High Speed Camera, containing a 1.5 m long strip of 35 mm film (Ilford HP5, 400 ASA). Shadow-light technique was used, having a Fischer Nanolite high frequency flash lamp (flash duration $O(10^{-8})$s) as light source. Two lenses focused its light to a beam of 7 mm diameter. Filming was done at magnification 3, with a field of view 6×8 mm^2 from the dust bed leading edge downstream (the x-y plane). The pulse generating the turbulent spot also triggered the camera system, which after a delay time of 60 - 75 ms (the time the disturbance needed to reach the particle bed) produced a flash series (frequency 2000 Hz, duration 20 - 35 ms). Simultaneously a hot-wire anemometer at x = 616, y $\simeq$ 3 mm recorded the longitudinal velocity-signal above the bed leading edge. This resulted in 75 - 80 frames showing the initiation of particle movement (the particles being at rest in the laminar flow at $U_\infty \simeq 7.0$ m/s) during passage of the spot. For each material at least 5 films were recorded. Analysis of the films was done using an x-y-discriminator (motion-analyser) connected to an Apple mini-computer. This produced tables and plots, containing as a function of time x- and y-coordinates and velocities of the entrained particles.

Results and discussion

Of the recorded number of films the major part showed particle movement. When looking at the bed after a spot passage, bare parts or sometimes a completely bare strip could be observed. Preferred entrainment existed near the leading edge. Some of the Durcal-10 and -15 fractions and for the Durcal-5 5 out of 10 films showed no motion at all. For these smaller sizes the bed roughness is small and apparently the forces induced by the spot are insufficient to lift particles from the strongly coherent bed.

Some preliminary films recorded at lower frame frequency and covering the entire spot-length (60 - 100 ms at y = 3 mm) indicate an increase in particle flux and entrained particle size near the end of the spot where close to the wall the ensemble-averaged longitudinal velocity shows a maximum (Ref. [4]). Sometimes so many particles were entrained, it was difficult to discern and trace individual particles from the film. About 170 trajectories were measured. Besides particle size was estimated from the films, but this must be considered very crude. The achieved accuracy in detemining the x-coordinate is ± 30 μm and for the y-coordinate ± 40 μm. Plate 1 shows an example of particle entrainment by the turbulent spot.

The initiation of particle movement nearly coincides with the arrival of the strong increase in u near the wall in the spot, or the negative peak in u'v'. Trajectories were almost 2-dimensional, and may be classified as sketched in Figure 3. About half of the number of particles are lifted according to direct entrainment. Whether there is rolling or impact first can not be seen. They are raised nearly vertically from the surface. Within 5 ms the trajectories are curved into the main flow direction, leaving the frame of view at a preferred angle 2 - 3°, velocities 0.1 - 0.5 U_∞ and a height up to 1 mm ($y/\delta_\ell \simeq 0.2$). There is a tendency for the smaller particles to achieve greater velocities and heights. An estimation of particle rotation gives 100 - 500 rev/s..

The next to largest group shows a movement close to the wall being either rolling (lumps of particles, $u \simeq 0,05\ U_\infty$) or entrainment at a height of order 0.1 - 0.2 mm. In some of these cases rolling lumps are lifted into the flow after impact on a wall irregularity. Part of the entrained lumps after reaching $y \approx 0.5$ mm broke into many smaller parts.

For the smaller Durcal fractions (Durcal-15,-10,-5) a great part of the bed (3 - 4 mm long parts), often starting from the leading edge, slided along the bed at a velocity $\simeq 0.05\ U_\infty$ accompanied by some local piling up of the bed. Piling continued until its top was about 0.2 mm above its environment. From then on, entrainment started from the top of the pile. For Durcal-5 this was the only type of particle movement observed.

Conclusions

The experiment confirms that structures in the boundary layer may be considered to be the driving force behind the entrainment process. This begins at arrival and ends after passage of the spot. The trend as predicted by the Shields-curve for the small particle-regime is confirmed. While the larger fraction films all showed movement, this did not occur for the smaller fractions. There is preferred entrainment near the bed-leading edge. The leading edge of the bed, being sometimes somewhat thicker and thus protruding from its environment (of order 0.1 mm) is the first irregularity the spot meets, resulting in greater probability of entrainment near that position. Several types of movement were observed, of which the most striking was a bulk-type motion of the small strongly coherent fraction. Comparing the particle velocities and heights, still accelerating when leaving the frame of view, to the spot trailing edge convection velocity one may conclude that, except for those very close to the wall, the entrained particles will stay within the turbulent spot.

List of Symbols

d	= particle diameter	[m]
g	= acceleration due to gravity	[m/s^2]
Re_d	= particle Reynolds number = $u_* d/\nu$	
U_∞	= free-stream velocity	[m/s]
u	= streamwise velocity	[m/s]
u_*	= wall shear velocity = $(\tau_o/\rho)^{\frac{1}{2}}$	[m/s]
v	= vertical velocity	[m/s]

x = streamwise coordinate [m]

y = vertical coordinate [m]

δ = boundary layer thickness [m]

ν = kinematic viscosity [m^2/s]

ρ = density of air [kg/m^3]

ρ_p = particle density [kg/m^3]

τ_o = wall shear stress [N/m^2]

References

1. Saffman, P.G.; Journ. of Fluid Mech.. vol. 22, part 2 (1965) 385-400
2. Cleaver, J.W.; Yates, B.; Journ. of Colloid and Interface Sci. Vol. 44, no. 3 (1973) 464-474.
3. Cantwell, B.J.; Ann. Rev. Fluid Mech. 13 (1981) 457-515.
4. Wygnanski, I.; Sokolov, M.; Friedman, D.; Journ. of Fluid Mech. Vol. 78, part 4 (1976) 785-819.
5. Cantwell, B.; Coles, D.; Dimotakis, P.; Journ. of Fluid Mech. Vol. 87, part 4 (1978) 641-672.
6. Van Atta, C.W.; Sokolov, M; Antonia, R.A.; Chambers, A.J.; The Phys. of Fluids,vol. 25, 3 (1982) 424-428.
7. Gad-el-Hak, M.; Blackwelder, R.F.; Riley, J.J.; Journ. of Fluid Mech. vol. 110 (1981) 73-95.
8. Head, M.R.; Bandyopadhyay, P.; Journ. of Fluid Mech. vol. 107 (1981) 297-338.

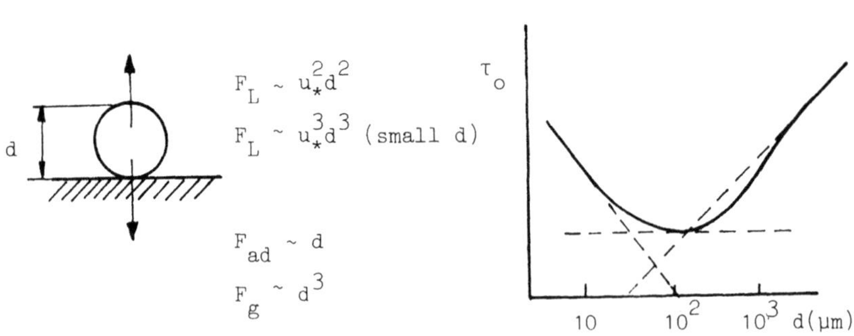

Fig. 1. Shields-diagram as predicted by the force balance on a single particle.

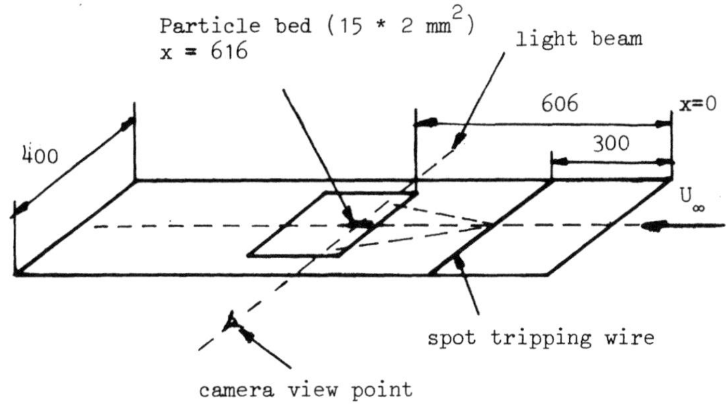

Fig. 2. Experimental set-up (flat plate in windtunnel).

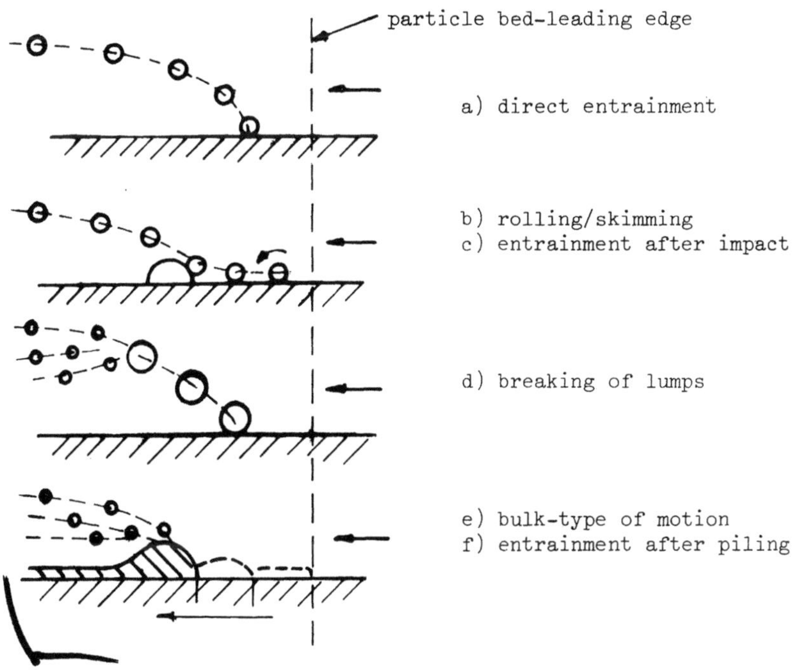

Fig. 3. Types of particle motion.

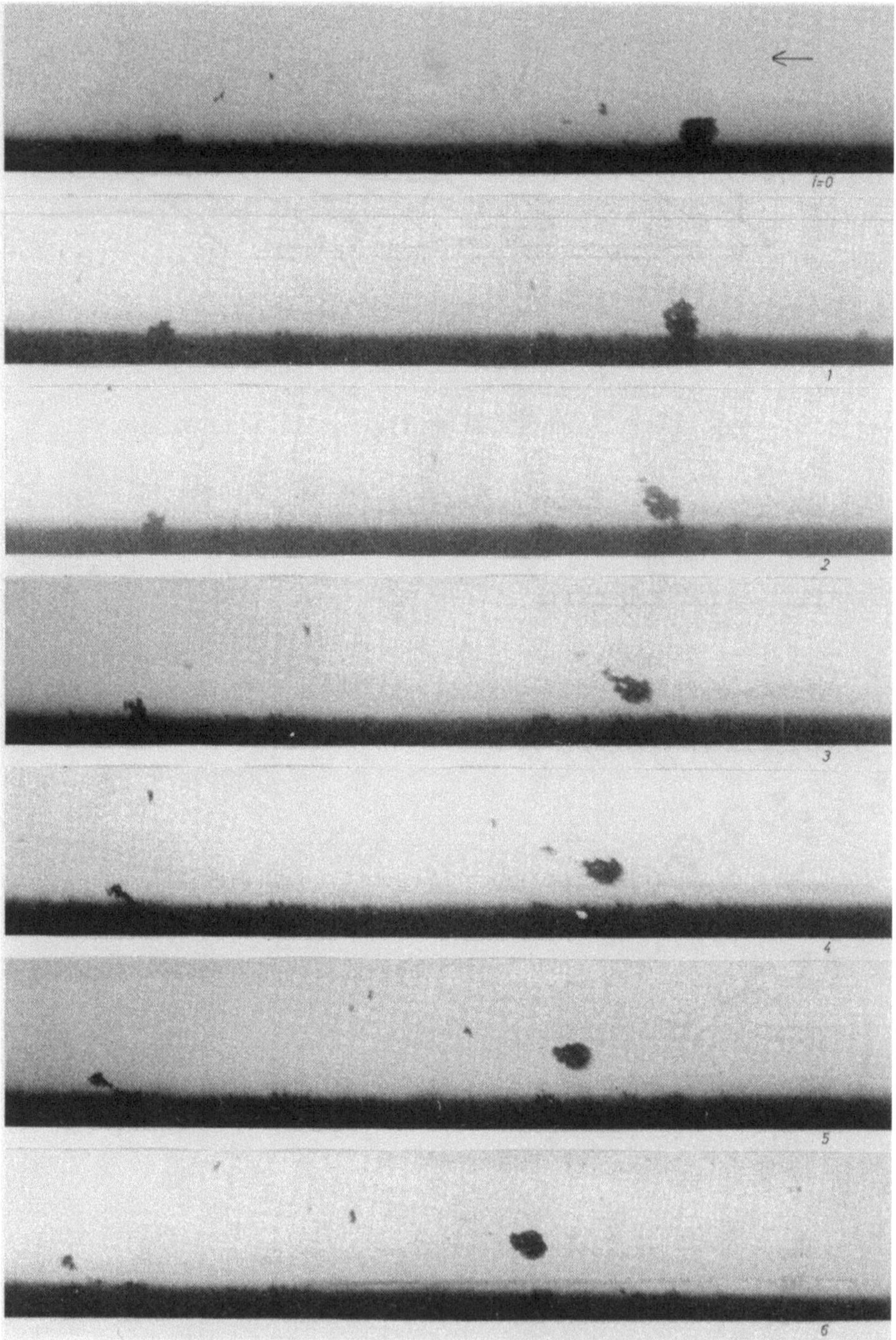

Plate 1. (for legend see next page)

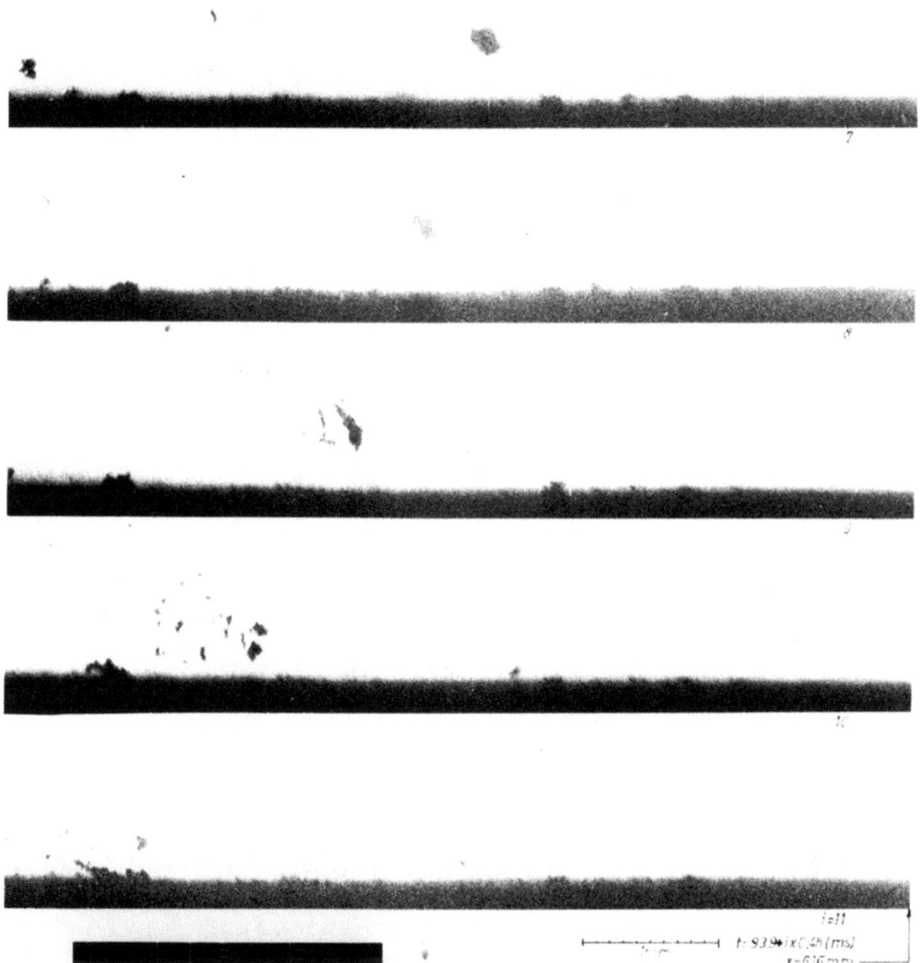

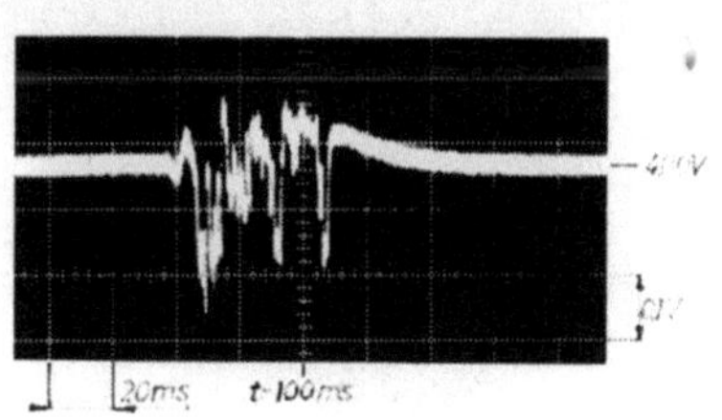

Plate 1. (continued).
Particle entrainment by the turbulent spot. Film 147.
U_∞ = 6.9 m/s; particle bed at x = 616, z = 10 mm; cement powder (ρ = 3.2 g/cm^3, d_{50} ≈ 65 µm).

On a Model for the Turbulent Re-Entrainment of Small Particles

M. W. REEKS

CEGB, Berkeley Nuclear Laboratories, Berkeley, Gloucestershire, U.K.

Abstract

A statistical model for the resuspension of small particles by a turbulent fluid is presented. The approach is similar to the escape of Brownian particles from a potential well. Particles are released from the well when they receive enough energy from the local interaction of the turbulent eddies to escape over a potential barrier formed from the average lift force and attractive adhesive surface forces. Using the harmonic approximation an expression is derived for the rate constant for long term resuspension for which we can assume most of the particles in the well are in quasi-equilibrium at the point of minimum potential. The rate constant is seen to depend upon:

(a) The timescale and intensity of the fluid induced lift force fluctuations.

(b) The natural frequency and depth of the potential well.

(c) The particle relaxation time based on a linearised drag law.

Introduction

Primarily because of its diversity and importance in environmental and industrial problems the re-entrainment of particles by a moving fluid has received great attention in the past (e.g. see Zimon (1969) and Corn (1966)). It naturally forms a significant part of the study of erosion, and in the dispersion and adsorption of contaminants by the atmosphere.

In the nuclear power industry various situations may result in the formation of radioactive particulate. Analysis of behaviour include assessment of the dispersion, deposition and subsequent re-entrainment of particles.

What we have to say here, however, is of fundamental regard to re-entrainment and not exclusively related to our own particular problem.

The large bulk of work that exists on the subject of particle removal from surfaces by aerodynamic forces has been mostly of an empirical nature (see e.g. Zimon, 1969). The general mechanisms seem well established and may be precisely stated. Fluid induced forces acting on a particle will be both tangential and normal to the particle substrate.

They are commonly referred to as drag and lift forces respectively, and cause the particle to either roll, slip or lift off the surface. This tendency will be counteracted by long range molecular forces acting between the particle and the substrate. In the study of re-entrainment, the study of adhesion and fluid induced drag, and lift forces are of equal importance.

Before continuing we would like to make a distinction between re-entrainment and resuspension. By re-entrainment we mean the collective process of particle removal from a surface by a moving fluid. By resuspension we refer exclusively to the removal by fluid lift forces and it is this aspect of re-entrainment to which this paper is devoted.

In particular we shall be concerned with the problem of long term resuspension. This contrasts with the short term or almost instantaneous resuspension that arises simply because the magnitude of the average lift force exceeds that of the surface adhesive force. Long term resuspension is more difficult to quantify since it is statistical in origin. The randomness of the event is intimately associated with the intrinsic turbulence of the resuspending flow. To be more specific, it depends upon the timescale and intensity of the fluctuations of the fluid lift force caused by the random shearing of the eddy motions close to the particle. This effect has been recognised before - most notably by Corn and Stein

They exposed glass spheres 10-100 μm in diameter on a nominally flat glass surface to an air flow of various velocities. The primary objective of the experiment was to investigate the relationship between aerodynamic drag and adhesion. However, they observed that for a range of particle sizes 5-40 μm at large air flows (~ 100 m/s) the number of particles removed from the surface increased with time. The process was merely recognised and no attempt was made to quantify it in terms of the controlling features of their experiment. In fact quite wrongly in our opinion they generally regarded random drag rather than lift as the principal mechanism for particle removal.

A closer examination of their results shows that the number of particles remaining on the surface decayed roughly exponentially with time. It suggests that we can associate with the resuspension a definite half life or more precisely that the probablity/unit time for the resuspension of a single particle is constant in time. It is to the calculation of this quantity, which we shall refer hereafter by the term 'rate constant', p, that this paper is primarily devoted.

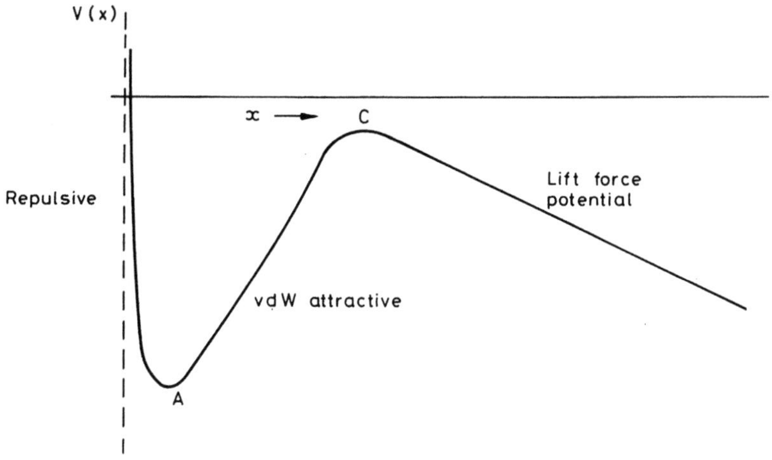

Fig. 1 Model of Resuspension

Statistical Theory

The treatment is similar in approach to the kinetic theory of the escape of particles from surfaces (Chandrasekhar, 1943; Dahneke, 1975). The effect of all the average forces acting normal to the surface can be conveniently represented by a potential diagram shown in Fig. 1. The horizontal axis represents two quantities depending upon whether the particle substrate is deformed or undeformed (rigid). For a smooth deformable particle on a smooth deformable surface, x, represents the distance of approach of the centre of the particle to the substrate surface from their undeformed state. For a rigid particle - rigid substrate x represents the minimum distance apart of the 2 surfaces. In equilibrium x $= x_0 \sim 3$Å for van der Waals molecular surface forces.

As long ago as 1934 Derjaguin (Derjaguin, 1934) showed that when surfaces are brought together under the influence of attractive forces they must deform in the contact region. The effect of this is to lower the equilibrium potential still further from that in its rigid state. The equilibrium finally obtained is a balance between the surface adhesive forces and normal Hertzian elastic forces modified by the action of the surface forces themselves. For the size of particles normally resuspended by turbulent flows (diam > 1 μm) the lowering of the potential can be substantial e.g. for a 20 μm diameter polystyrene particle on a smooth glass substrate the depth of the potential is a factor of 5 greater than that in an undeformed state, the effect increasing with increasing particle size. Fig. 1 shows two positions of equilibrium when the net force is zero.

(i) A stable equilibrium, A, where the attractive adhesive forces balance the repulsive elastic forces and average fluid lift force.

(ii) An unstable equilibrium, C, where the attractive force just balances the lift force alone. In this case the particle is in an undeformed state. For materials of low modulus of elasticity and short range surface forces, a particle may nominally be removed from the surface before this condition is ever attained. This is associated with the formation of a neck around the region of contact in the deformable solid which is suddenly broken when the mechanical energy is greater than the surface energy requirements, (Tabor, 1977). The particle relaxes instantaneously into an undeformed state out of range of the surface forces. This is the basis for particle removal from the surface in the adhesion model of Johnson, Kendall and Roberts (JKR), (1971).

For a smooth spherical particle of radius R on a smooth flat surface, the force required in this model to remove the particle from the surface (adhesive force) is given by

$$F_o = \frac{3}{2} \pi \gamma R \tag{1}$$

where γ is the surface energy/unit area. We note it does not depend upon the elastic properties of the 2 surface materials.

In the absence of the formation of a neck the contact area reduces continuously to zero with applied negative load (positive lift force), whence the particle is undeformed (Derjaguin, Muller and Toporov (DMT), 1975). From then on the potential energy is identical to that of the Van der Waals potential for a rigid particle on a smooth surface separated by a distance x at the point of contact i.e.

$$V(x) = - \frac{F_o x_o^2}{x} + \langle F_L \rangle \tag{2}$$

where F_o is the adhesive force given here by

$$F_o = 2 \pi \gamma R \tag{3}$$

and $\langle F_L \rangle$ is the average lift force which is assumed to be constant over the effective range of the adhesive forces. For illustration we have used the unretarded form of the Van der Waals potential. Clearly C in this instance is the point where

$$\frac{F_o x_o^2}{x^2} = \langle F_L \rangle$$

i.e.

$$x = \left(\frac{F_o}{\langle F_L \rangle} \right)^{1/2} x_o \qquad (4)$$

We shall suppose that particles that adhere to the surface lie in the well of the potential at A. They will leave the surface when they receive enough energy from the turbulent eddies that penetrate down to the particle to eject them out of the well. The fact that statistically most of the particles are assumed to be at A, means that their average potential energy must be very much less than Q, the potential height of C above A. In essence the particles are in almost equilibrium at A and their removal from the surface is long term i.e. on timescales very much greater than the periodicity of the well.

This is the approach used to consider the escape of Brownian particles from a potential well, and there is an obvious similarity. There are however, important fundamental differences between the 2 types of motion that must be reflected in the stochastic equations we may use to describe the average motion in the well.

(i) Equipartition of energy is not valid for particle motion in a turbulent fluid.

(ii) The relaxation time of a particle is not necessarily very much less than the timescale of the interacting turbulence i.e. the process is not necessarily Markovian as in Brownian motion.

(iii) The timescale of the turbulence, as opposed to that of the molecular motion, is likely to be very much greater than the periodicity of the well.

One direct consequence of (iii) is that the total energy of a particle may be drastically reduced by the well frequency from that in its free state, which in turn inhibits the ability of a particle to escape from the surface. In essence for small oscillations about the point of minimum potential energy, the system is extremely stiff with little transference of energy from the random forcing motion of the fluid to that of the particle in the well. In Brownian motion (B.M) the kinetic energy equals the average potential energy and is unaffected by the presence of the potential. Only the concentration is different from that of a free particle.

The behaviour of the particles at C is naturally of importance in determining the particle release rate from the surface. We suppose, as in the case of B.M., that the concentration way beyond C in the region of an average repulsive force is kept sufficiently small that it displaces the particles in the region of C from equilibrium. As a result there is a net current out of the well which we assume takes place under conditions in which the concentration at C normalised with respect to the number of particles in the well is constant. We shall further assume that the perturbation from equilibrium is only significant for small X about C, so that as with A we can use the harmonic approximation

$$V = -\frac{1}{2} m \omega^2 X^2 \tag{5}$$

Thus at A $\omega = \omega_A$ where $X = x$, and at C $\omega = i\omega_C$ where $X = x - X_C$, X_C being the position of C. ω_A and ω_C are necessarily positive constants and m is the mass of the particle.

Let us consider first the form of equilibrium solutions (zero net current) at A and C. Let us represent the lift force $F_L(t)$ by

$$F_L(t) = \langle F_L \rangle + \tilde{f}(t) \tag{6}$$

where $\langle F_L \rangle$ as before is the time/ensemble average lift force acting on the particle, and $\tilde{f}(t)$ the fluctuating component of F_L with zero mean. We shall further assume that $\tilde{f}(t)$ is stationary with an exponentially decaying auto-correlation

$$\langle \tilde{f}(t)\tilde{f}(t+s) \rangle = \langle \tilde{f}(o)^2 \rangle e^{-\gamma s} \tag{7}$$

Because the particles on the surface possess velocities considerably less than the focal fluid velocities we may linearise the equation of motion to a form equivalent to the Langevin equation. Thus using the harmonic forms for the potential, we have for the velocity v(t) of a particle at time t in the vicinity of either A or C

$$\frac{dv}{dt} = -\beta v - \omega^2 X + \frac{\tilde{f}(t)}{m} \tag{8}$$

$$\frac{dX}{dt} = v$$

where β^{-1} is the linearised relaxation time of the particle.

To solve for the concentration of the particles on the surface and ultimately the resuspension rate constant, we require a transport equation that embodies the important features of the motion described above. To this end we shall use a transport equation that is a direct counterpart of the Fokker-Planck equation for B.M. (Chadrasekhar, 1943). This describes the transport of the particle phase-space probability density $W(v,X,t)$ for a particle with velocity v and position X at time t. The equation of W for transport in a turbulent fluid is derived from the averaged particle Liouville equation using a closure scheme based on Lagrangian History Direct Interaction (Kraichnan, 1965). For a formal derivation of this equation we refer to Reeks 1980, we shall merely state it here and demonstrate its credibility. For particles undergoing motion of the form

$$\frac{dv}{dt} = -\beta v + K(X) + \tilde{f}(t) \tag{9}$$

$$\left(\frac{\partial}{\partial t} + v\frac{\partial}{\partial X} - \beta\frac{\partial}{\partial v} v + K(X)\frac{\partial}{\partial v}\right)W = \frac{\partial}{\partial v}\left(\alpha_1\frac{\partial W}{\partial v} + \alpha_2\frac{\partial W}{\partial X}\right) \tag{10}$$

Here α_1 and α_2 are in general functions of X and v as well as t. They are defined with reference to the averaged autonomous equation of motion of the system

$$\frac{du}{ds} = -\beta u + K(y) \tag{11}$$

$$\frac{dy}{ds} = u \tag{12}$$

We may define the solutions formally as

$$y = y(\Omega, Y|s); \qquad u = u(\Omega, Y/s) \tag{13}$$

where Ω and Y are the velocity and position of the particle at time zero.

This implies the inverse relation

$$\Omega = \Omega(u, y|s) \tag{14}$$

$$Y = Y(u, y|s) \tag{15}$$

found from equations (11) and (12) by replacing s by -s, with initial values (u, y). α_1 and α_2 are then given in terms of quantities

$$\frac{\partial y}{\partial \Omega}\{\Omega(v, X|s), Y(v, X|s)|s)\}; \quad \frac{\partial u}{\partial \Omega}\{\Omega(v, X|s), Y(v, X|s)|s\} \tag{16}$$

Explicitly

$$\alpha_1 = \frac{1}{m^2}\int_o^t ds \ \frac{\partial u}{\partial \Omega}(v, X, s) \ \langle\tilde{f}(o)\tilde{f}(s)\rangle \tag{17}$$

$$\alpha_2 = \frac{1}{m^2}\int_o^t ds \ . \ \frac{\partial y}{\partial \Omega}(v, X, s) \ \langle\tilde{f}(o)\tilde{f}(s)\rangle \tag{18}$$

If we define

$$\alpha_1 = \beta\mu; \quad \alpha_2 = (\beta\varepsilon-\mu) \tag{19}$$

then in cases where a_1 and a_2 are independent of x and v, in the limit of $t\to\infty$, μ and ε have a more transparent and important meaning; they are respectively the mean square velocity and spatial diffusion coefficient of the ensemble. For example, the motion of a free Stokes particle in a turbulent fluid has

$$\langle\tilde{f}(o)\tilde{f}(s)\rangle = m^2\beta^2 \langle u_f(o)u_f(s)\rangle \tag{20}$$

where $u_f(s)$ is the Lagrangian fluid velocity along a particle trajectory. Furthermore

$$\frac{\partial u}{\partial \Omega} = e^{-\beta s} \text{ and } \frac{\partial y}{\partial \Omega} = \frac{1}{\beta}(1-e^{-\beta s}) \tag{21}$$

so that

$$\mu(\infty) = \beta\int_o^\infty e^{-\beta s} \langle u_f(o)u_f(s)\rangle \, ds \tag{22}$$

$$\varepsilon(\infty) = \int_o^\infty \langle u_f(o)u_f(s)\rangle ds \tag{23}$$

which are precisely correct (Hinze, 1959; Reeks, 1977).

For motion of a particle given by equation (8) μ and ε also turn out to be independent of x and v. The solution to the subsidiary equation is

$$u = \frac{-(2\omega^2 Y + \beta\Omega)}{2\omega_1} e^{-\beta s/2} \sin\omega_1 s + \Omega e^{-\beta s/2} \cos\omega_1 s \tag{24}$$

$$y = \frac{(\beta Y + 2\Omega)}{2\omega_1} e^{-\beta s/2} \sin\omega_1 s + Y e^{-\beta s/2} \cos\omega_1 s \tag{25}$$

where $\omega_1{}^2 = \omega^2 - \beta^2/4$.
The solution is valid for both $\omega = (\omega_A, i\omega_C)$.
Using equations (17), (18) and (19) this yields

$$\mu(t) = \frac{1}{m^2\beta}\int_o^t ds\, e^{-\beta s/2} \left(\cos\omega_1 s - \frac{\beta}{2\omega_1}\sin\omega_1 s\right) \langle\tilde{f}(o)\tilde{f}(s)\rangle \tag{26}$$

$$\varepsilon(t)\ \frac{1}{m^2\beta}\int_o^t ds\, e^{-\beta s/2} \left(\frac{1}{\beta}\cos\omega_1 s + \frac{1}{2\omega_1}\sin\omega_1 s\right) \langle\tilde{f}(o)\tilde{f}(s)\rangle \tag{27}$$

When both $\mu(t)$ and $\varepsilon(t)$ have limiting forms, W at equilibrium has the form

$$W(v,X) = \text{const} \quad \exp\left[-\frac{1}{2}\left(\frac{v^2}{\mu(\infty)} + \frac{\omega^2 X^2}{\beta\varepsilon(\infty)}\right)\right] \tag{28}$$

$\mu(\infty)$ is again the particle mean square velocity associated with the ensemble, but it is also true of the local distribution at X. For $\omega = \omega_A$ we may reinterpret ε in terms of the average potential energy $\overline{PE}$ associated with the ensemble

$$m\beta\varepsilon = 2\ \overline{PE} \tag{29}$$

and that equation (28) represents a reliable approximation to the particles in the well so long as

$$\frac{Q}{m\beta\varepsilon} \gg 1$$

where Q is the height of the potential at C above A.
If we compare the spatial concentration $\rho(x)$ with that of the Boltzmann formula for B.M.

i.e.
$$\rho(x) \sim \exp - \left(\frac{1/2\ m\omega^2 X^2}{m\beta\varepsilon}\right) \qquad \text{Turbulent Motion} \tag{30}$$

$$\sim \exp - \left(\frac{1/2\ m\omega^2 X^2}{kT}\right) \qquad \text{B.M.} \tag{31}$$

we recognise that $m\beta\varepsilon$ is not equivalent to thermal energy kT i.e. it does not represent the turbulent energy of the particle except when both (β/ν) and $(\omega/\nu) \ll 1$.

For $t\to\infty$ (i.e. $t \gg \beta^{-1}$) the equilibrium values of μ and $\overline{PE}$ given above are identical to those obtained by averaging the equation of motion at equilibrium. For $t \gg \beta^{-1}$ the solution to equation (8) is

$$X(t) = \frac{1}{m\omega_1}\int_0^t \tilde{f}(s)\ e^{-\beta(t-s)} \sin \omega_1(t-s)\ ds \tag{32}$$

$$v(t) = \frac{1}{m}\int_0^t \tilde{f}(s)\ e^{-\beta(t-s)} \left(\cos \omega_1(t-s) - \frac{\beta}{2\omega_1} \sin \omega_1(t-s)\right) ds \tag{33}$$

Now multiplying equation (8) by v and averaging gives

$$\frac{1}{2}\frac{d}{dt}\overline{v^2} + \beta\ \overline{v^2} + \frac{1}{2}\omega^2 \frac{d}{dt}\overline{X^2} = \frac{\langle v(t)\tilde{f}(t)\rangle}{m} \tag{34}$$

so that at equilibrium we have

$$\overline{v^2} = \frac{1}{m\beta}\langle v(t)\tilde{f}(t)\rangle \tag{35}$$

Similarly multiplying equation (8) by X and averaging gives

$$\omega^2\overline{X^2} = \frac{\langle\tilde{f}(t)X(t)\rangle}{m} + \overline{v^2} \tag{36}$$

Substituting for X(t) and v(t) from equations (32) and (33) and using equation (29) gives expressions which are identical to equations (26) and (27).

Explicit Forms for $\mu(\infty)$ and $\varepsilon(\infty)$

Using the exponential decaying form $e^{-\nu t}$ for the lift force autocorrelation it is instructive to evaluate the equilibrium values for μ and ε. The integration is trivial. We obtain

$$\mu(\infty) = \frac{1}{\beta'(1+\beta'+\omega'^2} \frac{\langle\tilde{f}(o)^2\rangle}{m^2\nu^2} \tag{37}$$

$$m\beta\varepsilon(\infty) = \frac{\beta'+1}{\beta'(1+\beta'+\omega^2)} \frac{\langle\tilde{f}(o)^2\rangle}{m\ \nu^2} \tag{38}$$

In these formulae we have conveniently normalised both ω and β on the timescale ν of the lift force, so that here

$$\beta' = \beta/\nu; \quad \omega' = \omega/\nu \tag{39}$$

The case for $\omega^2 = -\omega_C{}^2$ i.e. equilibrium at C, is interesting. It would appear that when

$$\omega_C'^2 \geqslant 1 + \beta' \tag{40}$$

both $\mu(\infty)$ and $\varepsilon(\infty)$ are both negative, which is physically absurd. What in reality this condition expresses is that under these conditions $\mu(t)$ and $\varepsilon(t)$ do not tend to an asymptotic limit as $t\to\infty$. In other words it would appear that equilibrium solutions are not possible - the duration of the random lift is not sufficiently short to dampen out the accelerative motion away from C.

A closer examination would reveal that as $t\to\infty$ most of the particles are to be found at $X = \pm\infty$ about C, with extremely large velocities. It is clear however that the overall potential and that especially beyond A will not allow this to happen. In other words the approximation $V(x) = 1/2\ m\omega_C{}^2X^2$ about C is unreasonable and we must consider the entire potential. We would see this made manifest in the coefficients $\frac{\partial u}{\partial\Omega}$ and $\frac{\partial y}{\partial\Omega}$, which are controlled by the entire motion of the particle. Quite crudely the upper limit in the integrals in μ and ε are confined to values less than the typical time it takes a particle to travel from C to A. Equilibrium solutions are more difficult to obtain and we shall not attempt it here, though we will have more to say about it later on.

We shall now consider obtaining expressions for the rate constant for normalised frequencies at C such that

$$\omega_C{}'^2 \leqslant 1 + \beta' \tag{41}$$

Expressions for the Rate Constant

For convenience let us use subscript A to refer to quantities based on the harmonic approximation at A, and similarly subscript C for those at C. If we suppose that x is measured from A, and X_C refers to the position of C, then in the vicinity of C for constant current we have

$$v \frac{\partial W}{\partial X} + \omega_C^{\,2} X \frac{\partial W}{\partial v} - \beta v \frac{\partial W}{\partial v} - \beta W = \beta \mu_C \frac{\partial^2 W}{\partial v^2} + (\beta \varepsilon_C - \mu_C) \frac{\partial^2 W}{\partial v \partial X} \tag{42}$$

where $X = x - X_C$.

We consider

$$W(X,v) = \text{const} \quad \exp\left(-\frac{v^2}{2\mu_C}\right) \exp\left(\frac{\omega_C^{\,2} X^2}{2\beta\varepsilon_C}\right) . F(X,v) \tag{43}$$

so that for equilibrium (zero current) $F(X,v) = 1$.

For future reference we shall call the equilibrium distribution at $X = 0$, W_C.

Substituting in equation (42), we have for F

$$\frac{\beta\varepsilon_C}{\mu_C} v \frac{\partial F}{\partial X} + \left(\beta v + \frac{\mu_C \omega_C^{\,2} X}{\beta \varepsilon_C}\right) \frac{\partial F}{\partial v} = \beta \mu_C \frac{\partial^2 F}{\partial v^2} + (\beta\varepsilon_C - \mu_C) \frac{\partial^2 F}{\partial v \partial X} \tag{44}$$

We suppose F to be of the form $F(\eta)$ where

$$\eta = v - aX \tag{45}$$

This means

$$\left[\beta\left(1 - \frac{\varepsilon_C a}{\mu_C}\right) v + \frac{\mu_C \omega_C^{\,2} X}{\beta\varepsilon_C}\right] \frac{dF}{d\eta} = \left[\beta\mu_C - a(\beta\varepsilon_C - \mu_C)\right] \frac{d^2 F}{d\eta^2} \tag{46}$$

For F() to be a single function of η

$$\frac{\dfrac{\mu_C \omega_C^{\,2}}{\beta\varepsilon_C}}{\beta\left(1 - \dfrac{\varepsilon_C a}{\mu_C}\right)} = -a$$

or
$$a = \frac{\mu_C}{\beta\varepsilon_C}\left[\frac{\beta}{2} \pm \sqrt{\frac{\beta}{2}^2 + \omega_C^{\,2}}\right] \tag{47}$$

Thus
$$-\left(\frac{a\beta\varepsilon_C}{\mu_C} - \beta\right)\eta\,\frac{dF}{d\eta} = \left[\beta\mu - a(\beta\varepsilon_C - \mu)\right]\frac{d^2F}{d\eta^2} \tag{48}$$

We demand that for meaningful solutions

$$\begin{aligned} F &\to 1 \text{ for } X \to -\infty \\ &\to 0 \text{ for } X \to +\infty \end{aligned}$$

The solution to (48) satisfying these requirements is

$$F = \left(\frac{\sigma}{2\pi q}\right)^{1/2}\int_{-\infty}^{\eta} \exp -\left(\frac{\sigma}{2q}\,s^2\right) ds \tag{49}$$

where
$$\sigma = \frac{a\beta\varepsilon_C}{\mu_C} - \beta \tag{50}$$

$$q = \beta\mu_C - a(\beta\varepsilon_C - \mu_C). \tag{51}$$

In this respect we choose the positive root for a, and demand $q \geqslant 0$. Using the form for μ_C and ε_C given in (37) and (38) we find this is equivalent to

$$\omega_C'^{\,2} \leqslant 1 + \beta' \tag{41}$$

our initial requirement.

We have however a problem in normalising W in the vicinity of C to the number of particles at any one instant within the well. We make the reasonable assumption that if Q is the height of the potential barrier of C above A, then

$$W_C \simeq \rho_A \exp -\left(\frac{Q}{m\beta\varepsilon_A}\right)\cdot\frac{1}{\sqrt{2\pi\mu_C}}\exp -\left(\frac{v^2}{2\mu_C}\right) \tag{52}$$

where ρ_A is the spatial concentration of particles at A. Since most of the particles are in the vicinity of A, we can normalise ρ_A for one particle in the well using the harmonic approximation for V(x) around A i.e.

$$1 = \rho_A \int_{-\infty}^{\infty} \exp - \frac{\omega_A{}^2 x^2}{2\beta\varepsilon_A} \, dx \tag{53}$$

i.e

$$\rho_A = \frac{\omega_A}{\sqrt{2\pi\beta\varepsilon_A}} \tag{54}$$

so that for constant current at C we have

$$W(X_C,v) \simeq \cdot \frac{\omega_A}{2\pi} \frac{1}{\sqrt{\mu_C \beta\varepsilon_A}} \cdot \exp - \left(\frac{Q}{m\beta\varepsilon_A} + \frac{v^2}{2\mu_C}\right) \left(\frac{\sigma}{2\pi q}\right)^{1/2} \int_{-\infty}^{v} \exp - \left(\frac{\sigma s^2}{2q}\right) ds \tag{55}$$

The current at $x = X_C$, which is constant throughout the potential is

$$p = \int_{-\infty}^{\infty} vW(X_C,v) \; dv \tag{56}$$

which, after substituting for W, integrating and using the forms for σ and q, finally contracts to the simple form.

$$p = \frac{\omega_A}{2\pi\omega_C} \sqrt{\frac{\varepsilon_C}{\varepsilon_A}} \left[\sqrt{\frac{\beta}{2}^2 + \omega_C{}^2} - \frac{\beta}{2} \right] \exp - \left(\frac{Q}{m\beta\varepsilon_A}\right) \tag{57}$$

Since we have normalised ρ_A for 1 particle in the well at any one instant of time, p also represents the rate constant. Thus for quasi-equilibrium the number of particles N in the well is given by

$$N(t) = N_o e^{-pt}$$

Let us look at p for the 2 possible ranges of ω_C which can satisfy in-equality (41), namely

(a) $\beta' \ll 1$ $\omega_C'^2 < 1$

This embraces the condition on the scales for motion equivalent to B.M. i.e. $\beta' \ll 1$, $\omega_C'^2 \ll 1$, $\omega_A'^2 \ll 1$. Under these latter constraints

$$\varepsilon_C = \varepsilon_A \text{ and } \beta\varepsilon_A = \mu_A = \mu_C = \mu \tag{58}$$

giving the form

$$p = \frac{\omega_A}{2\pi\omega_C}\left[\sqrt{\left(\frac{\beta}{2}\right)^2 + \omega_C{}^2} - \frac{\beta}{2}\right] e^{-\frac{Q}{m\mu}} \tag{59}$$

which if we replace μ by $\frac{kT}{m}$ is identical to that obtained by Chandrasekhar (1943) and Kramers (1940) for B.M.

(b) $\omega_C'^2 \ll \beta'$, $\beta' \ll 1$

This case is interesting. It is the limit of extremely small particles, and outside the normal range of applicability of the assumptions made in B.M. However, we see that if $\omega_A'^2$ is also $\ll \beta'$, then again $\varepsilon_A(\infty) = \varepsilon_C(\infty) = \varepsilon(\infty)$ the local diffusion coefficient of a free particle which in this instance is the local diffusion coefficient of the fluid $\varepsilon_f(\infty)$. For p we have

$$p = \frac{\omega_A \omega_C}{2\pi\beta} \exp\left(-\frac{Q}{m\beta\varepsilon_f(\infty)}\right) \tag{60}$$

c.f. the B.M. formula for $\omega_C \ll \beta$. Furthermore in this limit we may assume for a sphere of radius R

$$m\beta = 6\pi R\mu_f$$

where μ_f is the dynamic viscosity.
In the case of Van der Waals surface forces

$$Q = 2\pi R\gamma x_o \tag{61}$$

$$\text{and } p \to \frac{\omega_A\omega_C}{2\pi\beta} e^{-\frac{1}{3}\frac{\gamma x_o}{\mu_f\varepsilon_f'(\infty)}} \tag{62}$$

i.e. $p \to 0$. The equilibrium spatial distribution within the well for constant μ and ε is of the form

$$\exp\left(-\frac{V(X)}{6\pi\mu_f R\varepsilon_f(\infty)}\right)$$

which for Van der Waals surface forces has the limiting form

$$\exp\left(-\frac{\gamma x_o{}^2}{3\mu_f\varepsilon_f(\infty)}\right)$$

The Transition Approximation

The condition corresponding to B.M. with $\beta \ll \mu_C$ gives

$$p = \frac{\omega_A}{2\pi} \cdot \exp - \left(\frac{Q}{m\beta\mu}\right) \tag{63}$$

which is commonly referred to as the transition approximation. It can be interpreted very simply. The distribution of particles immediately to the left of C (X < 0) is the equilibrium distribution

$$\frac{\omega_A}{\overline{2\pi\mu}} e^{-\frac{\omega}{m\beta\mu}} \cdot e^{-\frac{1}{2}\frac{\omega_C{}^2X^2}{\beta\mu}} \cdot \phi_C(v) \tag{64}$$

where $\phi_C(v)$ is the velocity distribution at C.
This distribution of particles propagates to the right of C by the action of the force $m\omega_C{}^2X$ alone, i.e.

$$\begin{aligned} W(X>0) &= \frac{\omega_A}{\sqrt{2\pi\mu}} e^{-\frac{Q}{m\beta\mu}} \quad \phi_C\,(v^2 - \omega^2X^2) \text{ for } v^2>\omega^2X^2 \\ &= 0 \qquad\qquad\qquad\qquad \text{for } v^2<\omega^2X^2 \end{aligned} \tag{65}$$

There are only positive velocities to the right of C, giving rise to a current of the form

$$p = \frac{\omega_A}{\sqrt{2\pi\mu}} e^{-\frac{Q}{m\beta\mu}} \cdot \frac{1}{\sqrt{2\pi\mu}} \int_{\omega X}^{\infty} v\, e^{-\frac{(v^2-\omega^2X^2)}{2\mu}} dv \tag{66}$$

$$= \frac{\omega_A}{2\pi} \exp - \frac{Q}{m\beta\mu} \tag{66}$$

Irrespective of the condition we have imposed on ω_C we may extend this approximation to the case where C is a discontinuity in the well and the particle leaves the surface because the mechanical energy is greater than the surface energy requirements (J.K.R. model). After escaping we assume a particle never returns to the surface. There are thus no negative particle velocities observed beyond C. For v < 0, and X > 0, $\frac{\partial y}{\partial \Omega}$ and $\frac{\partial u}{\partial W}$ are both zero, implying both α_1 and α_2 are zer. There can be no possible mechanism for the creation of negative velocities beyond C which is consistent with our boundary conditions. The discontinuity in force from finite to zero we can regard as so rapid that the equilibrium to the left of the transition is determined by that at A.

Using the formulae for ε_A and μ_A for $\omega_A'^2 \gg 1$, gives

$$p = \frac{\omega_A}{2\pi\sqrt{\beta'+1}} \exp - \left[\frac{\beta'}{\beta'+1} \cdot \left(\frac{m\omega_A{}^2 Q}{\langle\tilde{f}^2\rangle}\right)\right] \qquad (67)$$

This leads us more generally into the case $\omega_C'^2 \gg (1+\beta')$. If the values of both α_1 and α_2 are limiting, though not adequately described by the harmonic approximation, we can always assert that the transition approximation will apply so long as

$$\left(\frac{\omega_C}{\beta}\right)^2 \gg 1 \qquad (68)$$

The question then remains as to what we take for the equilibrium distribution C? If this distribution is determined by the limiting values of α_1 and α_2 around C, then particles emerging from the well must spend sufficient time around C to come to equilibrium with the new force at C. This in turn means

$$\omega_C \ll \beta$$

which is in contradiction to our original inequality i.e. the equilibrium distribution at C is more closely associated with ω_A than ω_C, and formula (67) is still valid.

Well Frequencies Associated with a Spherical Particle on a Smooth Flat Plate

As an example let us consider the well frequencies associated with a smooth spherical particle on a flat plate elastically deformed under the action of surface Van der Waals adhesive forces. The frequency ω_A will depend upon the particle mass m, the particle radius R, together with the surface energy/unit area of the interacting surface forces, and both the Elastic moduli E_i and Poisson's ratio ν_i of the 2 surface materials. With hindsight we shall clump E_i and ν_i into a single constant

$$K = \frac{4}{3}\left[\frac{1-\nu_1{}^2}{E_1} + \frac{1-\nu_2{}^2}{E_2}\right]^{-1} \qquad (69)$$

where K is the constant of elasticity in the Hertz Law of contact.

On purely dimensional grounds ω_A will be of the form

$$\omega_A \sim \gamma^{1/6} K^{1/3} R^{1/3} m^{-1/2} \qquad (70)$$

As an example for a 10 μm radius smooth glass sphere, with density 2.47 gm cm^{-3} on a smooth glass plate $\gamma \sim 0.15\ Jm^{-2}$, and $K \sim 5.78 \times 10^{10}\ N\ m^{-2}$,

$$\omega_A \sim 1.9 \times 10^7\ Hz$$

Let us compare this with the scale of turbulent motion at ground level in a neutrally stable air flow of 5 m/s for which the friction velocity, u_τ, over a smooth surface $\sim 16\ cms^{-1}$ (Sutton, 1953). Here

$$\nu \sim \frac{u_\tau^{\,2}}{\nu_f T^+} \qquad (71)$$

where ν_f is the kinematic viscosity and $T^+ \sim 10$ based on measurements of the bursting rate. This gives $\nu \sim 1.7 \times 10^3$ Hz. The value of β for a 10 μm radius glass particle in air based upon Stokes drag $\sim 3 \times 10^2$ Hz. The enormous disparity between ω_A and both β and ν means that for turbulent air flows normally encountered in practice

$$\omega_A'^{\,2} >> 1$$

and that for particles > 1 μm, $\frac{\omega_A}{\beta} >> 1$. For glass on glass. C is more likely to be associated with a discontinuity in the potential diagram, and formula (67) based on the transition approximation will represent the resuspension rate for $Q >> m\beta\varepsilon_A$.

For the case where the particle continuously reduces to an undeformed state with applied negative load (D.M.T) C is a balance between the average lift force $\langle F_L \rangle$ and Van der Waals attractive forces for a rigid sphere.

i.e.
$$X_C = \left(\frac{F_o}{\langle F_L \rangle}\right)^{1/2} x_o \qquad (4)$$

where F_o is the adhesive tearing off force given by

$$F_o = 2\ \pi\gamma R \qquad (72)$$

and

$$\omega_C^{\ 2} = \frac{4\pi\gamma R}{m\ x_o}\left(\frac{\langle F_L\rangle}{F_o}\right)^{3/2} \tag{73}$$

For the case of 10 μm radius steel spheres, of density 7.8 μm/cc on a steel plate $\gamma = 0.19\ \mathrm{Jm}^{-2}$, and exposed to the same air flow

$$\omega_C'^{\,2} \sim 8.6 \times 10^{15}\left(\frac{\langle F_L\rangle}{F_o}\right)^{3/2} \tag{74}$$

For the formula to apply $\left(\frac{\langle F_L\rangle}{F_o}\right) \leqslant 2 \times 10^{-11}$. We would not normally expect lift forces to be so exceedingly small and that in practice $\omega_C'^{\,2} \gg 1$, suggesting again use of the transition approximation of equation (67).

Summary and Conclusions

We have shown in this paper in what way the long term resuspension of small particles by a turbulent fluid depends upon:

(a) the timescale and intensity of the induced lift force fluctuations,

(b) the natural frequency and depth of the adhesive well potential,

(c) the particle inertial response to changes in fluid motion embodied in its relaxation time β^{-1}.

The significant feature of this analysis has been to demonstrate that for most practical problems the natural frequency of the well is very much greater than the typical frequency of the lift force fluctuations, and that this disparity significantly inhibits the ability of the particle to escape from the surface. For this situation the formula presented for the rate constant reduces to the transition approximation.

To obtain these formulae we have used a transport equation that is the direct counterpart of the Fokker Planck equation in Brownian Motion, but not limited in applicability to the relative range of timescales for which $\beta' \ll 1$ and $\omega' \ll 1$.

The object of future work will be to investigate the validity of the formula for the rate constant from existing experimental measurements.

Acknowledgement

This paper is published by permission of the Central Electricity Generating Board.

REFERENCES

1. Chandrasekhar, S., 1943, Stochastic Problems in Physics and Astronomy, Rev. Mod. Phys. 15(1), 1-89.

2. Corn, M., 1966, Aerosol Science, ed. C. N. Davies, chap. Xl, Acad. Press, London and New York.

3. Corn, M. and Stein, F., 1965, Amer. Ind. Hyg. Assoc. 26, 325-337.

4. Dahneke, B., 1975, J. Colloid, Int. Sci., 50(1), 89-107.

5. Derjaguin, B. V., 1934, Colloid 69, 155.

6. Derjaguin, B. V., Muller, V. M. and Toporov, Yu. P., 1975, J. Colloid Int. Sci., 53, 314-326.

7. Hinze, J. O., 1959, Turbulence, 352-364, McGraw-Hill, New York.

8. Johnson, K. L., Kendall, K. and Roberts, A. D., 1971, Proc. Roy. Soc. A, 324, 301-313.

9. Kraichnan, R. H., 1965, Phys. Fluids, 8, 575-597.

10. Kramers, H. A., 1940, Physica, 7, 284- .

11. Reeks, M. W., 1977, J. Fluid Mech. 83, 529-546.

12. Reeks, M. W., 1980, J. Fluid Mech. 97, 569-590.

13. Sutton, O. G., 1953, Micrometeorology, p 233, McGraw-Hill, New York.

14. Tabor, D., 1977, J. Colloid Int. Sci., 58(1), 2-12.

15. Zimon, A. D., 1969, Adhesion of Dust and Powder, Plenum Press, New York and London.

Observations on the Current Status of Field Experimentation on Heavy Gas Dispersion

J. McQUAID

Safety Engineering Laboratory
Health and Safety Executive
Red Hill, Sheffield S3 7HQ, UK

Summary

An ability to estimate the dispersion of heavy gases is an important requirement in the assessment of the hazards of flammable and toxic gas releases to the atmosphere. Methods of achieving such estimates, either by physical or mathematical modelling, require validation by reliable field data, especially at large scale. Field experiments on the dispersion of heavy gases started on a small scale around 1970 and culminated in the early 1980's in a number of large-scale programmes. The information gained from this work is briefly reviewed but the main purpose of this paper is to identify and discuss several recurrent features with implications for future experimentation.

1. Introduction

During the 1970's there was a rapid expansion of interest in the dispersion of heavy gases in the atmosphere. Several experimental programmes were undertaken both in the field and the laboratory, accompanied by the development of mathematical models for predicting dispersion. Large uncertainties in estimates of dispersion, particularly in the case of large-scale releases, persisted and were highlighted by Havens [1]. Moves to resolve the main outstanding issues were initiated in the UK and USA around 1980 and resulted in large-scale experimental programmes at Thorney Island and Maplin Sands in the UK and at China Lake in the USA. Although the full results of these programmes are not yet available for analysis, it is nevertheless useful at this stage to review what has been achieved and what still needs to be done.

A review of field experiments conducted up to 1978 was given in McQuaid [2] and a detailed description has more recently been published by Puttock et al [3]. Only a brief outline will therefore be given in this paper as an introduction to the topics sel-elected for discussion. Although the presentation will emphasise the scarcity of good data, at least prior to the three programmes mentioned above, two contributory factors should be borne in mind. Firstly, successful experimentation in this field is difficult to achieve and this is particularly so where the gas is released in the liquefied state. Secondly, many of the early experiments were mainly concerned with practical questions and the acquisition of dispersion data in sufficient detail for developing and validating predictive models was a secondary consideration. Although the respective experimental programmes are usually included in listings of dispersion experiments, it was really only in the later programmes that the dispersion aspects received the necessary attention.

2. Summary of Field Experiments

The reported experimental investigations have covered a variety of conditions, encompassing the time-dependency of the release, the material physical state and a land or water-based site. In this summary, the experiments will be classified according to the physical state of the heavy gas prior to release, whether as a liquefied gas under refrigeration or pressure, or as a gas at ambient pressure and temperature.

2.1. Refrigerated Liquefied Gas Experiments

US Bureau of Mines 1970-72 - These experiments, described by Burgess et al [4,5], consisted of quasi-instantaneous (maximum quantity $0.5m^3$) and continuous (maximum rate $1.3m^3/min$) spills of liquefied natural gas (LNG) on water. There were few measurements of concentration and it was likely that the site topography affected the results. Although the experiments were valuable in highlighting several features of heavy gas behaviour, the measurements were too limited in scope for current requirements.

Matagorda Bay Trials 1971 - These experiments were organised by ESSO and the American Petroleum Institute and reported by May et al [6]. The experiments were conducted at sea and 17 tests were performed in which quantities of LNG between 0.76 and $10.2m^3$ were spilled at a rate of about $19m^3/min$. The spill times ranged from about 6 to 35s. Concentrations were measured with 14 sensors located on two crosswind lines at sea level and 4 sensors at 3m height. The rate of evolution of vapour was very time-dependent. Although the concentration data are reasonably comprehensive, the uncertainty in the source conditions seems to have inhibited their use by mathematical modellers. The Maplin Sands experiments, to be discussed below, are very similar in concept and may now be taken as superseding the Matagorda Bay trials.

Gadila Jettison Tests 1973 - These tests were conducted in the Bay of Biscay and consisted of the discharge of LNG as a jet at high level from the stern of the Gadila, an LNG carrier. The tests were reported by Kneebone and Prew [7]. There were no direct measurements of concentration. However, the visible outline of the plume was taken to correspond to a concentration of 0.5%. Overhead photographic records, interpreted in this way, have been used by a number of modellers e.g. Britter [8], Cox and Carpenter [9], te Riele [10].

AGA Capistrano Tests 1974 - In these land-based tests, LNG was spilled into bunds up to 24m diameter. The tests were organised by the American Gas Association and carried out by the Battelle Institute. They were described by Duffy et al [11]. There was a total of 42 spills, of which 28 provided useful dispersion data from 36 gas sensors. The quantity spilled ranged from 0.4 to $51m^3$ and the spill time was between 20 and 30s. The time-dependent source condition is again a complicating factor but the data have nonetheless proved useful in the development of early models e.g. Cox and Roe [12].

China Lake 'Avocet' Series 1978 - These were the first of three series of tests at the US Naval Weapons Center test site at China Lake, California. They were conducted by the Lawrence Livermore National Laboratory for the US Department of Energy and were reported by Koopman et al [13]. The China Lake site consists of a shallow pond 58m diameter with a central spill point. Dispersion is mostly over land. In the Avocet series, 4 spills were carried out, each of $5m^3$ of LNG and with a spill time of about 1 minute. Their main purpose was the development of the facility and instrumentation for the subsequent larger-scale and more fully-instrumented tests in the Burro and Coyote series.

China Lake 'Burro' Series 1980 - In this series, 8 spills of $40m^3$ of LNG were conducted. Spill times ranged up to 3.5 minutes and the spills are therefore classified as continuous. The experimental conditions and the results obtained were described by Koopman et al [14]. Instrumentation was particularly comprehensive and included provision for measurement of turbulence and high-frequency concentration fluctuations in the cloud. Very useful data were obtained and detailed analyses have been reported by Koopman et al [15] and Rodean [16].

China Lake 'Coyote' Series 1981 - These tests were primarily oriented towards a study of rapid phase transition of LNG spilled on water and flammability of LNG vapour clouds. The test configuration was similar to that in the Burro series. The experiments are due to be reported in late 1983. Dispersion data are understood to be available.

Shell Maplin Sands Experiments 1980 - The site for the experiments was a large area of tidal sands on the north bank of the Thames estuary. The experiments have been described by Puttock et al [17, 18] and analysis of the data by Colenbrander and Puttock [19, 20]. A total of 34 spills was performed, comprising instantaneous and continuous spills of LNG and liquefied propane. A very extensive array of meteorological and gas concentration instruments was deployed, together with comprehensive photographic coverage. Although the detailed results have not yet been released (this is due in late 1983), all the indications are that the trials will provide a substantial contribution to the database on disperson of heavy gas clouds.

2.2 Pressurised Liquefied Gas Experiments

DGA Netherlands 1973 - One experiment was performed in this investigation (Buschmann [21]). It consisted of the release of 1 tonne of Refrigerant-12 (dichlorodifluoromethane). The pressurised liquid was discharged into a tank of hot water so that a gas cloud was generated rapidly. Dispersion was over land and the principal results consisted of the radius and height of the cloud as functions of time. The results of the experiment were used to verify the gravitational slumping model of van Ulden [22].

Nevada Ammonia Spills 1983 - These tests consisted of the release of about 20 tonnes of ammonia and were carried out by the Lawrence Livermore National Laboratory for a consortium led by the US Coast Guard. The test site was well instrumented and 4 successful trials were performed in August and September 1983. Details of the trials are not yet available.

HSE Water-Spray Barrier Trials 1981-82 - The objective of these trials was the evaluation of the performance of water-spray barriers as an aid to dispersion of heavy gas plumes. In the trials, conducted at HSE's Buxton site, carbon dioxide was released at a rate of 1 to 2kg/s and concentrations were measured with an array of up to 34 gas sensors. Each trial included a period of up to 3 minutes without the water-spray barrier in operation thus providing dispersion data on continuous heavy gas releases. The experiments are described by Moodie [23, 24].

2.3 Ambient Pressure and Temperature Gas Experiments

Porton Down Trials 1976-78 - These experiments were performed by the Chemical Defence Establishment for HSE and have been reported by Picknett [25]. In each experiment, a cloud of gas of $40m^3$ volume was released instantaneously from a cubical container. A total of 42 trials was performed in a variety of weather conditions and ground roughnesses and included releases on sloping as well as flat terrain. The gas used was a mixture of air and Refrigerant-12, allowing initial relative densities up to 4.2 to be achieved. The principal data available are photographic records from which the cloud geometry as a function of time is evaluated. The data have been used extensively in evaluations of predictive models e.g. Woodward et al [26], and in wind-tunnel simulation experiments by Hall et al [27].

Thorney Island Heavy Gas Dispersion Trials 1980-83 - These trials are similar to the Porton Down trials and involved the release of $2000m^3$ of gas. They were organised by a consortium led by HSE and performed by the National Maritime Institute. The are described in another paper in this volume (McQuaid [28]).

2.4 Conclusions

The requirements of an experimental database for the evaluation or development of predictive models are comprehensive data on concentration and turbulence in the cloud and on the meteorological conditions prevailing during each test. The investigations that provide the most satisfactory data in these regards are the China Lake Burro Series, Maplin Sands, Thorney Island and (most likely) the China Lake Coyote Series and the Nevada ammonia spill tests. To date, only the results of the Burro Series and Maplin Sands tests have been compared with model predictions and then only with the models of the respective investigators. The experimental results that are now becoming available will provide an extremely large database which seems likely to satisfy needs for some time to come. During the next few years, the subject development is likely to be concentrated on analyses of the existing data rather than the conduct of more experiments, at least at large scale. The remainder of this paper will discuss some themes of interest both for interpretation of the data and for the conduct of future investigations.

3. The Vapour Blanket Effect

Predictive models generally assume that the air flow above the cloud is describable in terms of the properties of the upwind approach flow. This assumption was discussed briefly in McQuaid [2]. Its validity has come into question as a result of observations during the Burro Series of LNG experiments at China Lake. Koopman et al [15] highlighted a marked difference of behaviour for one of the trials. They pointed out that in the Burro 8 trials the mean flow was observed to diverge around the cold cloud and was reduced significantly within the cloud. The ambient wind field was displaced upward by about 1.5m causing the wind speed within the cloud to drop essentially to zero. Koopman et al believed that this behaviour is likely to occur on larger spills under a variety of conditions and that the ability of large masses of cold, dense gas to displace the normal atmospheric flow has profound implications for hazard prediction from large accidental spills. It is relevant also to refer back to the earlier Matagorda Bay trials where a similar effect was observed. May et al [6] reported, for instantaneous spills, that the LNG vapour was not carried away by the wind as quickly as it was formed; a large and visible accumulation of vapour built up over the spill point. The effect was very dependent on wind speed, and was particularly pronounced at low wind speed.

This has become known as the 'vapour blanket effect' and it is interesting to consider the phenomenon in some detail and to extend the points that have been made. First, consider the reduction in velocity within the cloud and the upward displacement of the flow which results. Insight can be gained from studies that have been made of the effect of uniform injection of secondary fluid into a turbulent boundary layer on a flat plate. Evaporation from a pool of liquid belongs to that class of flows. It is known from this work that the boundary layer will separate or blow off at quite low values of the ratio of injection velocity to free stream velocity. Separation is characterised by a zero velocity gradient at the surface, an inflected velocity profile and a large value of the ratio of displacement to momentum thickness of the boundary layer. All of these effects are consistent with the description by Koopman et al for the Burro 8 test. Experimental studies (e.g. McQuaid [29]) have suggested that separation occurs at a value of the injection velocity to free-stream velocity ratio of around 0.012 to 0.014. These experiments were for air-to-air injection but the result is applicable to injected gases other than air if the density ratio is factored to the velocity ratio. Note that the result takes no account of buoyancy effects - the velocities in the application of original interest were too high for these to be important. There is a slight difficulty in applying the result to evaporation into the atmospheric surface layer since the relevant free-stream velocity is not obvious. However, it is not a critical difficulty and a velocity representative of the flow above the plume will be appropriate.

For an evaporating LNG pool, an estimate of the injection ratio can be prepared using the evaporation rate for LNG on water. Ermak et al [30] have used 4.2×10^{-4} m/s for the regression rate of the liquid surface i.e. the volumetric rate of evaporation per unit area of the pool. This figure indicates that blow off will occur at free-stream velocities below about 12m/s or in other words for all the LNG experiments carried out at China Lake and not just Burro 8. The effect will be greatest with Burro 8, which was conducted at very low wind speed. The conclusion is that the flow will be affected as described by Koopman et al and this would happen even if the vapour was neutrally buoyant. Although the effect as described is primarily over the evaporating pool itself, the boundary layer will take a time to adjust after it has passed the evaporation region. The experimental studies of McQuaid [29] have shown that the readjustment requires at least 30 to 40 boundary layer thicknesses of downstream development. The measurements on which Koopman et al based their discussion were carried out at 57m so that measurable effects would be expected, since the layer depth was only a few metres. The observations by Koopman et al are therefore entirely consistent with known behaviour.

But what happens to the structure of the ambient atmosphere above the plume? It might be expected that the flow over a large-scale plume would be affected since it is insulated from the influence of the ground shear. The turbulence energy that is advected from upstream will decay and this will manifest itself as a reduced entrainment rate through the top surface. The effect depends on the downwind extent of the region wherein buoyancy effects are important in the plume. It also depends on the lateral extent of the plume, governing the time-scale for lateral transport of turbulence kinetic energy from the sides towards the centre of the plume. The effect, if it is present, will therefore depend on the scale of the release. In the case of an instantaneously-formed cloud, with a limited extent in the wind direction, the air flow over the cloud is constantly replaced since the mean advection velocity of the cloud is substantially less than the wind speed above the cloud. There is therefore the possibility that a cloud would not be affected to the same degree as a plume. This would have consequences for the validity of the often-employed hypothesis that a plume can be modelled as a succession of slices, each of which has an entrainment rate through its top surface determined from a cloud model.

Some information on the effects on the ambient flow is available from small-scale experiments. These consisted of a study of a continuous release of carbon dioxide at floor level into a fully-developed channel flow (McQuaid [31]). The carbon dioxide was emitted from a line source and the layer formed had an average depth of about 0.05m. The dimensions of the channel were 0.3m (horizontal) and 0.9m (vertical). The development of the layer was studied up to a distance of about 100 layer depths from the source and the results included the intensities of the three components of the turbulent velocity within and above the layer. The Reynolds number based on layer depth was between 2×10^3 and 10^4. The Richardson number was about 17.5, using the definition of Puttock et al [3] i.e. $g'H_o/U_*^2$ where g' is the reduced gravitational acceleration, H_o the initial depth and U_* the friction velocity in the approach flow. The ratio of the mass fluxes of carbon dioxide and of the air flow in the channel was about 0.015. This was sufficiently high to introduce momentum transfer effects unrelated to the buoyancy effects, in the manner just described. Two comparison experiments were therefore performed. The first was with a neutrally-buoyant layer, in which the injected fluid was air at the same value of the mass flux ratio as in the experiment with carbon dioxide. The second comparison experiment measured the flow structure in the absence of injected fluid.

The intensities of the three turbulent velocity components are shown in Figs. 1a, b and c. The measurements shown were made at the final measuring station 4.83m, or about 100 layer depths, from the source. The upper edge of the layer is indicated in each figure and is defined as the height at which the concentration has fallen to 0.1% of the concentration at floor level at the same location. It can be seen from the figures that all three intensities in the density-stratified layer are substantially reduced below those in the two comparison experiments, as would be expected. The main point to note is that the effect persists well above the edge of the layer into the region where the mean concentration is negligible. The intensities near the floor in the neutrally-buoyant layer have returned to the levels present without any layer-forming fluid but the increased intensities, originating in the disturbance at the source, are still present further from the floor. The u-component intensity (Fig.1a) near the wall in the density-stratified layer has also recovered, reflecting the fact that the production of turbulence kinetic ener-

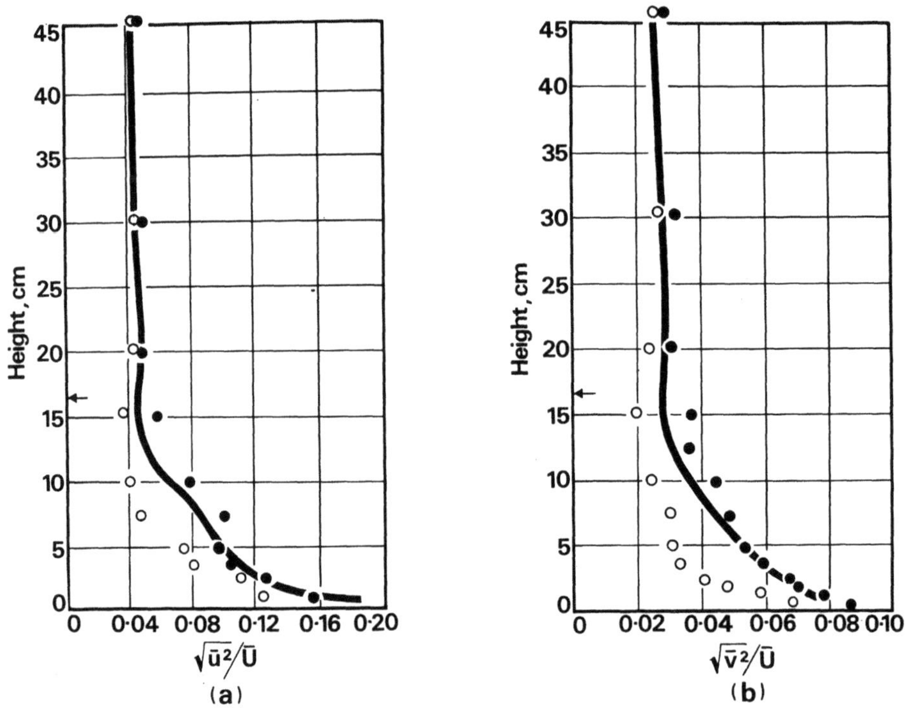

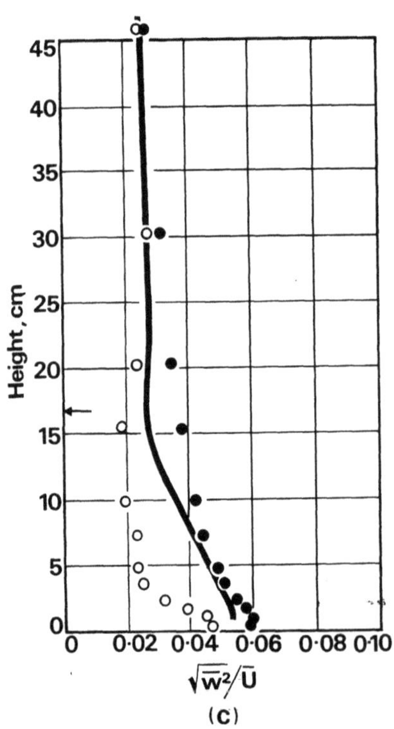

Fig.1 Distributions of Turbulent Intensity Components (from McQuaid [33]).

━━━ without layer

● neutrally buoyant layer

○ negatively buoyant layer

← edge of negatively buoyant layer

gy by the mean shear goes into the u-component only. Redistribution to the v and w-components has little effect by this stage. At the edge of the layer, the v and w-component intensities are a factor of 2 below those in the neutrally-buoyant layer.

These measurements show that there is a history effect which prevents the local conditions being described in terms of local quantities only. It follows that models based on k-theory will not be able to represent the turbulent transfer rates. The prescriptions for the eddy diffusivities of mass and momentum are generally dependent on the local gradient Richardson number. Similarly, box models with prescriptions for the entrainment rate which depend on the ambient structure in the absence of the layer will also be in error. It would need a more elaborate turbulence model with allowance for history effects, via transport equations for Reynolds stresses and a length scale, to describe the flow. There is certainly a need to be able to specify the conditions under which the simple, local-equilibrium models may not be reliable. However, it is highly desirable that some physical reasoning is applied before embarking on even larger scale field tests. Useful progress might be made through the analogy between a layer with a stable density stratification and a turbulent boundary layer on a concave surface. Strongly-stable layers can occur on such a surface due to the stabilising centrifugal force. There has been much effort on this topic relevant to the modelling of heavy gas plumes (see Bradshaw [32] for a comprehensive review).

4. Photographic Records as a Source of Quantitative Information

In some investigations, photographic records have provided the main information on the progress of the cloud e.g. the Gadila jettison tests and the Porton Down trials. In more recent investigations, the photographic coverage has been very extensive. Some records have also been obtained fortuitously at the scenes of accidental releases. Photography is a comparatively inexpensive experimental tool and it is worthwhile to review the uses made of the information, both in the heavy gas dispersion field and in dispersion studies generally.

The main quantitative application of the records has been the delineation of the cloud geometry, as defined by the visible boundaries of the cloud. The visibility of cold clouds is produced by the condensation of atmospheric water vapour. The edge of the visible cloud can be related to concentration if it is assumed that the local temperature at the cloud edge equals the dew point of the ambient air. However, Colenbrander and Puttock [19], in analysing the Maplin Sands results, have concluded that, for experiments over water, care is needed in selecting the relative humidity to be used in evaluating the concentration corresponding to the visible edge. They found that the measured temperature at the cloud edge was invariably higher than the dew point calculated from ambient air temperature and relative humidity. They attributed the effect to the presence of more water vapour in the cloud than can be accounted for by air entrainment. Their work suggests that the visible edge is an uncertain indicator of concentration in the absence of measurements of temperature or relative humidity near the cloud edge.

In the Porton Down trials, the released gas was marked with smoke and the cloud volume was inferred from plan and side view photographs. It was assumed that the concentration was uniform within the volume defined by the visible edge and that this volume contained all the released gas. The volume-averaged concentration was then obtained from continuity of species. This procedure is consistent with the assumptions in box models of heavy-gas dispersion. The concentration data obtained in this way, together with the data on cloud depth, area and position as functions of time have proved useful in the evaluation of box models and wind tunnel simulations. Brighton et al [33] have analysed some of the overhead photographic records obtained in the Thorney Island trials in a similar way. They determined the path of the cloud centroid and the area and position of the cloud as functions of time. The direction of movement of the cloud and its mean advection velocity have been compared with meteorological measurements whilst the rate of growth of cloud area has been compared with the predictions of gravity-spreading models. Although the analysis of the photographic records from these trials is still at an early stage, the indications are that the results will provide a valuable supplement to the data from fixed instrumentation.

Hall et al [27], in wind-tunnel simulations of the Porton Down trials, have qualitatively compared photographic records in the respective experiments, with scaling of the time from release based on equality of Froude number. The side-by-side comparisons of the photographs showed strong visual similarities and, taken in conjunction with quantitative comparisons, it was possible to conclude that the modelling technique was valid.

Image analysis techniques, in which the optical density distribution of the image is determined instrumentally, have been applied by Puttock et al [18] to still photographs from the Maplin Sands trials and by Riethmuller [34] to the video film records from the Thorney Island trials. No attempt was made to relate the optical density to any integrated measure of the concentration distribution along the viewing direction. This is a major problem, due to the difficulty in separating the contributions from the light attenuated by the marker particles in the viewing direction (marker opacity) and the light scattered into that direction by particles in the neighbourhood (marker brightness). Developments in this field have been reviewed by Gifford [35] while Lilienfeld et al [36] have measured opacity directly using the polarized component of skylight. In both the Maplin Sands and Thorney Island experiments, the fixed instrument masts provide time-synchronised vertical distributions of concentration at discrete points on the overhead photographic images. This information may be of value in calibrating the optical density distribution in terms of some measure of the integrated concentration.

In the Thorney Island trials, a supplementary study was performed to assess the utility of remote-sensing using still cameras (Leck [37]). Twelve remotely-operated cameras were distributed around the perimeter of a 350m radius circle at a height of 1.5m and pointing at a fixed point in the path of the cloud. During the early stages after release, when the smoke-marked cloud is dense and effectively opaque, the images, which will be processed by computer, will identify the cloud edge as seen from each of the twelve camera positions. During the later stages of release when the cloud is semi-transparent, it is hoped that details of the internal structure can be determined. This technique, known as optical tomography, has been used successfully by Santoro et al [38].

Although the acquisition of photographic records is comparatively inexpensive, the subsequent processing to extract quantitative information is time-consuming, especially if performed visually. Standard computerised packages for image processing are now readily available and provide scope for realising the full potential of photographic remote sensing. The advantages are considerable through lessening the reliance on a fixed instrument array which necessarily has to have built-in redundancy to cope with wind direction variations. The remote-sensing technique is very suitable for studying the interaction of clouds with obstacles where the rational deployment of gas sensors is a particularly difficult problem.

5. Control and Definition of Source Conditions

Accidental releases of heavy gas to the atmosphere can occur in a variety of ways. Some of the field experiments described in Section 2 have simulated what can happen in an accident, for example, spillage of cryogenic liquids. Such experiments are inevitably subject to the influence of a large number of variables. This causes difficulties in experimental control and definition of the source conditions for dispersion. At a previous IUTAM Symposium, Lumley [39], in the closing discussion, stated 'Many of our experiments are too complicated. We should try to isolate phenomena. Combine only after investigating separately'. This was the philosophy adopted in the HSE programmes at Porton Down and Thorney Island, while still maintaining an essential connection with the sequence of events observed in many accidental releases. An examination of the evidence available suggested that catastrophic failure of a container is followed very quickly by the formation of a heavy gas cloud around the failed container. This cloud then slumps under gravity and disperses. This division of events into formation and dispersion phases, although an admittedly simplified picture, was adopted for the experimental programme. The physical processes occurring in the formation phase are reviewed in another paper in this volume (Jagger [40]).

For passive releases, it is known that in the far field the cloud will have forgotten what happened at the source. The same would be expected to be true also for releases which are initially heavier than air. However, for releases of heavy flammable gas, the near field is the region of importance and evidence is accumulating that the source conditions can have a dominating influence. Puttock et al [3] report that, in the Maplin Sands tests, the behaviour of LNG plumes was noticeably dependent on the height above the water at which the LNG was discharged. Indeed, in one experiment in which the LNG was discharged below the water level, the resulting plume was buoyant and passed over the sensors at 40m. Experimental spills of LNG on water are susceptible to the rapid phase transition (RPT) phenomenon and when this occurs markedly different plume behaviour is observed (Koopman et al [15]). In a study reported elsewhere in this volume, Fay [41], describes very significent differences in behaviour between experiments by Hall [42] and Meroney and Lohmeyer [43]. He found that the entrainment velocity differed by up to a factor of 10 between the two experiments. He attributed the difference to effects arising from the different degrees of initial mixing induced by the methods of releasing the gas clouds. He further found that in the experiments of Meroney and Lohmeyer, in which the initial mixing was particularly vigorous, the entrainment velocity was independent of the ambient flow conditions. The hypothesis that there is a period after release when atmospheric turbulence is of little effect and that the dispersion is dominated by the initial gravity spreading has also been put forward by Rottman et al [44].

The use of a fixed-volume release configuration, as at Porton Down and Thorney Island, achieves experimental control of geometry, composition and timing of release. However, it has posed a number of problems for mathematical modellers such as the initial motion of the cloud on exposure to the wind and the effects of wind shear and of the container wake on the initial motion. Some of the physical processes involved in the motion of an initially stationary cloud suddenly released have been studied by Rottman and Simpson [45] and Chatwin [46].

With releases of liquefied gas, the main problem is the definition of the source strength. In the China Lake Burro series, for example, the diameter of the liquid pool is quoted only for the Burro 9 test and is so qualified as to be of questionable value. In the Maplin Sands experiments, no details have been given of the source conditions although the dispersion results have been published. It is general practice to assume an evaporation rate (which is subject to much uncertainty) and to combine this with the rate of release of liquid to define the source area. This is as good as one can do in the absence of direct measurements of source geometry and evaporation rate. Such measurements are difficult to perform. It would be useful, however, to have a sensitivity analysis whenever indirect experimental results are used in comparisons with model predictions. Otherwise the source condition becomes a disposable constant of unknown effect.

6. Time Averaging of, and Variability between, Results of Experiments

Heavy gas dispersion models predict 'mean' concentrations but modellers are remarkably reticent about what they mean by 'mean'. The presentation of mean values of statistical parameters of concentration-time records has therefore been approached in different ways in the absence of any consensus. The problem of interpretation was first highlighted in the original Bureau of Mines tests in which very high intermittency was observed. It has been a feature in the China Lake and Maplin Sands tests and it will also be a feature of the Thorney Island trials, as will be illustrated below.

For steady-state continuous releases, the non-stationarity of the concentration-time record is a result of plume meander. For an instantaneous release, there is also, of course, the fact that the limited duration of cloud passage makes for highly non-stationary records. The subject of plume meander is discussed in detail by Pasquill [47] for neutrally buoyant plumes. In passive dispersion prescriptions the dispersion coefficients are based on 10 minute averages of the wind direction records so that the effects of meander are largely smoothed out. This cannot be done, or is at least inappropriate, for dispersion of flammable gas clouds. A 10-minute average concentration is of limited relevance to the assessment of the hazard of such clouds. The meandering results in the short time-average concentration being larger, by perhaps an order of magnitude, than the 10 minute-average concentration at the same point. Burgess et al [5] attempted to take account of this variability by specifying a peak-to-mean concentration ratio derived from their experiments. Although there are theoretical objections to this, its simplicity appeals to hazard analysts and it is a much-used device. In the analysis of the Maplin Sands experiments (Puttock et al [3]), the observed maximum concentration at a sensor position was equated to the centreline concentration of the instantaneous plume. This value, for which the averaging time was about 3s, was used in comparing the results with model predictions. In any one experiment, there was a relatively small number of sensors over which a plume passed so that the information that can be extracted on the decay of the centreline concentration of the instantaneous plume is limited. The properties of instantaneous plumes have recently been studied in some detail, for neutrally buoyant plumes in a wind tunnel, by Fackrell and Robins [48]. A similar study for heavy gas plumes would be very valuable. The presentation of the China Lake data for model comparisons was approached in a quite different way (Koopman et al [15]). They used the wind velocity data and an atmospheric transport code to compute the trajectories of markers released successively at the gas source. The positions of the markers over the duration of an experiment define the plume centreline. The concentrations are presented as 10s moving averages, this time being chosen to preserve plume meander (Ermak et al [30]).

For the instantaneous case, three representative examples of the records from the Thorney Island trials have been chosen for illustration and are shown in Figs. 2a,b and c. The records in Figs. 2a and b were obtained near the source and that in Fig.2c in the far field. The experiments are characterised by a pronounced cloud front, especially near the source. The passage of this front, and of secondary fronts at later times, is clearly evident in Figs.2a and b. The problem of selecting an averaging time which does not smooth out fluctuations due to the spatial structure of the cloud will be appreciated from these records. An analysis has been performed to determine an averaging time which would not give unacceptable attenuation of the peaks of the records (Nussey et al [49]). An unacceptable attenuation was defined as one that exceeded the accuracy of measurement. It was found that an averaging time of more than about 1s was not acceptable. This averaging time happens to correspond approximately to the 1 Hz frequency response of the gas sensors. Clearly this averaging time cannot

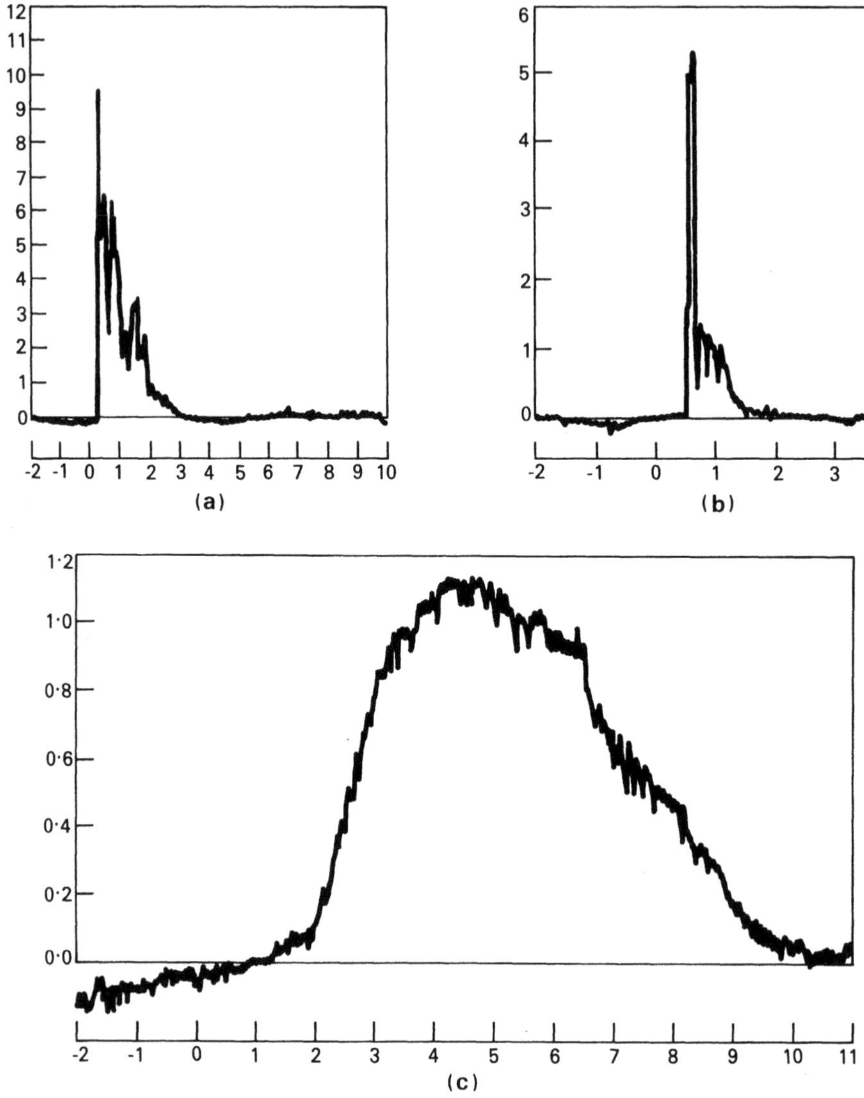

Fig.2 Examples of concentration - time records in a heavy gas cloud.

Ordinate - concentration, volume per cent
Abscissa - time, secs x 10^{-2}

be used as the basis for calculation of turbulent properties according to Reynolds decomposition. Yet if a longer time average is taken, unacceptable attenuation of the records is introduced. It should be remembered also that output of a gas sensor with a 1 Hz frequency response is already attenuated. In experiments reported by Storebo et al [50], the output of an ion-concentration sensor with a frequency response in excess of 100 Hz was filtered at successively lower cut-off frequencies. The sensor was positioned at 10m downwind of a source of ionised air. It was found that attenuation increased rapidly below a cut-off frequency of 10 Hz. Several gas sensors with a frequency response of 10 Hz were deployed in the Thorney Island trials and it is hoped that the data from them will be of value in resolving the problem discussed above.

The (almost) universal attention that has been given by experimenters to the presentation of results as time averages results from their perception of the demands of modellers. Chatwin [51] has argued that this demand, if it exists, is misplaced and attention should instead be directed to statistics of ensembles of experiments. Encouragement for this alternative approach has been provided by the investigations of Hall et al [27] and Meroney and Lohmeyer [43] which each included repetitions of experiments. The variability between repetitions was found to decrease as the initial Richardson number increased. This is consistent with the conclusion of Fay [41] that there is an initial phase where dispersion is dominated by the conditions of release and these are of course reproduced in each repetition. A further aspect, which distinguishes heavy gas from passive releases, is the inertia of the cloud which will attenuate the variability resulting from the randomness of the ambient turbulence. There is therefore some hope that dispersion of heavy gas clouds may not exhibit the same degree of variability as neutrally buoyant clouds.

7. Organisation and Reporting of Experiments

The major programmes that have been undertaken in recent years point to a number of conclusions regarding organisation and reporting. Some of these are obvious but nonetheless worth stating. The characteristics of large-scale field experiments that demand particular attention are the inability to control the flow conditions and the large amount of data generated in each experiment.

Regarding the organisation of an investigation, the following are the main ingredients for success:

- a good site. Apart from being flat and unobstructed within the tolerances laid down by the requirement of horizontal homogeneity of the flow, it should be secure but not too remote and the weather history should be available for a number of years. The weather characteristics should be examined to ensure a reasonable probability of occurrence of the desired weather conditions during the period of the year planned for the trials. A weather forecasting service, on both a daily and a weekly basis, should be available.

- a good design of experiment. The ability to control the governing parameters (other than the weather) is important so that hypotheses can be tested rigorously. Otherwise, to quote Lumley [39] again, 'We end up having two or three things going on at once; its very hard to tell what is influencing what, and in what way'.

- a developed gas-sensing technology. Experience in all the major programmes has shown that it is rarely possible to buy suitable sensors off the shelf. It has been a feature that in-house technology, with full technical back-up service, is a prerequisite.

- adequate finance. Good field experiments cannot be done cheaply. The cost of a large-scale experimental programme is measured in millions of pounds.

- a good central organisation. It needs to have expertise, both managerial and technical. It must have authority to take the necessary decisions.

- a good trials team. Experience in similar programmes and continuity of staffing are essential.

- careful preplanning. Wind tunnel and computer modelling and preliminary tests to try out the engineering of the experimental design all help to maximise the probability of success. Such measures are assuredly cost effective in relation to the cost of abortive effort in the field.

- continuous monitoring. During the execution of the trials

a variety of technical questions will arise. Resolving them, and considering the interactions between them, is preferably carried out by a small group independent of the trials team.

The reporting of a major investigation should take account of the needs of users of the data. These are discussed at length in a recent report of an exercise by Kline et al [52] on the evaluation of available data on complex turbulent flows. They concluded that assessment by individuals with respect to their own work is not sufficient. The trust-worthiness of the data needs to be accepted by consensus of the research community and this can nowadays only be the case if the database is computer readable and widely accessible. The procedures for validating the data need to be described and estimates of uncertainty provided.

8. Discussion

The information so far available on the recent investigations gives encouragement to the view that a reliable database on heavy gas dispersion is now available. There is a need to ensure accessibility to it in order to derive full benefit from the effort and expense. There is a strong argument for some organised analysis, perhaps in the manner of the exercise described by Kline et al [52]. Buoyancy - influenced flows were excluded from that exercise because it was felt (in 1980) that there was an absence of good data - a view that hopefully can now be denied.

9. References

1. Havens, J.A.; An assessment of predictability of LNG vapour dispersion from catastrophic spills onto water. Paper C2, Proc. 5th Int. Symp. on the Transport of Dangerous Goods by Sea and Inland Waterways. Hamburg, April 1978.
2. McQuaid, J.; Dispersion of heavier-than-air gases in the atmospere: review of research and progress report on HSE activities. HSL Tech. Paper 8, Health and Safety Executive, Sheffield 1980.
3. Puttock, J.S.; Blackmore, D.R.; Colenbrander, G.W.: Field experiments on dense gas dispersion. J.Haz.Matls. 6 (1982) 13-41.
4. Burgess, D.S.; Murphy, J.N.; Zabetakis, M.G.: Hazards associated with the spillage of liquified natural gas on water. Rep. No.RI 7448, US Bureau of Mines 1970.
5. Burgess, D.S.; Biordi, J.; Murphy, J.N.: Hazards of spillage of LNG onto water. Rep. No. 4177, Pittsburgh Mining and

Safety Research Center, US Bureau of Mines 1972.

6. May, W.G.; Feldbauer, G.F.; Haigh, J.J.; McQueen, W.; Whipp, R.H.: Spills of LNG onto water: vaporisation and downwind drift of combustible mixtures. Rep. No. EE61E-72 Esso Engineering Ltd. 1972.

7. Kneebone, A.; Prew, L.R.: Shipboard jettison tests of LNG onto the sea. Proc. 4th Int. LNG Conference, Algiers 1974.

8. Britter, R.E.: The ground level extent of a negatively buoyant plume in a turbulent boundary layer. Atm.Env.14(1980)779-785.

9. Cox, R.A.; Carpenter, R.J.: Further development of a dense vapour cloud dispersion model for hazard analysis. Heavy Gas and Risk Assessment(S. Hartwig, Ed.). Dordrecht: D. Reidel 1980.

10. te Riele, P.H.M.: Atmospheric dispersion of heavy gas emitted at or near ground level. Proc. 2nd Int.Symp. on Loss Prevention and Safety Promotion in the Process Industries, Heidelberg 1977.

11. Duffy, A.R.; Gideon, D.N.; Putnam, A.A.: Dispersion and radiation experiments. LNG Safety Program Interim Report on Phase II Work, AGA Project Is-3-1. Battelle Columbus Laboratories 1974.

12. Cox, R.A.; Roe, D.R.: A model of the dispersion of dense vapour clouds. Proc. 2nd Int.Symp. on Loss Prevention and Safety Promotion in the Process Industries. Heidelberg 1977.

13. Koopman, R.P.; Bowman, B.R.; Ermak, D.L.: Data and calculations on 5 m^3 LNG spill tests. Rep. No.DoE/EV-0085, Liquefied Gaseous Fuels Safety and Environmental Control Assessment Program. Washington: Dept. of Energy 1980.

14. Koopman, R.P.; Baker, J.; Cederwall, R.T.; Goldwire, H.C.; Hogan, W.J.; Kampinen, L.M.; Kiefer, R.D.; McLure, J.W.; McRae, T.G.; Morgan, D.L.; Morris, L.K.; Spann, M.W.; Lind, C.D.: Burro series data report LLNL/NWC 1980 LNG spill tests. Lawrence Livermore Laboratory 1982.

15. Koopman, R.P.; Cederwall, R.T.; Ermak, D.L.; Goldwire, H.C.; Hogan, W.J.; McClure, J.W.; McRae, T.G.; Morgan, D.L.; Rodean, H.C.; Shinn, J.H.: Analysis of Burro series 40 m^3 LNG spill experiments. J.Haz.Matls. 6 (1982) 43-83.

16. Rodean, H.C.: Effects of a spill of LNG on mean flow and turbulence under low wind speed, slightly stable atmospheric conditions. Proc. IUTAM Symp. on Atmospheric Dispersion of Heavy Gases and Small Particles.Berlin: Springer-Verlag 1984.

17. Puttock, J.S.; Colenbrander, G.W.; Blackmore, D.R.: Maplin Sands experiments 1980: Dispersion results from continuous releases of refrigerated liquid propane and LNG. Proc. NATO/ CCMS 13th Int. Tech. Meeting on Air Pollution and Modelling. Isle of Embiez Sept. 1982.

18. Puttock, J.S.; Colenbrander, G.W.; Blackmore, D.R.: Maplin Sands experiments 1980: Dispersion results from continuous releases of refrigerated liquid propane. Heavy Gas and Risk Assessment (S. Hartwig, Ed.). Dordrecht: D. Reidel 1982.

19. Colenbrander, G.W.; Puttock, J.S.: Dense gas dispersion

behaviour: experimental observations and model development. Proc. 4th Int.Symp. on Loss Prevention and Safety Promotion in the Process Industries. Harrogate 1983.

20. Colenbrander, G.W.; Puttock, J.S.: Maplin Sands experiments 1980: interpretation and modelling of liquefied gas spills onto the sea. Proc. IUTAM Symp. on Atmospheric Dispersion of Heavy Gases and Small Particles. Berlin: Springer-Verlag 1984.

21. Buschmann, C.H.: Experiments on the dispersion of heavy gases and abatement of chlorine clouds. Proc. Symp. on Transport of Hazardous Cargoes by Sea. Jacksonville 1975.

22. Van Ulden, A.P.: On the spreading of a heavy gas released near the ground. Proc. 1st Int.Symp. on Loss Prevention and Safety Promotion in the Process Industries. Delft 1974.

23. Moodie, K.: Experimental assessment of a full-scale water spray barrier for dispersing dense gases. Proc. I.Chem.E.Symp. on Containment and Dispersion of Gases by Water Sprays. Manchester 1981.

24. Moodie, K.: An experimental assessment of water spray barriers for dispersing clouds of heavy gas. To be presented to I.Chem.E.Symp. on Heavy Gas Releases - Dispersion and Control. Utrecht 1984.

25. Picknett, R.G: Dispersion of dense-gas puffs released in the atmosphere at ground level. Atm.Env. 15 (1981) 509-525.

26. Woodward, J.L.; Havens, J.A.; McBride, W.C.; Taft, J.R.: A comparison with experimental data of several models for dispersion of heavy vapour clouds. J.Haz.Matls. 6 (1982) 161-180.

27. Hall, D.J.; Hollis, E.J.; Ishaq, H.: A wind tunnel model of the Porton dense gas spill field trials. Rep. No. LR 394(AP), Warren Spring Laboratory. Stevenage 1982.

28. McQuaid, J.: Large-scale experiments on the dispersion of heavy gas clouds. Proc. IUTAM Symp. on Atmospheric Dispersion of Heavy Gases and Small Particles. Berlin: Springer-Verlag 1984.

29. McQuaid, J.: Experiments on incompressible turbulent boundary layers with distributed injection. Aero.Res.Council Rep. and Memo. No. 3549. London: HMSO 1968.

30. Ermak, D.L.; Chan, S.T.; Morgan, D.L.; Morris, L.K.: A comparison of dense gas dispersion model simulations with Burro Series LNG spill test results. J.Haz.Matls. 6 (1982) 129-160.

31. McQuaid, J.: Some experiments on the structure of stably-stratified shear flows. Saf. in Mines Res.Est.Tech. Paper P21. Sheffield 1976.

32. Bradshaw, P.: Effects of streamline curvature on turbulent flow. AGARDograph No. 169. North Atlantic Treaty Organisation. Neuilly sur Seine 1973.

33. Brighton, P.W.M.; Prince, A.J.; Webber, D.M.: Determination of cloud area and path from visual and concentration records. To be presented to Symp. on Heavy Gas Dispersion Trials at Thorney Island. Sheffield 1984.

34. Riethmuller, M.: Private communication 1983.

35. Gifford, F.A.: Smoke as a quantitative atmospheric diffusion tracer. Atm. Env. 14 (1980) 119-1121.

36. Lilienfeld, P.A.; Woker, G.; Stern, R.; McVay, L.: Passive remote smoke plume opacity sensing: a technique. Appl. Optics 20 (1981) 800-806.

37. Leck, M.J.: Private communication 1983.

38. Santoro, R.J.; Smerjian, H.G.; Emmerman, P.J.; Goulard, R.: Optical tomography for flow field diagnostics. Int. J.Heat Mass Transfer 24 (1981) 1139-1150.

39. Lumley, J.L.: Closing discussion. Structure of Complex Tubulent Shear Flow (R. Dumas and L. Fulachiar, Eds.). Berlin: Springer-Verlag 1983.

40. Jagger, S.F.: Formation of heavy gas clouds. Proc. IUTAM Symp. on Atmospheric Dispersion of Heavy Gases and Small Particles. Berlin: Springer-Verlag 1984.

41. Fay, J.A.: Experimental observations of entrainment rates in dense dispersion tests. Proc. IUTAM Symp. on Atmospheric Dispersion of Heavy Gases and Small Particles. Berlin: Springer-Verlag 1984.

42. Hall, D.J.: Further experiments on a model of an escape of heavy gas. Rep. No. LR 312 (AP), Warren Spring Laboratory. Stenvenage 1979.

43. Meroney, R.N.; Lohmeyer, A.: Statistical characteristics of instantaneous dense gas clouds released in an atmospheric boundary layer wind tunnel. Proc. IUTAM Symp. on Atmospheric Dispersion of Heavy Gases and Small Particles. Berlin: Springer-Verlag 1984.

44. Rottman, J.W.; Hunt, J.C.R.; Britter, R.E.: Some physical processes involved in the dispersion of dense gases. Proc. IUTAM Symp. on Atmospheric Dispersion of Heavy Gases and Small Particles. Berlin: Springer-Verlag 1984.

45. Rottman, J.W.; Simpson, J.E.: The initial development of gravity currents from fixed-volume releases of heavy fluids. Proc. IUTAM Symp. on Atmospheric Dispersion of Heavy Gases and Small Particles. Berlin: Springer-Verlag 1984.

46. Chatwin, P.C.: The incorporation of wind shear effects into box models of heavy gas dispersion. Proc. IUTAM Symp. on Atmospheric Dispersion of Heavy Gases and Small Particles. Berlin: Springer-Verlag 1984.

47. Pasquill, F.: Atmospheric Diffusion. 2nd Edition. Chichester: Ellis Horwood 1974.

48. Fackrell, J.E.; Robins, A.G.: The effect of source size on concentration fluctuations in plumes. Boundary-Layer Met. 22 (1982) 335-350.

49. Nussey, C.; Davies, J.K.W.: The effect of averaging time on the statistical properties of sensor records. To be presented to Symp. on Heavy Gas Dispersion Trials at Thorney Island. Sheffield 1984.

50. Storebo, P.B.; Lillegraven, A.; Honnashagen, K.; Bjorvatten, T.; Jones, C.D.; van Buijtenen, C.J.P.: Tracer experiments

with turbulently dispersed air ions. FFI/Rapport - 81/3006. Norwegian Defence Research Establishment 1981.

51. Chatwin, P.C.: The use of statistics in describing and predicting the effects of dispersing gas clouds. J.Haz.Matls. 6 (1982) 213-230.

52. Kline, S.J.; Cantwell, B.J.; Lilley, G.M.(Eds.): The 1980/81 AFOSR - HTTM - Stanford Conference on Complex Turbulent Flows: Comparison of Computation and Experiment. Thermosciences Divn., Mech.Eng.Dept., Stanford University 1981.

A Probabilistic Model for Dosage

D. J. RIDE

Chemical Defence Establishment,
Porton Down, Salisbury, U.K.

Summary

A modified binomial pdf is proposed for use in the prediction of dosages and by implication for concentrations. Although based on an idealised cloud configuration it can readily be extended to provide a good description of the properties of real clouds. It is shown that even for the simplest cloud a minimum of three parameters is necessary to define a two-parameter pdf which describes fluctuations realistically, and that a useful, independent set of such parameters for a real cloud consists of the first two central moments of the concentration field and the pulse repetition rate. The inadequacy of the log-normal pdf is described and attributed to its inability to handle zero values and to the lack of control which may be exercised over the specification of its skewness. The model proposed here can reconcile the differing time periods of responders to the contamination in the cloud, a feature which exists independently of the use of the binomial pdf. An experimental technique using ionised air molecules is referenced which is ideally suited to exploring and defining aspects of the model and for similar studies.

For many years the models used for predicting atmospheric concentrations of contaminants have been those which produce time-averaged values. These models are appropriate in those cases where the effects of the contaminants depend on their accumulation over a time period which is long compared with the time scales of the fluctuations in concentration which occur as a result of atmospheric turbulence. However, some potential atmospheric contaminants are so toxic that harmful effects may be achieved by exposure times which are much shorter than these time scales; for these contaminants, models which supply only expected values of concentration possess inadequate prediction capabilities. Forecasting the fluctuating concentration field is also important in considerations of the explosive and flammable properties of clouds, odour nuisance, and line-of-sight probabilities through obscuring clouds. A further, related problem is concerned with the relationship between the response time of a measuring instrument and the description of a fluctuating scalar field which it provides.

Parameters which have been used to characterise fluctuating concentrations are: peak-to-mean ratio, variance, intermittency (the fraction of the time that measurable concentrations are absent), and the pulse repetition rate. Under certain conditions some of these parameters are functionally related one to another, a point which will be examined in more depth below.

Instead of considering concentration as the fluctuating variable, attention will be directed to the dosage. Formally the dosage is the integral of the (time dependent) concentration with respect to the time over some period of interest. Its importance arises from the fact that nearly all responders (including the human lung) possess a response time, τ say. Thus the concentration which they perceive is given by the expression

$$\frac{1}{\tau}\int_{T}^{T+\tau} C(t).dt$$

that is, a time-averaged concentration.

Responders which are not time dependent may have a volumetric response or a linear distance response. In all cases an integral term arises, the existence of which determines that interest should be concentrated on the practically important quantity, dosage (in its broadest sense), rather than the instantaneous description of the concentration field. A satisfactory model of the fluctuations would yield a probability density function (pdf) of the dosage acquired over some specified time period and its distribution in space. The value of employing pdf's has been described by Chatwin (1).

Csanady (2) has examined this problem and produced a probabilistic model for dosage based on the assumption that the pdf for fluctuations conforms to a log-normal distribution. He gives a plausible explanation for the occurrence of a log-normal distribution; however, there is no sound theoretical basis for its existence, and its shape is approximated only in some regions of the cloud. Csanady notes this last point as a critical assumption in the application of his model. Additionally, the log-normal distribution suffers from serious practical drawbacks: it cannot described the occurrence of concentrations whose indicated value is zero, and it is limited to positively skewed data. Nevertheless, its usage is widespread, and one aim of this paper is to demonstrate that other pdf's are worth exploring.

The approach to the construction of a simple prediction model for fluctuations adopted here was chosen after reviewing the techniques which have led to the development of the most widely used atmospheric prediction models. The most commonly used mean concentration model for neutrally buoyant plumes and puffs remains the simple Gaussian one, whilst the elementary concept of a cylindrical mass continuously deforming under the force of gravity, originally introduced by Van Ulden (3), continues as a useful basis for prediction models of a dispersing volume of heavy gas. Continuing in the tradition of these pragmatic advances, the simplest possible geometric description of a cloud which gives rise to fluctuations was sought. A model which uses the smallest number of parameters is a cloud composed of identical, uniformly randomly distributed spheres of radius r, each containing the same uniform concentration of contaminant, C_* (Figure 1). The space surrounding the spheres within the boundary of the cloud is free from contaminant. This configuration reflects the discrete nature of contaminant found in real clouds. Three parameters are necessary to achieve a description of pdf's in such a cloud:

the mean concentration, $\bar{C}$
the peak-to mean ratio of concentration, $C_*/\bar{C}$
and the mean number of pulses of concentration experienced, $\bar{m}$.

The following identities for the model cloud may be simply determined:

$$C_*/\bar{C} = (1 - \gamma)^{-1}$$

where γ is the intermittency.

$$(\sigma/\bar{C})^2 = (C_*/\bar{C}) - 1$$

where σ is the standard deviation of the instantaneous concentration field.

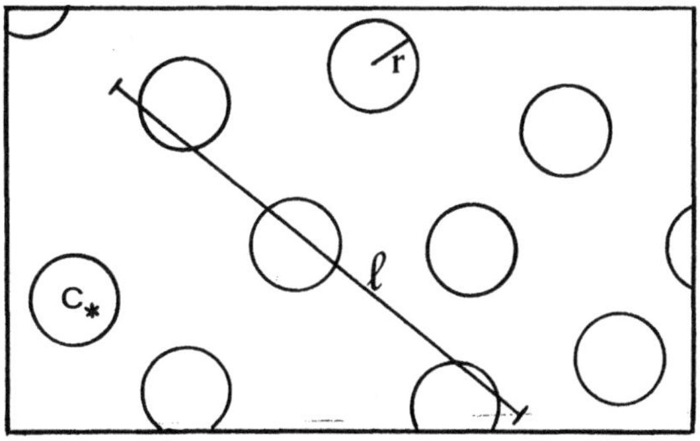

Fig.1. The basic model cloud

If the time of exposure to the cloud is t (= ℓ/u, where ℓ is the path length and u the velocity of traverse), then the mean dosage in the model cloud is given by

$$\bar{D} = \bar{C}.t$$

An equivalent expression for D_*, the maximal dosage, may be written

$$D_* = C(t).t$$

where C(t) is the maximal possible concentration averaged over time t, that is, $C(t)/\bar{C}$ is the peak-to-mean ratio observed for a response time of t.

For the model cloud the peak-to-mean ratio is as complete a specification of the characteristics of fluctuation as the variance. The same is not true of real clouds. The value of the peak-to-mean ratio for a real cloud is very difficult to measure as the peak is likely to be attained at only a few, isolated places. On the other hand, the variance is much easier to measure reliably, is therefore robust and, statistically, is a much more representative parameter than the peak-to-mean ratio.

The identity

$$(C_*/\bar{C}) - 1 = (\sigma/\bar{C})^2$$

which exists for the simple model cloud, will not hold when more than one size of sphere of contaminant is present. Rodean and Cederwall (4) have examined the relationship between the two measures obtained during the Burro series of LNG spills and have discovered an excellent linear relationship between the logarithms of $(C(t)/\bar{C}) - 1$ and $\sigma(t)/\bar{C}$, where $\sigma(t)$ is the standard deviation of fluctuations measured with a response time of t.

$$(C(t)/\bar{C}) - 1 = k.(\sigma(t)/\bar{C})^{\lambda}$$

Rodean and Cederwall also found that the constant of proportionality k was relatively insensitive to experimental conditions.

If a path of length ℓ is taken through the interior of the cloud, traversed with a velocity u, and m spheres are cut, then the dosage D achieved is given by $D = m\alpha rC_*/u$,

and similarly, the maximal dosage D_* is given by $D_* = n\alpha rC_*/u$,

where n is the greatest number of spheres that it is possible to encounter on the path of length ℓ, and αr is the mean path length through a sphere of radius r. It follows that the mean dosage $\bar{D}$ is obtained from

$$\bar{D} = \bar{m}\alpha r C_*/u.$$

These relationships yield $m = (D/\bar{D}).\bar{m}$

$$n = (D_*/\bar{D}).\bar{m}$$

Now, if p is the probability of encountering an isolated sphere in the region traversed by the path, and individual encounters can be regarded as independent events, then the probability of encountering m spheres is given by the binomial pdf. However, m and n will generally take non-integer values as a result of their determination from the equations above, so gamma functions must be used in place of the factorials and the pdf thus transformed from a discrete to a continuous one. It is also necessary to normalise by the area A under the curve between m = 0 and m = n.

$$P(m) = \frac{1}{A}\cdot\frac{\Gamma(1+n)}{\Gamma(1+m).\Gamma(1+n-m)}\cdot p^m.(1-p)^{n-m} \qquad \text{for } 0\leqslant m\leqslant n$$

$$P(m) = 0 \qquad \text{for } m<0,\ m>n$$

The pdf for normalised dosage is simply $dm/d(D/\bar{D}).P(m)$, which is $\bar{m}.P(m)$.

Sampling theory suggests that $\sigma(t)$ is proportional to the minus one-half power of the number of events, which is itself proportional to the product ut.

Using these relationsnips it is possible to construct an expression for C(t) which satisfies the asymptotic requirements

$$\lim_{t\to 0} C(t) = C_* \quad \text{and} \quad \lim_{t\to 00} C(t) = \bar{C}$$

thus

$$C(t) = \bar{C}.\left[k.(\sigma/\bar{C})^{\lambda}\left(\frac{\beta}{ut+\beta}\right)^{\lambda/2} + 1\right]$$

where β is a constant - for given cloud parameters - with the dimension of length. Note the subsidiary relationship

$$\sigma(t)/\bar{C} = (\sigma/\bar{C})\left(\frac{\beta}{ut+\beta}\right)^{\frac{1}{2}}$$

Data have been published by C D Jones (5) which enable the determination of k, λ and β and the testing of the goodness of fit of the expressions involving them. They relate to an instrument which measures concentrations of unipolar ions generated by a corona discharge. Measurements were made at four distances downwind from a continuous ion source operating for 48 minutes out of doors. The data were filtered to simulate five instruments with responses between 30 Hz and 0.3 Hz. Measurements included mean concentration, standard deviation of concentration, peak-to-mean ratios and intermittency. Using these data, a least-squares technique was employed to determine k and λ for a best fit of $C(t)/\bar{C}$ to the observed results (Table 1). λ showed a definite correlation with distance (correlation coefficient = 0.98) but a satisfactory fit was obtained by using a constant value of $\lambda = 1.5$, it being obvious that the trend could not be sustained indefinitely. No trend with distance, and little variation, was noted for k, consistent with the findings of Rodean and Cederwall; its best average value was 11.0. Similar techniques were used to determine β. Since β increases with distance and has dimension of length, a simple linear regression was chosen, $\beta = 0.41x$, where x is the distance downwind from the source. The data were consistent with a zero value at the source, which means that - conveniently - the only singularity occurs there. Calculated values of the fluctuations were obtained using this formulation for β and compared with measured values (Table 2). The fit may be seen as very good. Less satisfactory is the fit of the calculated values of intermittency; however, the observed values do not behave as expected in that there is no consistent decrease with increasing response time. This point has not been satisfactorily explained and more confirmatory data are awaited.

RANGE	RESPONSE Hz	$C(t)/\bar{C}$	
		observed	calculated
2 m	30	31.90	47.38
	10	25.60	33.76
	3	18.40	17.87
	1	13.90	10.39
	0.3	6.74	5.73
5 m	30	62.70	87.68
	10	58.90	72.72
	3	43.40	46.06
	1	22.20	31.81
	0.3	10.60	18.83
10 m	30	90.40	63.97
	10	79.00	53.65
	3	48.00	37.88
	1	28.50	24.31
	0.3	12.20	14.92
15 m	30	42.70	33.76
	10	34.10	30.95
	3	19.20	23.26
	1	13.70	17.87
	0.3	10.60	14.21

Table 1. The peak-to-mean ratio for various sampling times calculated as a function of the intensity of fluctuations compared with Jones's data.

RANGE x	RESPONSE Hz	INTENSITY $\sigma(t)/\bar{C}$ observed	INTENSITY $\sigma(t)/\bar{C}$ calculated	INTERMITTENCY γ observed	INTERMITTENCY γ calculated
2 m	Infinite ($\sigma/\bar{C}$)		2·57		
	Unfiltered	2·99		0·85	
	30	2·61	2·34	0·83	0·85
	10	2·07	2·02	0·81	0·80
	3	1·33	1·47	0·79	0·69
	1	0·99	0·96	0·79	0·48
	0·3	0·57	0·56	0·87	0·24
5m	Infinite ($\sigma/\bar{C}$)		3·85		
	Unfiltered	4·21		0·90	
	30	3·96	3·70	0·88	0·93
	10	3·49	3·45	0·83	0·92
	3	2·56	2·86	0·85	0·89
	1	1·96	2·08	0·81	0·81
	0·3	1·38	1·27	0·83	0·62
10 m	Infinite ($\sigma/\bar{C}$)		2·80		
	Unfiltered	3·42		0·84	
	30	3·20	2·75	0·84	0·88
	10	2·84	2·65	0·85	0·88
	3	2·24	2·36	0·84	0·85
	1	1·65	1·88	0·84	0·78
	0·3	1·17	1·25	0·85	0·61
15m	Infinite ($\sigma/\bar{C}$)		1·98		
	Unfiltered	2·18		0·71	
	30	2·07	1·95	0·73	0·79
	10	1·95	1·90	0·74	0·78
	3	1·60	1·76	0·72	0·76
	1	1·33	1·47	0·69	0·68
	0·3	1·13	1·03	0·71	0·51

Table 2. Calculated values of the intensity of fluctuations and intermittency compared with Jones's data.

This experimental justification of the functional form of C(t) has revealed a useful method by which values of a concentration statistic appropriate to the response time of one receptor may be related to that of a second.

Reverting to the discussion of the parameters of the continuous binomial distribution, we may now write, by substitution,

$$n = \left[k.(\sigma/\bar{C})^{\lambda}\left(\frac{\beta}{ut+\beta}\right)^{\lambda/2} + 1\right].\bar{m}$$

It is convenient to leave the prescription of m in terms of the normalised dosage, ie. $m = (D/\bar{D}).\bar{m}$

Values of $\bar{m}$ and A are easily tabulated for various values of n and p using a digital computer. The values of p and A may be read off opposite the observed value of $\bar{m}$ in the computed tables.

A physical interpretation of the parameter $\bar{m}$ involves both the density of the discrete elements of contaminated air composing the cloud and the threshold concentration to which the responder reacts. In the absence of extensive data and a comprehensive analysis, there appears to be no practical alternative to measuring $\bar{m}$ in individual cases at present. This is a straightforward procedure and was performed by Jones in the experiments described above. His results contain a table giving the total number of pulses over the 48 minutes for each of four downwind distances and for each of nine levels of threshold concentration.

The appropriate nature of the binomial pdf may be judged by comparing published results of Birch, Brown, Dodson and Thomas (6) with such pdf's with parameters chosen to reproduce the shape of the results (Figure 2). In all but one instance the match is very good.

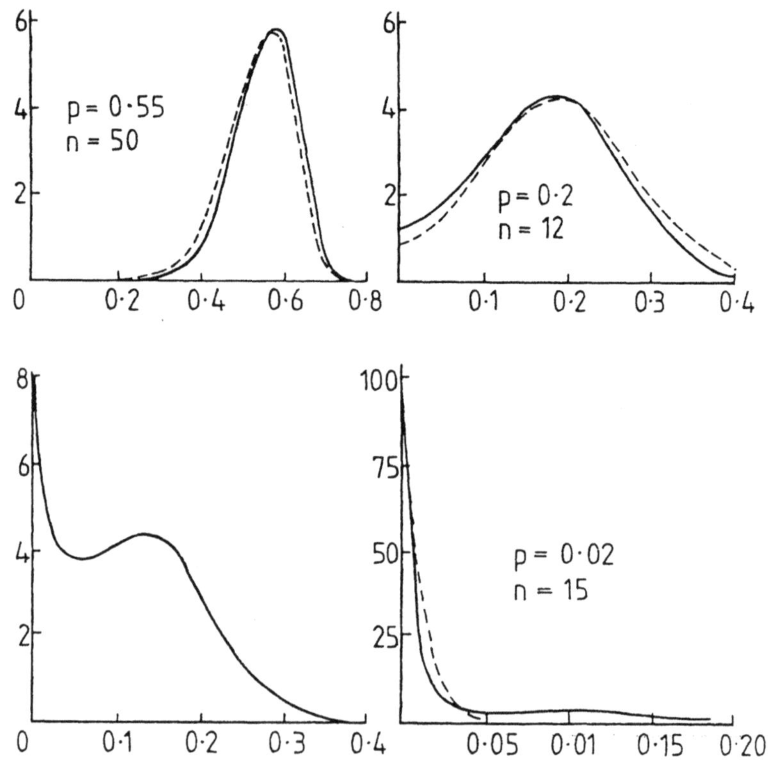

Fig.2. Concentration (horizontal axis) v. probability density (data of Birch et al) compared with binomial pdf's (shown with broken lines). Indicated concentrations are in reality dosages scaled by instrumental response times.

A set of pdf's calculated from Jones's data by using the model is shown in Figure 3. The probability of detecting low dosages is seen, as expected, to be far higher for fast response receptors. For these the curves are highly positively skewed and, with the usual manipulation of zero dosages, can be approximated by log-normal functions. The same is not true for the receptors with long sampling times. The 0-1 Hz curve is nearly Gaussian in shape and the variance has decreased markedly. These curves imply that the spectrum of toxic effects in a population subjected to a given mean dosage changes markedly depending on the time period over which the dosage is acquired.

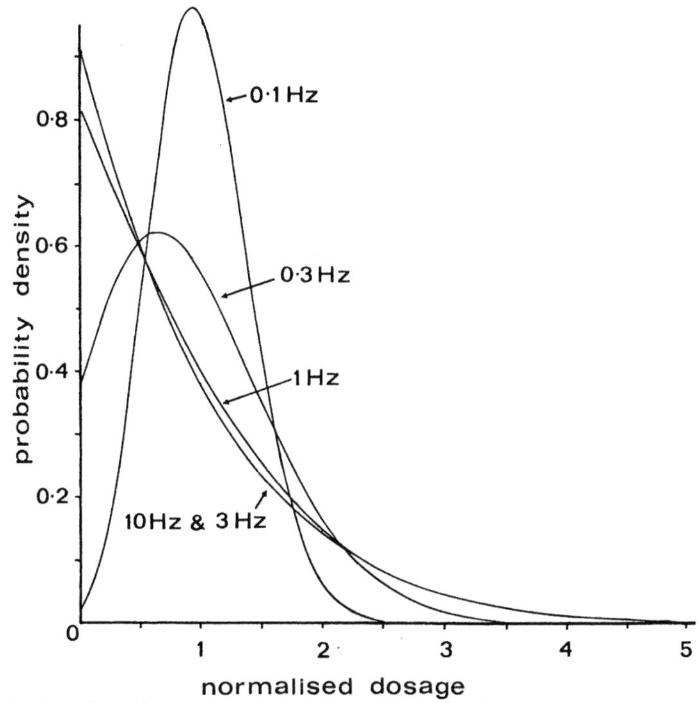

Fig.3. Calculated pdf's of normalised dosage (indicated concentration) for several response times. Jones's data, 15 m.

The model has been developed for a homogeneous region of the cloud. Real clouds are not homogeneous as a rule. The trials reported by Ramsdell and Hinds (7), for instance, show an increase in the intensity of turbulence away from the centre of a cloud. If a cloud may be approximated by two homogeneous regions with dosage probability density functions P_1 and P_2, based on the C, $\sigma/\bar{C}$ and $\bar{m}$ parameters for each region, then the probability that the dosage lies between D and D + ΔD is

$$(\Delta D/2)^2 \int_0^{\Phi} P_1(\phi).P_2(D - \phi).d\phi$$

Extensions to more than two regions can be made in similar fashion. However, since in Jones's experiments the statistics of the cloud reflect discontinuities arising from the internal cloud structure and from the meandering of the plume, it appears that a simple one-region model utilising gross cloud statistics can yield useful results of - at least - peak-to-mean ratios and intensities of fluctuations. This apparent robustness must derive from a dominant statistical mechanism of the cloud which is insensitive to moments of the discrete element distribution higher than the first. Whilst Richardson (8) was arguing from a Lagrangian standpoint, it would seem that his distance-neighbour function (defined as the mean number of neighbours per length, as a function of their distance apart) is a promising candidate for this dominant statistic.

References

1. Chatwin, P.C.: The use of statistics in describing and predicting the effects of dispersing gas clouds. J.Haz.Mat. 6(1982)213-230.
2. Csanady, G.T.: Dosage probabilities and area coverage from instantaneous point sources on ground level. Atm.Envir. 3(1969)25-46.
3. Van Ulder, A.P.: On the spreading of a heavy gas released near the ground. Loss prevention Symp.(1974) Elsevier Press.
4. Rodean, H.C.: Analysis of turbulent wind-velocity and gas concentration fluctuations during the Burro series 40-m^3 LNG spill experiments. Lawrence Livermore Lab. Report UCRL 53353 (1982).
5. Jones, C.D: On the structure of instantaneous plumes in the atmosphere. J.Haz.Mat. 7(1982)87-112.
6. Birch, A.D.; Brown, D.R.; Dodson, M G.; Thomas, J.R.: The turbulent concentration field of a methane jet. J.Fluid Mech. 88(3)(1978)431-449.
7. Ramsdell, J.V.; Hinds, W.T.: Concentration fluctuations and peak-to-mean concentration ratios in plumes from a ground-level continuous point source. Atm.Envir. 5(1979)483-495.
8. Richardson, L.F.: Atmospheric diffusion shown on a distance-neighbour graph. Proc.Roy. Soc. A 110(1926)709-737.

Maplin Sands Experiments 1980: Interpretation and Modelling of Liquefied Gas Spills onto the Sea

G.W. COLENBRANDER* and J.S. PUTTOCK**

* Koninklijke/Shell-Laboratorium, Amsterdam (Shell Research B.V.)
** Thornton Research Centre (Shell Research Ltd.)

Summary

In 1980, Shell Research performed a major series of controlled spills of LNG and refrigerated propane on the sea at Maplin Sands in the South of England. Results of both continuous and instantaneous dispersion trials are presented. Comparisons are made with predictions from the dense gas dispersion model HEGADAS, showing conservative model results for LNG.

An expression for heat transfer from the sea surface to the cold cloud has been added to the model. With the heat transfer included, the model is sensitive to the release of latent heat from water vapour transferred into the cloud. This effect could have enhanced the dispersion in the Maplin Sands LNG spills, since the water content of the clouds appears to have been much higher than could be explained by entrainment of ambient air alone. Laboratory experiments have been performed using liquid nitrogen to study the pick-up of water by cold gas clouds.

Comparison of four instantaneous Maplin Sands spills with model predictions provides information on cloud shape, dispersion distance and advection velocity.

1. Introduction

The Maplin Sands experiments were performed in the summer of 1980. The aim was to study the dispersion and combustion of releases of dense flammable gases. For this purpose 34 spills of liquefied gases onto the sea were performed. The gases used were refrigerated liquid propane and liquefied natural gas (LNG), in quantities up to about twenty cubic metres. The use of these two gases enabled us to study the dispersion of a simple dense gas (propane) and the more complicated behaviour of LNG with the additional effects of significant heat transfer from the water surface and possible transition to buoyancy.

Release of the liquid was either continuous or instantaneous. Continuous spills consisted in the release of liquid at a steady rate from the end of a pipe near the water surface. For instantaneous spills the liquid was poured into an open-topped insulated octagonal barge, 12.5 m across, which was then submerged, creating the spill as water flowed in to displace the liquefied gas.

Cross-correlations of velocity fluctuations measured by sonic anemometers give a typical value for u_*/U_{10} of 0.034, consistent with the observed wind velocity profile and with the smooth mud or sea surface upwind. Nearly all the experiments, details of which have been published elsewhere[1,2,3], were conducted under conditions of near-neutral atmospheric stability.

Results of simulations with the HEGADAS II model show, on the whole, good agreement with the experimental propane data[1], but are found to be conservative with respect to LNG dispersion distances. This prompted us to investigate the total heat input into LNG clouds more closely. The heat input into the cloud arises from the sea surface and the surrounding air, as well as from latent heat release of water (vapour) in the cloud. In Section 2 the results of the dispersion trials are examined with respect to the water content of the cloud. Incorporation of a higher cloud humidity results in a far better agreement with experiments.

Section 3 of this paper describes laboratory experiments on heat and mass transfer into cold, dense vapour clouds dispersing over a water surface. The experimental results are used to quantify the heat transfer due to latent heat release of water (vapour) to the cloud.

Results of the analyses of the four instantaneous spills (two propane and two LNG) are presented in Section 4. Aspects considered are the cloud shape, the dispersion distance and, for propane spills, the advection velocity.

2. Analysis of Maplin Sands Continuous LNG Trials

The experimental results of the continuous LNG spills have been compared with the predictions of the dense-gas dispersion model HEGADAS II. The model[4] assumes, crosswind, a flat-topped concentration profile with Gaussian edges. The flat part can be eroded so that the profile becomes completely Gaussian far downwind. Gravity-spreading can increase the width of the plume. Combination of the assumed vertical concentration profile with a power-law velocity profile gives an advection velocity which is a function of the height parameter S_z. The vertical entrainment is a function of the bulk Richardson number. For stably stratified conditions this function has been arranged so as to be compatible with published two-layer laboratory experiments and wind-tunnel data. The entrainment function for unstable stratification (which in the case of LNG can occur even in neutral ambient conditions if the plume becomes slightly buoyant) was chosen such as to be compatible with the results of the Prairie Grass tracer dispersion experiments[3].

Since the heat transfer from the sea surface to the cold gas cloud is likely to be significant for LNG, expressions for heat transfer have been incorporated into the model, for testing against the observations. We used standard expressions for free and forced convection heat transfer, taking the larger of the two values[3].

Table I gives data on the spills in which the LNG was released above and close to the water surface. In this table, values of the initial Richardson number Ri_o, as defined in Ref. 2, are given. This allows comparison, between the spills, of the magnitude of density effects, which increases with increasing Ri_o.

The results confirmed the model predictions for spills at high wind speeds, where density effects were unimportant. However, for other spills the model without heat transfer consistently overpredicted the LFL distance, the difference roughly increasing with initial Richardson number. With heat transfer taken into account, the predictions were better for spills where the measured ambient relative humidity was high, but for several other spills the change had comparatively little effect.

Attempts have been made to improve agreement between the results of model simulations and experimental data. For this purpose the data yielding quantitative information on cloud humidity are being closely analysed. There is distinct evidence that the moisture content of the cloud is higher than could be explained by entrainment of air, using the measured relative humidity at 10 m height given in Table I. For example, temperature measurements at 0.7 m above the water surface during trial 29 indicate that there the relative humidity was 60-73 %, i.e. considerably larger than that measured at a height of 10 m (52 %). Confirmation as to the higher water content of the cloud is obtained from cloud visibility data. Visibility of the cloud is due to condensation of water vapour. Hence a cloud ceases to be visible when its temperature rises above the dew point. Temperature measurements in visible clouds showed the dew point to be higher than that derived from air temperature and humidity measured at 10 m height.

Moisture content has a significant influence on calculated LFL distances. Model runs show that, when combined with heat transfer from the water surface, high levels of ambient humidity cause the plume to become buoyant over an extended downwind distance, resulting in enhanced vertical entrainment. HEGADAS simulations for trial 29 and 39 may serve as illustrations.

For these runs we used relative humidities of 89 % and 79 %, respectively, instead of the values at 10 m height of 52 % and 63 % indicated in Table I. The higher values were chosen because they give calculated temperatures and concentrations at the moment the cloud ceases to be visible that are in agreement with the measurements.

The use of these higher relative humidity values brings the calculated LFL distances down from 245 m to 140 m for trial 29 and from 270 m to 135 m for trial 39, in excellent agreement with the observed distances. For both trials the calculated visible cloud length is 380 m. This agrees well with the cloud length of 325 m observed in trial 29, but not with the much smaller visibility length of 210 m of trial 39.

There are various mechanisms by which the water (vapour) content of the cloud may reach a high level. For instance, a high humidity of the air close to the water surface, or water pick-up during the evaporation and dispersion phases of the LNG spill. Boyle and Kneebone[5], for instance, have measured substantial water pick-up during the LNG pool boiling experiments. We have performed laboratory experiments to investigate the water (vapour) content of a refrigerated gas cloud evaporating and dispersing on a water surface. In addition, these experiments have been set up to check the heat transfer relations incorporated in HEGADAS and to assess the air entrainment due to convective currents into the cloud.

3. Laboratory Experiments on Heat and Mass Transfer into Cold, Dense Vapour Clouds Dispersing over a Water Surface

In the experiments a steady stream of liquid nitrogen (-196 ^{o}C) was released into a 1 m wide and 20 cm deep tray filled with water up to about 3 cm below the brim. The nitrogen, floating on the water surface, evaporated and the cold, dense vapour entered a perspex box, the flow being homogeneously distributed over the width of the box (see Fig. 1). The box contained a water layer 20 cm deep, over the surface of which the vapour was flowing. The nitrogen left the box through an opening over its entire width. The opening was provided with a flow restriction consisting of vertical bars with a triangular cross-section. The length and width of the box were equal (1 m), the height was 60 cm. On top a lid was placed to prevent air circulating in the room from entering the box. Holes in the lid allowed the air to flow in and out freely.

At locations A, B and C (25, 50 and 75 cm from the box entrance) temperatures and concentrations in the vapour stream were measured at, respectively, 3, 2 and 1 cm above the water surface. A mass spectrometer was used for measuring the concentrations of nitrogen, oxygen and water, and determining air entrainment and water (vapour) pick-up into the nitrogen vapour stream. Additionally, the water temperature in the box was measured, as well as a vertical temperature profile in the vapour stream at location B (using nine thermocouples).

The stream of liquid nitrogen was generated by displacement from a dewar vessel by means of pressurized nitrogen gas. The flow rate was determined from the decrease in weight of the dewar vessel. After about 5 minutes a fairly steady flow rate was established.

Six experimental runs were done, at the same nitrogen release rate. In all runs the temperature was measured simultaneously at each of the locations. However, only one concentration could be measured at a time. So the concentration probe remained at the same place during an experiment and was installed at another location for the next run.

Data on Temperature Profile and Air and Water Concentrations in the Cloud

In Fig. 2 a typical temperature profile at location B is shown, measured 405 s after the start of the nitrogen release. The temperature shows hardly any variation with height between 2 and 5 cm above the water surface. From the temperature profiles we estimated the height of the nitrogen vapour stream to be 7 cm.

Fig. 3a is a typical example of the course of the temperature as a function of time from the start of the release, at locations A, B and C. During the period of steady nitrogen flow ($t > 300$ s) the temperatures do not fluctuate strongly. The maximum variation in measured temperatures between the different runs was ± 5 ^{o}C. At $t = 200$ s a peak is seen in all three curves. At that point in time it was observed in all experiments that a solid ice layer had formed on parts of the water surface in the evaporation tray. The water in this tray was heated by three coils; a stirrer was installed in the middle of the tray, where no ice formation occurred.

In Figs. 3b and c the measured air and water concentrations at 2 cm above the water surface are shown. For $t > 300$ s the air concentration levels are fairly steady, showing an increase with distance of only about 1 % air per 25 cm.

During the period of steady flow the measured water concentrations are not very stable. It should be noted that the response of the mass spectrometer to changes in water concentration is much slower (response time ≈1 min) than for oxygen and nitrogen.

However, the increase of the water concentration with distance is significant and cannot be explained at all by the entrainment of moist air. For, as we have just shown, the air concentrations are very low and the air contains only about 1.4 % water (compare the water concentrations measured at $t = 0$ in pure air). So there must have been transport of water from the water surface into the cold vapour stream.

Heat Transfer Relations

Temperature and water concentration data measured 400 s after the start of the nitrogen release were analysed. Averaged over all experiments, the nitrogen flow rate at that point of time was 94 ± 4 kg/h and the temperature T_g in the bulk of the cloud at location B was 183 ± 4 K. Since we estimated the height of the cloud halfway along the box at 7 cm, we find a bulk velocity U of the nitrogen vapour stream of 0.2 m/s; our estimate of the friction velocity is $u_* = 0.01$ m/s. The mean water temperature T_w in the box was found to be 290.6 K, with a variation of ±2.3 K between the different experiments.

To calculate the heat transfer from the water surface to the vapour stream we used the relations[3] as incorporated in HEGADAS for forced[6] and natural[7] convection heat fluxes:

$$Q_{H,f} = \tfrac{1}{2}\left(\frac{\alpha_T}{\nu}\right)^{2/3} C_f\, \rho\, C_p u\, (T_s - T_g)$$

and $$Q_{H,n} = 0.14\left[\frac{\alpha_T^2\, \zeta\, g}{\nu}\, (\rho C_p)^3\right]^{1/3} (T_s - T_g)^{4/3}$$

where C_f is the friction factor, u the gas velocity, T_s and T_g the temperatures of substrate and gas and g the gravitational acceleration. The expressions contain the following gas properties: α_T the thermal diffusivity, ν the kinematic viscosity, ρ the density and C_p the specific heat. The parameter $\zeta = -\frac{1}{\rho}\left(\frac{\partial\rho}{\partial T}\right)_p$, so for an ideal gas $\zeta = 1/T$. The friction factor $C_f = 2\left(\frac{u_*}{u}\right)^2$, where u_* is the friction velocity. For methane $\frac{\alpha_T}{\nu} = 1.35$, but for other gases this ratio is only slightly different.

Using these expressions, we find for the conditions at location B: $Q_{H,f} = 130\ W/m^2$ and $Q_{H,n} = 1000\ W/m^2$. Thus the heat transfer by natural convection is found to be an order of magnitude larger than that by forced convection. We follow the same calculation procedure as in HEGADAS, where the heat flux from the substrate to the cloud is given by:

$$Q_H = \max(Q_{H,f}, Q_{H,n}).$$

It is easily found that the gas temperature increase with distance x due to heat transfer from the water surface is given by:

$$\frac{dT_g}{dx} = \frac{b\ Q_H}{E\ C_p}$$

where E is the nitrogen flow rate (in kg/s) and b the width (in m) of the box.

Using the calculated heat flux at point B we find: $\frac{dT_g}{dx} = 37\ ^oC/m$. The measured increase of the air concentration of 4 % air per metre gives rise to a heating rate of 5 $^oC/m$. So the heat transfer from the water surface and the ambient air together should give a temperature increase in the bulk of the nitrogen stream of 42 $^oC/m$.

As an average over the six experimental runs we found a temperature increase between locations A and C of 37 ^oC (minimum 32 ^oC, maximum 40 ^oC), or 74 $^oC/m$. This is significantly larger than in the calculated figure of 42 $^oC/m$.

We believe that this discrepancy can be explained by looking at the measured increase in water concentration between locations A and C. At t = 400 s and 2 cm above the water surface this increase was found to be 0.01 mole water per mole nitrogen. If we assume that this water releases its sensible and latent heat, then the increase in water content over a distance of 0.5 m gives rise to a temperature increase of 19 ^oC. The calculated temperature increase rate then becomes 80 $^{o}C/m$ instead of 42 $^{o}C/m$. Compared with the measured figure of 74 $^{o}C/m$, this is considered satisfactory.

These laboratory experiments indicate that the observed high water content of the clouds in the Maplin Sands experiments may be partly explained by transfer from the water surface to the cloud. But a high air humidity close to the surface and water pick-up during pool boiling will have contributed as well; further investigations are needed to quantify these contributions. The results of the experiments indicate that significant heat transfer into the cloud may occur, associated with water pick-up from the surface during the dispersion phase. (Approximately 40 % of the total heat input was due to latent heat release from water (vapour) trapped in the cloud.)

4. Analysis of Maplin Sands Instantaneous Spills

Four instantaneous spills provided useful data at Maplin, two LNG and two propane spills (see Table II). The LNG spills were performed under similar conditions: No. 22 (12 m^3) in 5.5 m/s wind, and No. 23 (8.5 m^3) in 6.6 m/s wind. Both were ignited, although in spill 23 this was at a late stage and only a small patch of the cloud burned. We shall describe spill 22 in some detail.

The visible outline of the cloud, in Fig. 4a, shows the elongation due to the length of time the liquid takes to evaporate from the surface of the sea. We normally model the liquid as spreading under gravity until a minimum thickness of 2 mm is reached, with the evaporation rate proportional to the pool area. A constant liquid regression rate of 2 x 10^{-4} m/s (0.085 $kg.m^{-2}.s^{-1}$), derived from a continuous LNG spill, is used. Once the minimum thickness is reached, the pool is assumed to break up, the subsequent evaporation causing a reduction in the total area rather than any decrease in the thickness of the liquid layer.

Use of the pool model as described, however, gives a maximum pool diameter which is much larger than the approximate 36 m observed, and consequently too short an evaporation time. The minimum thickness has to be set to 6 mm to agree with the observations. With this done, the time-dependent HEGADAS model, including surface heat transfer, has been run to simulate the gas dispersion; ground-level contours of 3 % concentration are shown in Fig. 4b. The concentration level which corresponds to the visible edge of the cloud is difficult to assess, since it is strongly dependent on the water content of the cloud, about which there is some uncertainty, as described above. The value of 3 % is the concentration measured at 250 m from the source when the visible edge of the cloud arrived.

The model results are shown in an alternative form in Fig. 5. This plots the progress of the 5 % and 15 % contours along the centre line of the cloud, as a function of time. 5 % and 15 % are the lower and upper flammability limits (LFL, UFL) for methane. Also plotted is the upwind and downwind travel of the flame, following ignition at 180 m from the source. The flame travelling upwind is interesting in that it met the rear of the flammable cloud, moving away from the source, at 140 s, forty metres from the source. This observation shows the period of significant gas evaporation to be slightly overestimated.

The downwind-moving part of the flame was extinguished at 240 m. This is consistent with the measurement of gas concentration at 250 m from the source and 1 m from the surface, which was between 4 % and 5 % at this time, and indeed for the whole period shown by the dashed line in Fig. 5. Thus the downwind distance to LFL is overpredicted by the model, just as for continuous LNG spills at low relative humidity. And, as for continuous spills, this overprediction can be removed by increasing the relative humidity input to the model, in this case from 62 % to 75 %.

The first instantaneous propane release was spill 60, performed in a very low wind. Unfortunately, the data collection system failed totally on this occasion. However, the wind speed, about 1.2 m/s, can be obtained from several separate upwind anemometers, and useful data can be derived from the photographs. Fig. 6 shows a number of outlines of the expanding circular cloud.

The wind speed is too low for this spill to be simulated by the current version of HEGADAS, which does not include longitudinal gravity spreading. The model is currently being extended for very low wind speeds by the addition of an initial phase for an instantaneously emitted gas cloud. In this phase the cloud is represented as a cylinder, and air entrainment associated with the gravity spreading is included, in addition to the existing mechanisms of dilution.

For propane spill 63, the wind speed was greater, 3.4 m/s. The most striking feature of this spill (Fig. 7) is that the cloud was not elongated in the wind direction, in contrast to spill 22. The wind speed was only about 40 % lower than in spill 22, and the evaporation behaviour of the liquids is similar. In fact, the cloud was wider than it was long, if "length" is measured along the wind direction.

The main difference between the propane and LNG spills is that propane remains denser than air, even if warmed up to ambient temperature. This, combined with the larger quantity and lower wind speed, caused the cloud to remain very low, the bulk of it being well below one metre height in the early stages. It spread in all directions as a strong gravity current with a prominent raised head at the front. Gas concentration measurements clearly show the passage of the head (e.g. Fig. 8), and also some increase in concentration as the rear of the cloud passed. The bulk of the gas in the gravity current behind the head was probably below the lowest gas sensor.

The reduced "length" of the cloud suggests that a head-wind has a stronger effect on a gravity current than a tail-wind. Some laboratory experiments on this effect have been reported by Simpson and Britter[8]. However, the ratio of wind speed to speed of the head in the laboratory was smaller than in spill 63 and a difference between the two directions was not discernible in laboratory data, but a small difference is present for Simpson and Britter's theoretical curves. In any case, as the upwind gravity spread is arrested and then reversed, there is a change to a saline-wedge flow, which was not analysed by Simpson and Britter.

Cloud shapes predicted by the time-dependent HEGADAS model for spill 63 are shown in Fig. 7b. In the early stages the predicted cloud is elongated because of the time taken to evaporate the liquid. But later this is counteracted by the lack of allowance for longitudinal gravity-spreading, producing a roughly circular cloud shape. The simulation might also be improved here by use of the cylindrical box front-end for the model.

Gas concentrations were measured, generally at three heights, at nine locations. However, the wind direction was well outside the planned "window", 112° from the array axis; so the cloud did not reach any far-field sensors. There are no measurements beyond 180 m from the source. It is difficult to deduce ground-level concentrations from readings of sensors which were probably above the bulk of the very low cloud. And in the later stages of dispersion, when the cloud was higher, there were no measurements. However, it will be interesting to compare the near-field observations with those from Thorney Island[9] to assess the effect of the totally different initial conditions, in one case dropping from a great height and in the other remaining low.

One feature which emerges from use of the cylindrical box front-end is the low advection velocity of the cloud. Initially, when using a box simulation of the early stages of dispersion, we set the cloud velocity equal to the average of the ambient wind velocity over the height of the cloud, assuming a logarithmic profile. However, this greatly overestimates the speed of the cloud. We have found that setting the cloud velocity to 60 % of the ambient gives a good fit to the observations. Similar results are obtained for spill 60, although the ambient wind was less accurately measured in this case. This finding compares well with the laboratory measurements of Simpson and Britter[8] showing that the effect of an ambient flow on the motion of a gravity current was to change its velocity by an amount equal to 62 % of the ambient flow velocity.

5. Conclusions

1. For LNG, the Maplin Sands data confirmed the model predictions for spills at high wind speeds, where density effects were unimportant. However, for other spills the model without heat transfer consistently overpredicted the LFL distance, the discrepancy roughly increasing with initial Richardson number. With heat transfer taken into account, the predictions were better for spills where the measured ambient relative humidity was high, but in several other spills the change had comparatively little effect.

   Large moisture contents of the clouds were found, which may be explained by a combination of high humidity of the air close to the water surface and water pick-up from the sea surface. Model runs have shown that high levels of humidity in clouds give rise to enhanced dispersion when the water is trapped as vapour at ambient temperature in the cold cloud and there releases its sensible and latent heat.

2. Laboratory experiments with liquid nitrogen evaporating from water showed that the water content of the cold vapour stream increases as the stream flows over a water surface, thus substantiating the findings from the Maplin Sands LNG spills. The results of these experiments support the validity of the natural-convection heat-transfer relation used in HEGADAS. They indicate that a substantial part of the total heat transfer from the water surface into the cloud is due to heat release from water picked up from the surface during the dispersion phase.

3. The two analysed instantaneous LNG spills of the Maplin Sands programme produced very elongated clouds in wind direction. This shows the importance of taking into account the finite pool evaporation time for moderate and high wind speeds. The downwind distance to LFL is overpredicted by the model. Just as for continuous LNG releases at low humidity this overprediction can be removed by increasing the relative humidity input to the model.

   The two instantaneous propane spills were performed in low winds, showing nearly circular shape of the clouds and very strong effects of gravity-spreading. Visual data show that the advection velocity during the early phases of cloud development was significantly lower than the ambient wind speed averaged over the height of the cloud as calculated with a box simulation of the early stages of dispersion.

References

1. Puttock, J.S., Colenbrander, G.W. and Blackmore, D.R., "Maplin Sands experiments 1980: Dispersion results from continuous releases of refrigerated liquid propane", Heavy Gas and Risk Assessment II, Frankfurt, May 1982 (publ. D. Reidel, Dordrecht).
2. Puttock, J.S., Colenbrander, G.W. and Blackmore, D.R., "Maplin Sands experiments 1980: Dispersion results from continuous releases of refrigerated liquid propane and LNG", NATO/CCMS 13th Int. Techn. Meeting on Air Pollution Modelling and its Application, Isle des Embiez, Sept. 1982.
3. Colenbrander, G.W. and Puttock, J.S., "Dense gas dispersion behaviour: Experimental observations and model developments", 4th Intl. Symp. on Loss Prevention and Safety Promotion in the Process Industries, Harrogate, England, Sept. 1983.
4. Colenbrander, G.W., "A mathematical model for the transient behaviour of dense vapour clouds", 3rd Intl. Symp. on Loss Prevention and Safety Promotion in the Process Industries, Basle, Sept. 1980.
5. Boyle, G.J. and Kneebone, A., "Laboratory investigations into the characteristics of LNG spills on water. Evaporation, spreading and vapour dispersion", Shell Research Ltd. Report to the API (1973).
6. Holman, J.P., "Heat Transfer", 5th ed., McGraw-Hill, 1981.
   or
   Welty, J.R., "Engineering Heat Transfer", John Wiley & Sons, 1974.
7. McAdams, W.H., "Heat Transmission", McGraw-Hill, 1954.
8. Simpson, J.E. and Britter, R.E., "A laboratory model of an atmospheric mesofront", Quart. J. Roy. Met. Soc., 106 (1980) 485-500.
9. McQuaid, J., "Large-scale experiments on the dispersion of heavy gas clouds", I.U.T.A.M. Symp. on Atmospheric Dispersion of Heavy Gases and Small Particles, Delft, Aug. 1983.

Table I

Details of continuous LNG spills

Spill no.	Rate, m^3/min	Wind speed, m/s	Relative humidity at 10 m, %	Ri_o	Observed LFL distance peak, z = 0.9 m, m	Predicted LFL distance, mean z = 0 m	
						HEGADAS II	Incl. heat transfer
27	3.2	5.5	53	16	190 ± 20	260	235
29	4.1	7.4	52	7	140 ± 15	250	245
34	3.0	8.6	72	4	150 ± 20	195	185
35	3.8	9.8	63	3	175 ± 25	205	205
39	4.7	4.1	63[a]	50	130 ± 20	355	270
56	2.5	5.1	83[a]	17	110 ± 30[b]	210	115

Notes: (a) Derived from observations at meteorological station 3 km away.
(b) Plume only briefly over sensors.
Lowest sensors at z = 0.6 m.

Table II

Details of instantaneous spills

Spill no.	Spilled liquefied gas	Quantity, m^3	Wind speed, m/s	Relative humidity at 10 m, %
22	LNG	12 ± 0.6	5.5	62
23	LNG	8.5 ± 0.6	6.6	66
60	propane	27 ± 6	1.2	79[a]
63	propane	17 ± 5	3.4	93

Note: (a) Derived from observations at meteorological station 3 km away.

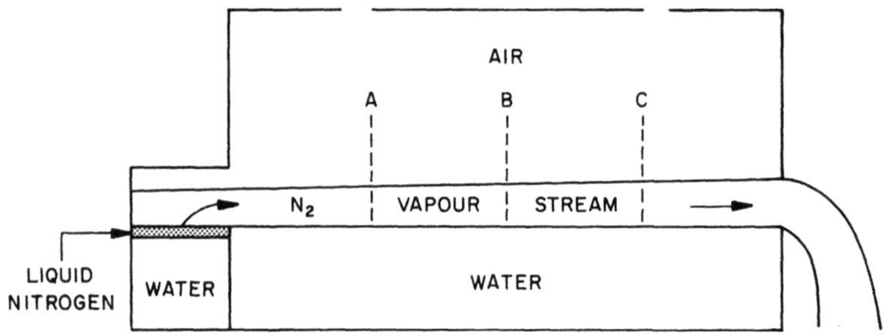

Fig. 1. Schematic representation of the experimental set-up

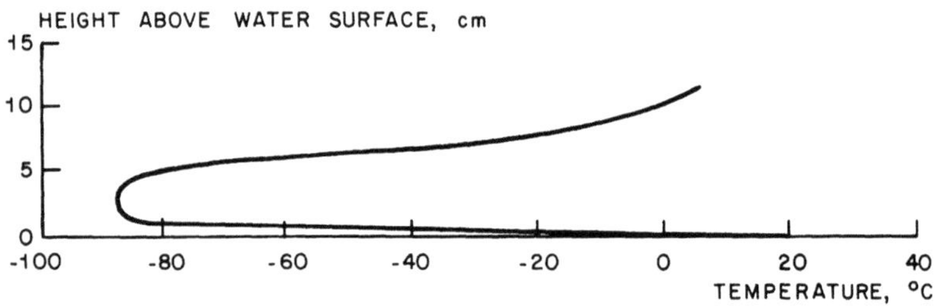

Fig. 2. Typical vertical temperature profile measured at location B at t = 405 s

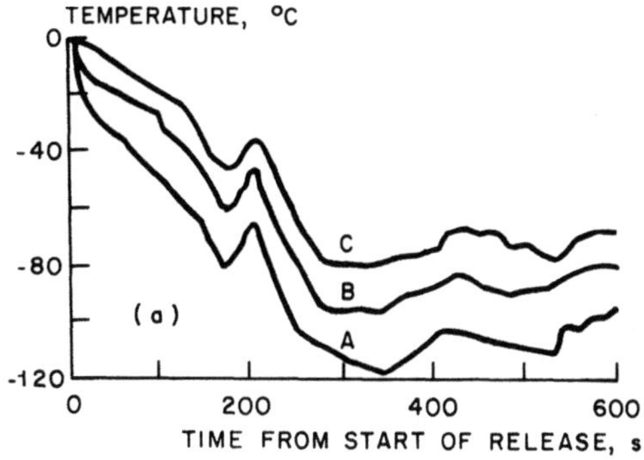

Fig. 3a. Temperatures as functions of time, measured at location A (height above water surface: 3 cm), B (height: 2 cm) and C (height: 1 cm)

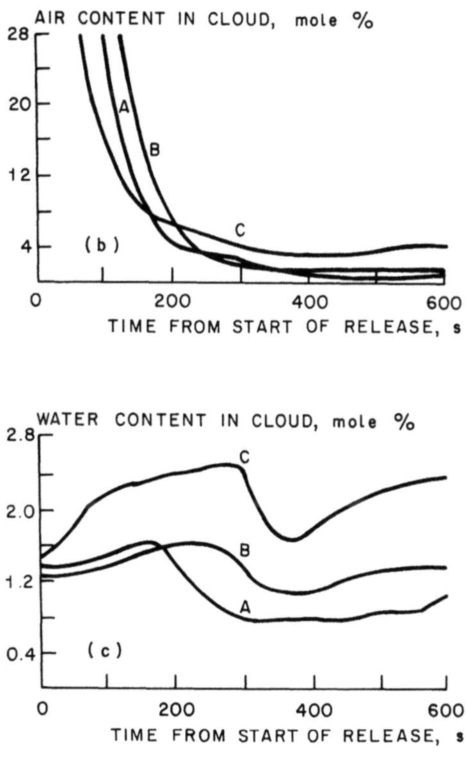

Fig. 3b and 3c. Air concentrations (b) and water concentrations (c), measured as functions of time at locations A, B and C at 2 cm above the water surface

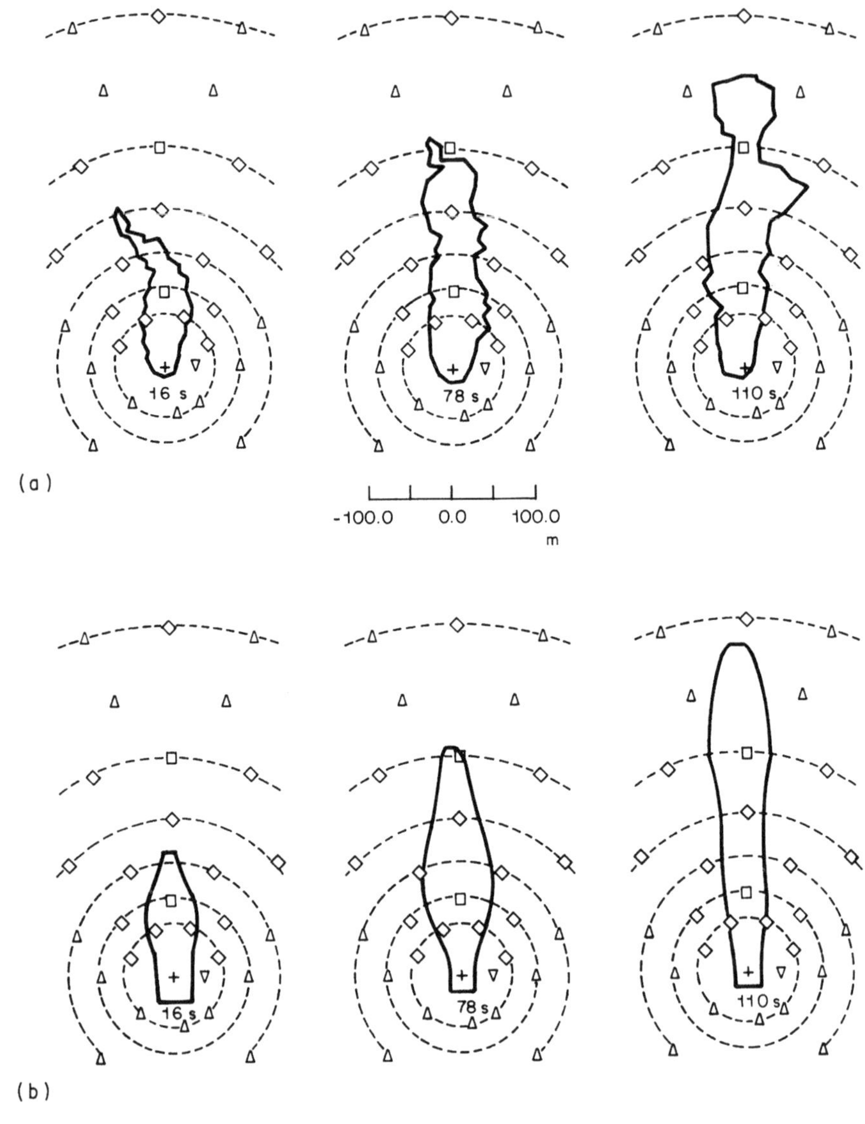

Fig. 4.(a) Outline of the developing visible cloud in LNG spill 22. Times indicated in seconds from an arbitrary start. The spill start was at 847 s. Various symbols indicate measurement stations

(b) Predictions from the HEGADAS model of the 3 % concentration contour, corresponding to the visible limit, at same times as (a) above. Indicated times measured from spill start

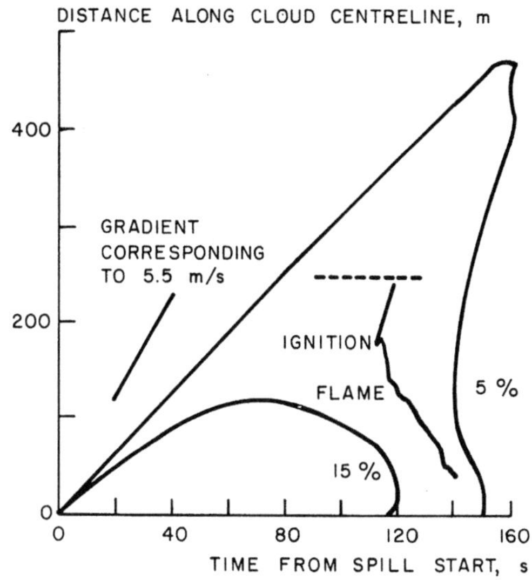

Fig. 5. The progress of 5 % and 15 % concentration contours along the centre line of the cloud in spill 22, as predicted by HEGADAS. The region between these curves corresponds to flammable gas. The progress of the flame upwind and downwind is also shown. The dashed line indicates the period during which the concentration measured at 250 m from the spill point, 1 m from the surface, was between 4 % and 5 %

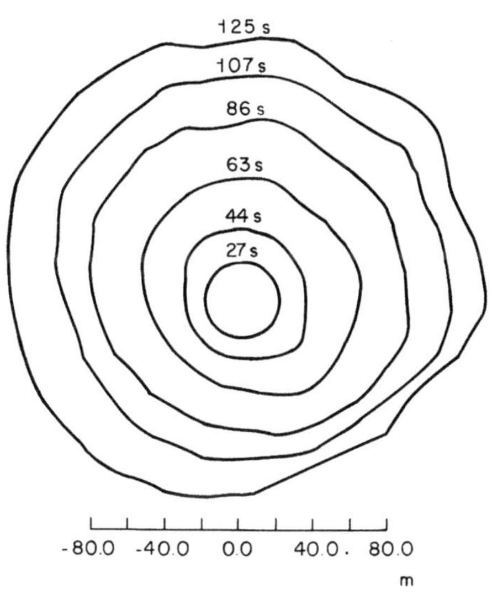

Fig. 6. The spreading cloud from propane spill 60. Times are measured from start of spill

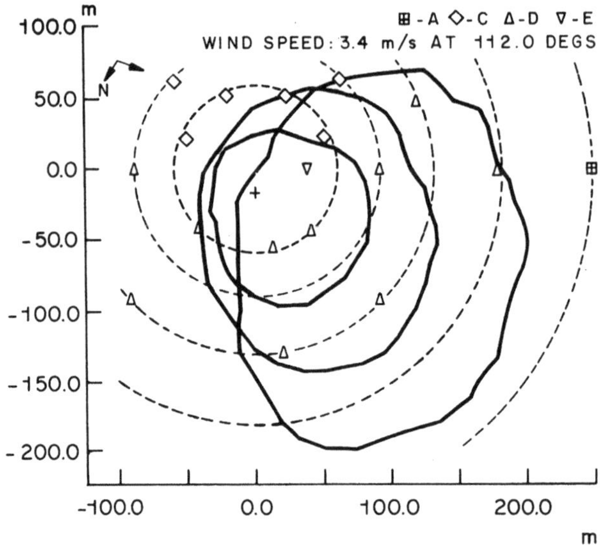

Fig. 7a. The outline of the cloud in spill 63 at times 41 s, 65 s and 94 s from the spill start

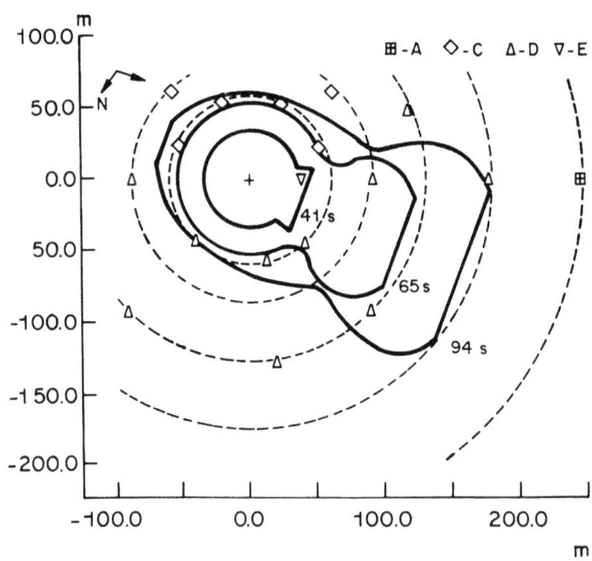

Fig. 7b. The predicted cloud outline (3 % contour) from HEGADAS at the same times as Fig. 7a

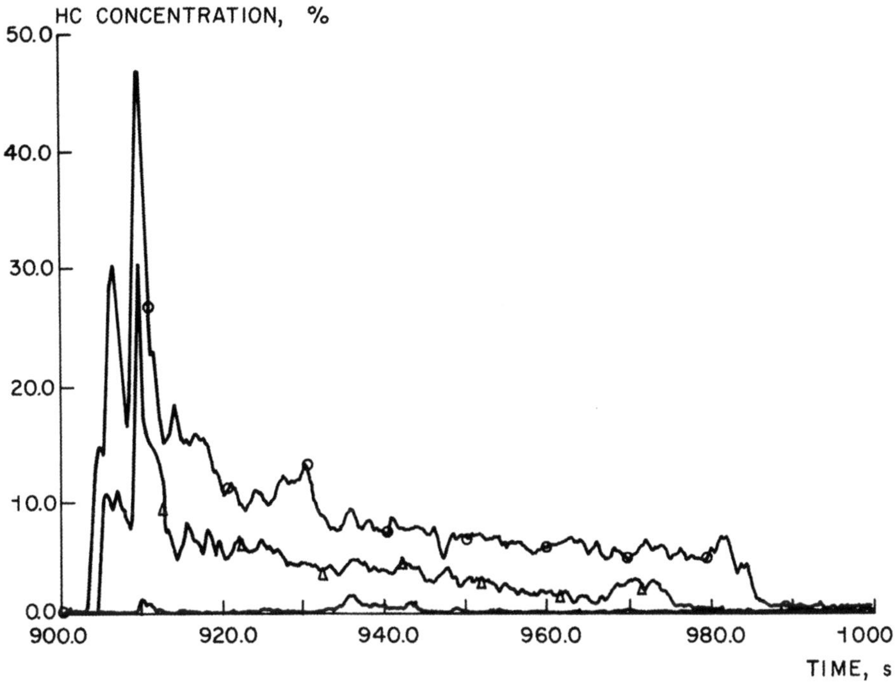

Fig. 8. Gas concentrations measured in spill 63 at 57 m from the spill point, 45° from the centre line, at 0.6 m, 1.3 m and 2.3 m heights. The gas sensor response has been accelerated, using the time derivative of the signals. The time of the spill is 857 s on this scale

Monitoring the Atmospheric Diffusion of Puffs and Plumes with Bipolar Space Charge at Small Scale in Wind Tunnels

B W. BOREHAM and J.K. HARVEY

Department of Aeronautics
Imperial College of Science and Technology
London SW7 2BY
England

Introduction

The use of charged particles as markers to simulate the transport of neutrally buoyant contaminants within puffs and plumes within the atmosphere is an established technique [1]. In this paper the use of a bipolar space charge method is discussed. The main advantage of this method over the use of unipolar space charge is avoidance of the difficulties encountered with the self repulsion of unipolar ions, although recombination now becomes an effect that must be allowed for.

It is often desirable in simulation studies to know the time dependent distribution of pollutants and, if tests are done at reduced scale, a rapidly responding monitoring instrument is needed to resolve the time detail. A new Langmuir-type flux probe has been developed that is capable of such monitoring [2]. This probe has several advantages over the aspiration-type concentration probes normally used for measuring the mean concentration of particles. The flux probe does not require the same prior knowledge of the velocity field for accurate operation. Aspiration-probes require that the pumping speed be matched at all times to the approach flow velocity which may well be fluctuating, otherwise the sampling volume will be undefined, leading to erroneous results in any but the simplest of flow fields. The present device, being passive, rather than dynamic, overcomes this problem and the two-dimensional design-symmetry reduces the alignment problem and alleviates the dependence of results on alignment accuracy.

A theoretical model of the probe behaviour has been developed and good agreement is obtained between the calculated and observed operating characteristics. Probe response time is the order of 1 m sec or less. Several probes have been developed and the total system has been used to study concentration fields in the 4' × 4.5' and 3' × 3' cross-section wind tunnels at Imperial College. The probe design is similar to the familiar single electrode Langmuir probe used for low pressure measurements and consists of two concentric cylindrical electrodes. The inner, small diameter electrode is solid and functions as the collector, whilst the outer electrode consists of a thin wire mesh or grid and serves to define the sampling volume, which remains constant. The outer grid is earthed and an attractive potential is applied to the inner electrode. The current flow to the collector from the gas sample is monitored to provide ion flux readings. Concentration, if also required, may be obtained using separate velocity field measurements. Hence a concentration reading, based on the known constant sampling volume can be made.

Probe Development

The probe current-voltage characteristic can be calculated in order to obtain the required bias voltage at saturation and the probe response time. The characteristic will be influenced by the following considerations:

(1) The applied bias field.

(2) Collision between the ions and the gas.

(3) Space charge effects consisting of the following:

  (i) Electrostatic (Debye) attraction between oppositely charged particles within the probe.

  (ii) Electrostatic repulsion between like-charged particles.

  (iii) Electrostatic repulsion due to the collector current flow.

At pressures very much less than atmospheric the applied bias field dominates and ions can be captured easily in distances of the order of 1 mm. At atmospheric pressure, however, collisions will dominate and may be strong enough to prevent particle capture unless large attractive and/or retarding electric fields are applied. Both types of field are possible with the cylindrical geometry detector, aligned perpendicularly to the flow. (With the attractive potential applied to the collector the field experienced by the collected particles will be attractive in the half of the detector that is upstream of the collector and retarding in the half that is downstream of the collector, as shown in figure 1).

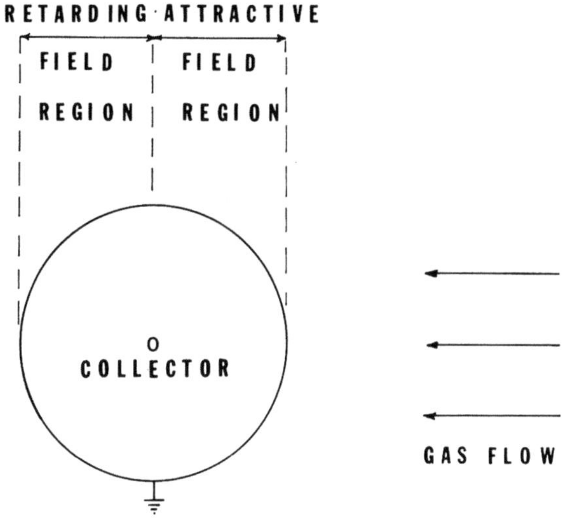

PROBE FIELD SCHEMATIC
Figure 1

Two limiting modes of collision dominated probe behaviour are identified and separately treated. These are:

(a) The mode in which the gas is assumed to be fully disturbed on passing through the outer grid. In this mode collisions occur between the charged particles and a randomly distributed gas flow velocity field in which the particles (accelerated by the applied field) lose energy to the neutral gas particles. The mean direction of motion of the charged particles is that imposed by the bias field and the particles can be captured by the attractive field upstream of the collector.

(b) In this mode the gas flow remains undisturbed by the grid and the gas-particle collisions carry the particles downstream with the gas flow. Particle capture is now more difficult and requires a combination of the attractive field upstream of the collector and the retarding field downstream of the collector.

For each of these cases the detector current-voltage characteristic equation is derived in full elsewhere [2] and only the results are summarised here.

Several different probes were used with collection cross-sections ranging from 9 to 49 mm^2. Operation was in both the disturbed and undisturbed flow regimes. Figure 2 compares the measured and calculated characteristics for a 48 mm^2 cross-section probe located 5 cm from the source exit and wind tunnel flow velocity of 5 ms^{-1}. Good agreement is obtained with the calculation underpredicting the measured values. The analytical model has so far assumed that the flow streamlines terminate on the probe collection surface. The charge loss caused by the streamlines carrying ions around the inner electrode has therefore not been included. This may be accounted for with an expression of the form

$$(i/i_o)_{S.E.} = \left(1 - \frac{\beta}{U} \frac{U}{\langle U_r \rangle} \frac{1}{R_o} \frac{1}{\lambda}\right) (i/i_o)_{calc}$$

where $(i/i_o)_{calc}$ is the previously calculated normalised current

$(i/i_o)_{S.E.}$ is the semi-empirical normalised current

$\langle U_r \rangle$ is the mean ion velocity in the radial direction

β is an experimentally determined constant.

Applying this correction to the calculations of figure 2 the excellent agreement of figure 3 is obtained.

Operation in the undisturbed flow regime was obtained by replacing the wire mesh outer grid with a coiled wire outer grid and the agreement between theory and measurement for a 9 mm^2 collection area probe of this type is shown in figure 4.

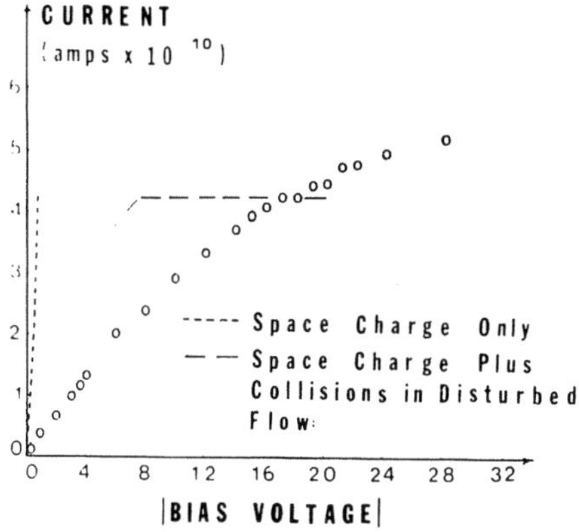

Figure 2

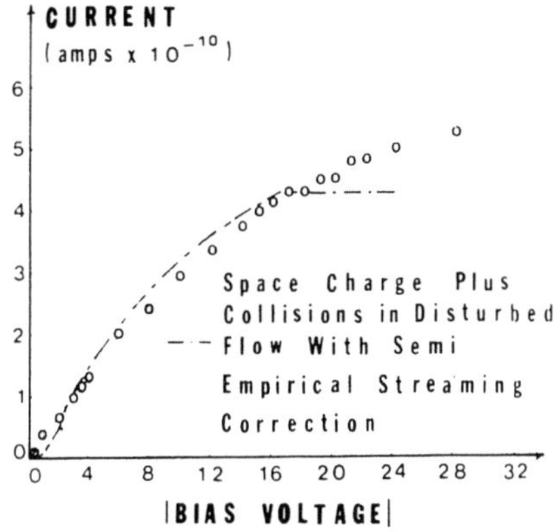

Figure 3

Wind Tunnel Diffusion Studies

Mean flux cross-section measurements were obtained at several points varying from 5 cm to 1.8 m downstream of the ion source, using both bipolar and unipolar sources. Air speed was measured with a built-in differential pressure manometer system. The bipolar source consisted of a 30 mm × 10 mm Nickel-63 foil rolled into a cylindrical shape and fitted into a hollow aerodynamically shaped holder. The foil activity was 15 mCi and the bipolar ions were expelled from the source by a current of air along the axis, the velocity of which was matched to the wind tunnel free stream flow velocity. A small mesh wire grid fitted in the source upstream of the foil ensured source mixing and introduced small-scale turbulence into the source flow.

The unipolar source was similar to that used by Jones [1] and consisted of a co-axial electrode system producing a point discharge. The discharge filled the space between the electrodes with unipolar ions that were again expelled from the apparatus by a current of air along the axis, as for the bipolar source.

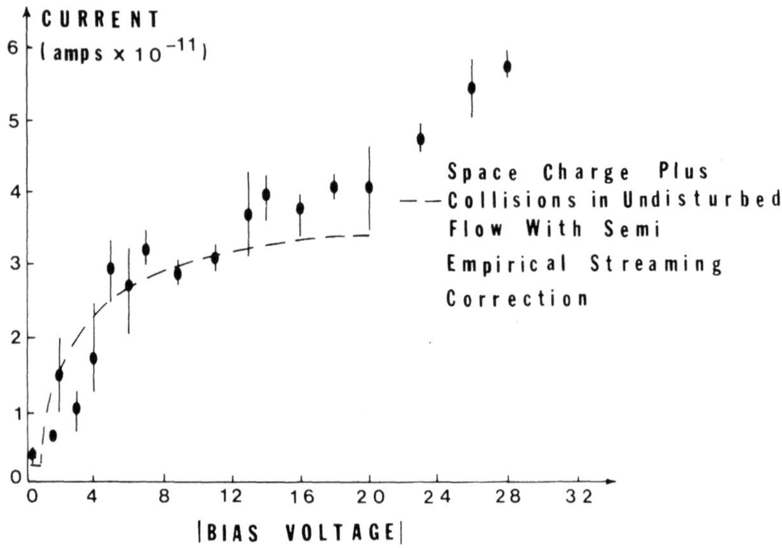

Figure 4

Most measurements were obtained with relatively small ambient turbulence levels (∿0.2%), although the mixing of the ionised plume and the boundary layer shed from the outer surface of the ion-generator into the main flow result in higher turbulence levels within the ion plume. Some results were obtained with a turbulence grid inserted into the wind tunnel flow to increase the general turbulence level and also with a simulated model atmosphere (1/125 scale) at an equivalent height of 75 m.

The results were compared with the model for the plume suggested by Pasquill [3], modified for ion recombination. The concentration χ is given by:

$$\chi(x,y,z) = \frac{Q}{2\pi U\sigma_y\sigma_z}\exp\left[-\tfrac{1}{2}\,\frac{y^2}{\sigma_y{}^2} + \frac{z^2}{\sigma_z{}^2}\right]$$

where $$\frac{\sigma}{x} = \frac{1}{\sqrt{2}}\,C\,x^{-n/2}$$

or, for a constant vertical velocity profile,

$$\sigma_y \equiv \sigma_y \equiv \sigma = C'x^m$$

$Q = N\,UA_o/\varepsilon$ is the particle flux

where N is the particle density

A_o is the effective source area

ε is the charge per particle.

For charge loss by recombination

$$N = \left(\frac{1}{N_o} + \frac{\alpha}{U}X\right)^{-1}$$

where X is the distance downstream from the effective source
α is the recombination coefficient
N_o is the source particle density.

Hence the mean flux that is measured by the probe (i.e. the flux integrated over the probe collection area) is

$$\overline{F(X,Y,Z)} = \left[\frac{UA_o\left(\frac{1}{N_o} + \frac{\alpha}{U}X\right)^{-1}}{2\pi\sigma^2}\right]\frac{1}{w\ell}\int_{Z-\ell/2}^{Z+\ell/2} e^{-z^2/2\sigma^2}\,dz \int_{Y-w/2}^{Y+w/2} e^{-y^2/2\sigma^2}\,dy$$

The required values of N_o and A_o may be obtained from measurement of total current in the plume as a function of downstream distance and σ is obtained by iteration.

Figure 5 compares the flux cross-section obtained with a 48 mm^2 area probe located 0.5 m downstream of the source exit with that calculated for $\sigma = C'x^m$ where $C' = 1.66 \times 10^{-4}$, $m = 0.9$, $N_o = 2.98 \times 10^{12}$ particles/m^3 and $A_o = 3.95 \times 10^{-4}$ m in a 10 ms^{-1} flow in the 4' × 4.5' wind tunnel. Each experimental point refers to 100 sample taken over a sample time of 200 ms. Figure 6 is the corresponding cross-section obtained in a unipolar flow under identical experimental conditions. It is seen that the motion of the unipolar ions is dominated by their mutual repulsion [1].

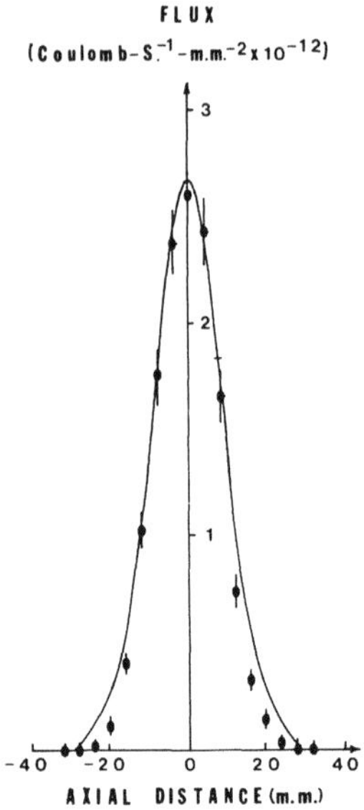

Figure 5

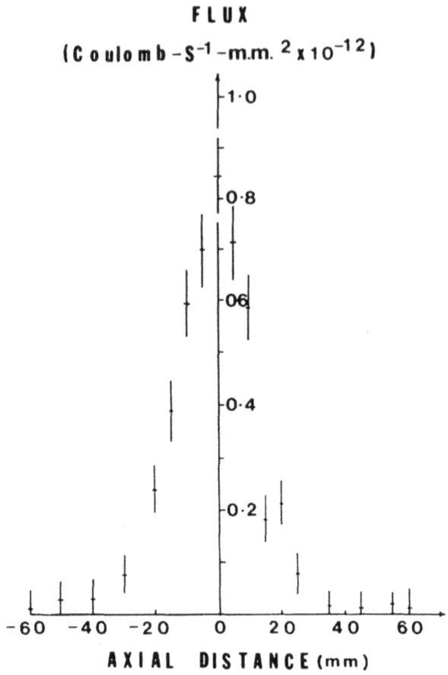

Figure 6

Figure 7 illustrates the effect of increased turbulence (produced by inserting the turbulence grid into the wind tunnel flow) on the bipolar source flow at a distance of 0.3 m from the source exit.

Figure 8 is an example of the time dependent flux signal obtained within the shear layer at the edge of a plume from a unipolar source and a flow of 10 ms^{-1}. Figure 9 compares the bipolar source time dependent signal in the centre of the plume (9a) and at the edge of the plume within the shear layer for the model atmosphere flow at a flow velocity of 4.26 ms^{-1}.

Conclusion

We conclude that a new technique has been successfully developed and applied to atmospheric dispersion.

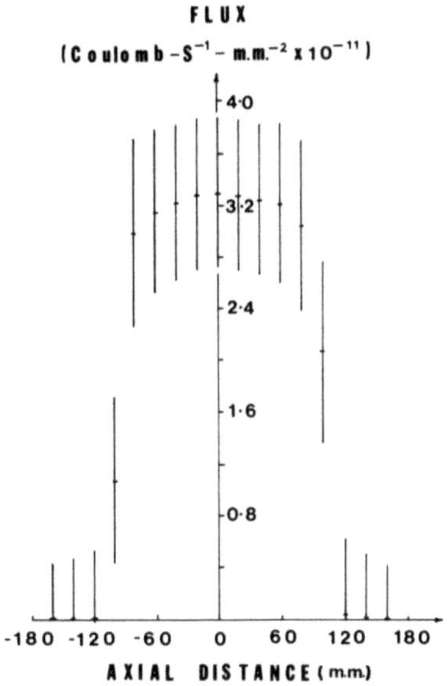

Figure 7

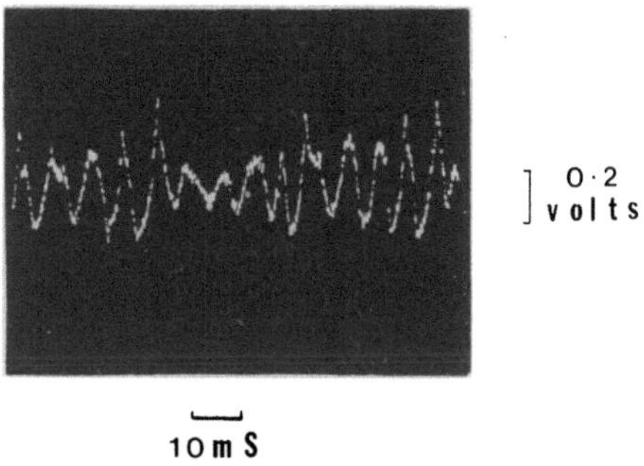

Figure 8

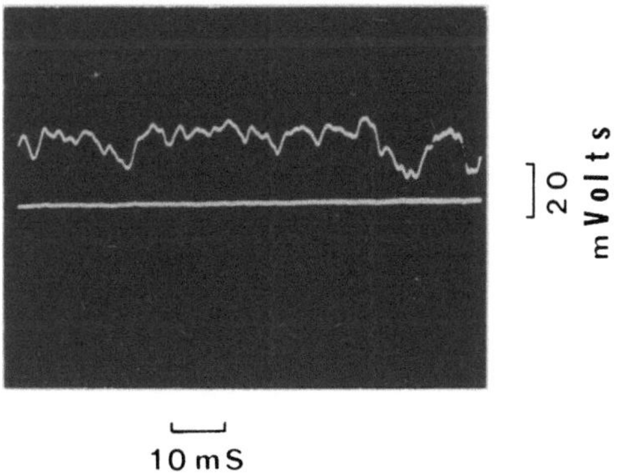

Figure 9a

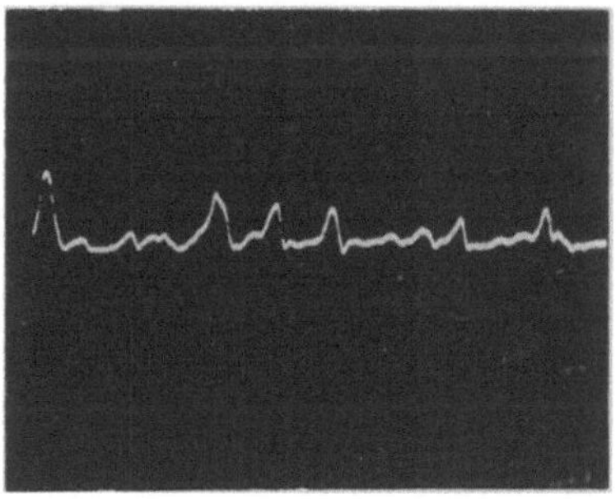

Figure 9b

References

1. Jones, C.D. and Hutchinson, W.C.A.; Plumes of electric space charge in the lower atmosphere. Journal of Atmospheric and Terrestial Physics, Vol. 38, pp 485-494, 1976.

2. Boreham, B.W.; A new fast response probe for diffusion modelling in atmospheric air. Submitted for publication.

3. Pasquill, F.; Atmospheric diffusion. John Wiley, 2nd Ed, p 351, 1974.

Experimental and Theoretical Studies in Heavy Gas Dispersion.

Part I. Experiments

K. Emblem*, P.Å. Krogstad
The Foundation of Scientific and Industrial Research at the Norwegian Institute of Technology

and T.K. Fanneløp**
Division of Aero- and Gas Dynamics
Norwegian Institute of Technology
Trondheim, Norway

ABSTRACT

This report describes an experimental study of heavy gas dispersion in a 24 m long and 2.4 m wide duct in a laboratory. The test gas used was cold nitrogen which was boiling off from a liquid pool. The gas was released both directly while boiling, giving a semicontinuous source, and instantaneously by blocking the duct until the boiling process was completed. In the case of a semicontinuous release the cloud spreads over a free water surface in four tests and over an isolating styrofoam surface also in four tests. Another four tests were made with instantaneously released gas spreading over the styrofoam surface. Measurements of temperatures, velocities and concentrations were made at different downstream positions in all tests. A simplified box model theory is used for comparison with the data from the experiments. This paper includes data for the entrainment parameters, observations concerning the leading edge velocity and the heat budget of the cloud as well as the velocity distribution inside the cloud and comparisons of data for instantaneously- and semicontinuously released clouds.

1. INTRODUCTION

Spreading and dispersion of flammable and toxic gases are of major concern in risk assessments for industrial plants and in other activities requiring handling of heavy gases.
The physical effect which most directly affects the gas dispersion, is the entrainment of air into the cloud. This entrainment will take different forms during a spreading process. In the early stage following the release the entrainment is thought to be influenced mainly by frontal mixing and interfacial shear with the surrounding air. At late times the dispersion is believed to be dominated by atmospheric turbulence. This report describes a series of experiments which have been undertaken to study entrainment in the early "gravitational" phase. The aim of the study has been to obtain experimental information under controlled laboratory conditions in the absence of atmospheric effects.

---

* Present adress: The Research Center, Norsk Hydro Corp.
**Now with the Swiss Federal Institute of Technology, ETHZ.

The main results described in this paper relates to the heat budget in the cloud, the leading edge velocity, the velocity-, concentration- and temperature distributions inside the cloud and the overall entrainment rate of air into it.
The first experiments in this study started three years ago and some results concerning experimental technique, overall flowfield and experimental data for cloud height, temperature profiles, leading edge velocities etc. have already been published (ref.1,2).

2. THE EXPERIMENT

The tests were conducted in a 24 m long, 2.4 m wide and 1.6 meters deep channel, designed and constructed for the heavy gas experiments (fig.1). The walls of the channel are made of clear plastic draped over a wooden frame. The bottom of the channel is covered with a 0.15 meters deep water layer. During some of the experiments the water surface was open and allowed heat exchange between the water and the cold, spreading cloud. In other experiments the water was covered with isolating styrofoam plates to eliminate this effect.
Cold nitrogen gas boiling off from a pool of liquid nitrogen released at the up-stream end of the channel was used for the tests. The boiling process from liquid nitrogen on water is quite vigorous. The phase change for a volume of 8 litre LN_2, boiling in the film boiling regime, requires about 40 seconds resulting in a semi continuous gas source. The major part of the gas release is, however, completed during the first 30 seconds,which is comparable to the time required for the gas to spread the lenght of the channel. The effects of the source conditions have been studied earlier and is described briefly in ref.2.
Tests of instantaneous releases of gas have also been conducted. A vertical plate was used to block the gas at 2.4 m from the upstream end of the channel and was removed suddenly when the boiling process was completed. The gas released in this way was slightly mixed with air during the boiling process and measurements indicated a mean initial concentration of about 85% (vol).

2.1 Instrumentation

Instruments for the measurement of physical data related to the spreading cloud were mounted on three vertical columns (fig.1). At these locations both velocities, concentrations and temperatures were measured continuously at the vertical positions shown.
Additional temperature measurements were made in between the main instrumentation columns giving more accurate information on the position of the leading edge of the cloud during the spreading process.
Temperature measurements were made by means of thermocouples with a very short response time. Gas concentrations were measured with paramagnetic analysers of a special design giving rapid response. The analysers were located in the channel to minimise the time delay between probe and analyser.
Velocities were measured using the pulsed-wire technique. The

geometry of the probe used is shown in fig.2 and consists of an upstream transmitter and two receivers mounted perpendicular to each other. The transmitter was mounted parallell to the bottom of the channel facing the oncoming cloud. The transmitter was heated to 200-300°C for about 2 ms at a constant rate of about 4 pulses pr. second thus generating a series of thermal tracers which were convected downstream to the receivers. Because of the short duration of the pulse and small volume of the transmitter (the transmitter was a 3 mm long tungsten wire of 5 μm diameter) the bouyancy effect of the heat pulse was considered negligible. The receivers were mounted in a plane perpendicular to the transmitter as indicated in fig 2. Further, this plane was oriented normal to the bottom of the channel, and by proper combination of the signals from the receivers the velocity components along the channel axis and in the vertical direction could both be determined. The workable velocity range with this technique depends on the probe geometry and for the probes used velocities down to 15 cm/s could be measured. Unfortunately the probe geometry also limits the flow angle that can be measured. The maximum flow angle that could be measured with the probes used in the experiments was ± 15 deg. Due to this restriction a large number of measurements had to be descarded. Although the data logging system was run at a maximum speed, only about 2 readings pr. probe pr. second could be taken due to the overall large number of sampling points and probes. This corresponds roughly to velocity readings at 20 cm intervals if one makes the approximate transformation $\Delta X = U_{LE}\Delta t$.

In the vertical direction the probes were arranged to give maximum resolution across the height of the main cloud. However, this meant that no measurements were possible near the top of the gravity current head. Further description of the measurement system is found in refs. 1 and 3.

3. ANALYTICAL BOX MODEL SOLUTION WITH TOP- AND FRONT ENTRAINMENT

A spreading heavy gas cloud does not appear to have a uniform height. A dominating feature is a pronounced gravity current head of approximately twice the height of the main part of the cloud. However, this and other nonhomogenuous features are neglected in the present analytical model which is adapted to the experimental set up, with heavy gas spreading two-dimensionally in a channel. An extension to the case of radial spreading is straight forward. According to the box-model theory presented earlier (ref.2) the solution for the leading edge position of a spreading heavy gas cloud from a time dependent source given by

$$V_g = V_o t^q \tag{1}$$

has been found to be:

$$X_{LE} = \left(\frac{3}{2+q}\right)^{2/3} K^{2/3} t^{\left(\frac{2+q}{3}\right)}, \quad K = \left(k \frac{\Delta\rho}{\rho} g V_o\right)^{\frac{1}{2}} \tag{2}$$

The growth in cloud volume due to air entrainment has been expressed by Fay (ref.4) as:

$$\frac{dV_E}{dt} = (\alpha_1 X_{LE} + \alpha_2 H)\,\frac{dX_{LE}}{dt} \tag{3}$$

The time-dependent source also contributes to the rate of volume growth, so the total time derivative of the cloud volume is given by:

$$\frac{dV}{dt} = \frac{d(V_E + V_G)}{dt} = (\alpha_1 X_{LE}+\alpha_2 H)\frac{dX_{LE}}{dt} + qV_o t^{(q-1)} \tag{4}$$

From earlier studies of isothermal mixing (refs.5 and 6) it is known that the entrainment of air does not influence the leading edge velocity and thus the front position for a given release at a specific time is independent of (3). This is also a good approximation in the present releases where the heat capacity and molecular weight of the gases mixed are nearly identical and where the ambient air had very low humidity. It is therefore assumed that the air entrainment affects only the cloud height. The overall volume expansion of the cloud is given by:

$$\frac{dv}{dt} = X_{LE}\frac{dH}{dt} + H\frac{dx_{LE}}{dt} \tag{5}$$

Equation (1), (4) and (5) now gives a complete set of equations to solve for the cloud height and thus the mean gas concentration as a function of time:

$$H(t) = H_o\left(\frac{t}{t_o}\right)^{-(1-\alpha_2)R} + t^{-(1-\alpha_2)R}\ \frac{\alpha_1}{2-\alpha_2}\left(\frac{K}{R}\right)^{2/3}\left(t^{(2-K_2)R} - t_o^{(2-K_2)R}\right)$$

$$+\ \frac{\left(\frac{R}{K}\right)^{2/3} qV_o}{q-K_2R}\left(t^{(q-K_2R)} - t_o^{(q-\alpha_2 R)}\right) \tag{6}$$

$$R = \frac{q+2}{3}$$

For instantaneous releases t_o should be calculated from equation (1). For time dependent sources $t_o = H_o = 0$. A time dependent release for which the supply stops after a limited period can also be calculated from equation (6) by dividing the process into two separate phases, a boiling and spreading phase (with $q > 0$), and a pure spreading phase with $q = 0$ and values of V_o, t_o and H_o which match the q-values in each phase. For instantaneous releases or time dependent releases with some initially released gas present, equation (1) is somewhat modified:

$$X_{LE} = X_o^{3/2} + \frac{K}{R}\left(t^{\frac{q+2}{3}} - t_o^{\frac{q+2}{3}}\right)^{2/3} \tag{7}$$

where $x_o = (K/R)^{2/3} t_o^R$ with the appropriate q-value for $t < t_o$.

The lack of homogeneity in the actual cloud must be taken into account before comparisons are made between the calculated curves and the test figures. Using the equations above, estimates of top and front entrainment coefficients (α_1 and α_2 respectively) can be obtained by curve fit to the experimental data.
The simplified theory does not include the effect of temperature on the density of the gases. This had to be accounted for by the use of a mean cloud temperature which represents an average in time and space of the test data.

4. RESULTS AND DISCUSSION

4.1. Heat transfer into the cloud.

The main sources for heat supply to the cloud arise from the entrained air which includes water vapor which condenses, from water mist generated during the vigorous boiling process and heat transfer from the bottom surface. Some of the present experiments were performed by isolating the water surface to quantify the amount and effect of heat supply from the free water surface.

The results reveal that this heat transfer was of little importance in the early stages of the spreading process. As can be observed from fig.3, the temperature readings close to the water surface are influenced to some extent, but in the bulk part of the cloud there is little temperature change. The temperature rise in the lower part results in a slight increase in cloud height, but the influence on the entrainment of air into the cloud should be small or negligible. This is also confirmed by figure 4 which shows the concentration distributions for the same leading edge position in the case of continuously released gas with and without insulation at the bottom surface. This does not contradict the previous finding of convective rolls (ref.1). The convective rolls appeared at a later stage in the spreading process, long after the leading edge had passed. The experiments therefore indicate that in the early stages of a dispersion process the warm layer produced by heat transfer near the bottom does not break through the heavy, cold layer above, and thus does not influence the gas temperature nor the gas concentration significantly.
Nitrogen and air have nearly identical physical properties. Without sacrificing accuracy the values for heat capacity and molecular weight can be considered the same for the two gases. This assumption leads to a linear relation between temperature and concentration as nitrogen is mixed adiabatically with dry air. This relation is shown in figure 5. Here the initial temperature of the pure nitrogen vapor is set equal to -145°C which is the temperature measured immediately above the boiling pool. This corresponds to a superheating of 50°C. Measured values from the experiments are also plotted in the figure. Both concentration- and temperature data plotted in fig. 5 are mean values averaged over the cloud volume. As can be observed from the figure the heat supply from sources other than entrained air must be considerable. The scale to

the right indicates the initial vapor temperature which corresponds to the measured data under the assumption that heat is transfered only from the entrained air. The air humidity in the laboratory is far too low to explain the very high values of cloud temperature shown in the figure 5. The remaining source for heat supply appears to be the water transferred to the cloud during the film boiling process. This was confirmed by the presence of large amounts of water fog in the cloud at the far end of the channel. Here the gas temperature was close to ambient. If only humidity from the entrained air were present, the cloud would have been invisible as the temperature was approximately 15°C above the dew point for the mixture.

The considerations of the heat budget and this particular observation leads to the conclusion that the transfer of water to the cloud represents a major heat source.

4.2 Leading edge velocity.

The leading edge position is tracked by nine thermocouples located at different positions downstream in the channel. From the time registrations the mean leading edge velocity in each space interval between the thermocouples is determined. The results are plotted in figure 6. The figure indicates an oscillation of the leading edge velocity both for semicontinuous and instantaneous releases with an amplitude of about 20% of the mean value. This phenomenon was present in all 12 tests, but the frequency of the oscillations was slightly different for the different release types. Similar oscillations have also been observed in the earlier experiments (ref.2).

When the measured results are compared with the box model calculations, good correspondence with the mean data is found when using relatively low values for the constant (k) in the equation for the leading edge velocity. The best curve fits correspond to the values k=0.8 and k=1.2 for semicontinuous and initial release respectively. (See fig. 6.) The calculations consider only isothermal dispersion of the cloud. This does not correspond to the experimental release and to make a meaningful comparison with the data a mean cloud temperature has been used in the calculations. This temperature is adjusted for the time averaged external heat supply depicted in figure 5. The mean cloud temperature is about 250°K for the instantaneous releases and 253°K for semicontinuous releases at the times when the leading edge has travelled a distance of 21 m.

The calculations have been divided into two phases: A boiling and spreading phase and a pure spreading phase. Both phases give results which agree well with the experimental data.

4.3 Velocity distribution inside the cloud.

A plot of the velocity vectors measured at the position 14.6m are shown in fig. 7 for instantaneous releases and the corresponding shape of the cloud is also indicated. From previous unpublished flow visualizations it has been found that smoke injected at the bottom of the current head requires bout 13.5 s at this station to complete a full turn around the vortex head and reappear at the front. From this one can estimate the speed of circulation around the head contour to be about 0.7 of the front velocity. The flow angle in the immediate neighbourhood of the front can therefore be as high as 35 degrees, too large to be measured by the present velocity probes. This also explains the scarcity of velocity data in this region. However, in regions of smaller flow angles measurements are available that clearly indicate the upward motion in the forward part of the front. Also the strong downward motion behind the head is clearly indicated in the neck region. No averaging of data was possible since the flow was very unsteady. The scatter in the data is believed to be the result of strong turbulence rather than experimental uncertainty.

The measurements indicate that the common assumption of a "top hat" type velocity distribution is physically correct at least downstream of the current head as very small changes in magnitude was measured across the cloud, (fig.8).

An inspection of unprocessed data from the lowest velocity probes reveals that the flow angle continuously shifts from above the upper measurement limit through zero flow angle to below the lower limit and back again. This is compatible with the turbulent nature of the flow. However, at later times (at a considerable distance downstream of the leading edge) quite slow transitions from one limit to another could be observed, and after each shift the flow would retain a constant flow angle for many seconds. It is believed that these occurences are caused by the crossing of streamwise vortices most likely generated as convection rolls.

4.4 Concentration and temperature measurements

The five concentration meters give a cross representation of the concentration distribution within the cloud at each point in time.

Additional information concerning the concentration distribution is also derived from the thermocouples located in the vicinity of each concentration meter. Figure 4 depicts some concentration profiles. The feature of special interest in these profiles is that the concentration is roughly uniform all over the cloud in the case of the instantaneous release. Another interesting detail is that the gas concentrations seem to reach the same value at the far end of the channel both for instantaneous and semicontinuous releases. Finally as described above, there are no significant differences in the concentration profiles between the semicontinuous releases with and without an insulated test surface. Also the temperature profiles show a striking uniformity over the cloud for instantaneous releases.

When the gas had been in contact with the water surface for some time thermal convection cells which broke through the gas layer were observed.

4.5 Entrainment parameters

The box model solution presented herein can be used to obtain a best curve fit to the experimental data by systematically changing the entrainment parameters for top and front entrainment. A sensitivity test for each of the entrainment parameters is given in figure 9 to indicate the relative importance of each in relation to the gas concentration. The top entrainment is clearly the dominating parameter and the figure indicates that the front entrainment influences the concentration only slightly. Figure 10 shows the calculated "best fit" concentrations compared to the measured data. The corresponding entrainment parameters are $\alpha_1 = 0.023$ and $\alpha_2 = 0.05$ for instantaneous release and $\alpha_1 = 0.008$ and $\alpha_2 = 0.08$ for semicontinuous releases.

4.6 Discussion

Several experiments have produced suprisingly high temperatures in gas clouds emanating from cryogenic liquid boiling on water. Also the large amount of mist formed during such boiling processes have been noted without any definite conclusions, (refs. 7 and 8). The observations made in the present study concerning heat transfer from the open water in particular with regard to mist correspond directly to these observations. But the mechanisms by which mist is produced is unknown and represent a field for further research.

The oscillation of the leading edge velocity was in the early phases of these experiments assumed to be related to the boiling process. When the oscillations also appeared for instantaneous gas releases this hypothesis had to be discarded. Some new tentative ideas have been tested, but the problem remains unsolved.

The measured velocity field inside the cloud confirms the visually observed vortex movement of the gravity current head and also supports the use of top hat velocity profiles in theoretical models for the remaining part of the cloud. The finding described in ref. 2 of a velocity maximum overshoot at the low positions in the neck region is also confirmed.

The onset of convective currents in the later stage of the spreading process is probably important only in dispersion situations where the gas comes to rest due to terrain effects, barriers or during extreme calm weather situations.

The values of the top entrainment parameter found in these experiments is relatively high as compared to the range of suitable values recommended by Fay (ref.4) i.e. $10^{-2} - 10^{-4}$. The overall Richardson number (Ri) of the present clouds is in the range of 1.5 - 5 during the spreading process. When compared to the experimental data described by Turner (ref.9) for this range of Ri, the entrainment parameters found seem on the other hand to be somewhat low. The entrainment parameters found are therefore not in conflict with known experimental information.

The difference in the top entrainment coefficients for the instantaneously and semicontinuously released clouds is in qualitative agreement with the information given by Turner (ref.9). In the early release phase Ri is approximately three times larger for the semicontinuously released clouds than for the instantaneously released clouds. Referring to Turner, the entrainment velocity should therefore be higher in the case of the instantaneously released gas, which is also found. The difference observed between the values of the constant (k) used in the caluculation of the leading-edge velocity for the instantaneous and semicontinuous releases has yet to be explained.

REFERENCES

1 P.Å.Krogstad and K.Emblem:"Division of Heavy Gas in Air, Preliminary experimental study. "SINTEF report no. STF A80005, Trondheim 1980

2 K.Emblem and T.K.Fanneløp:"Entrainment Mechanisms of Air in Heavy Gas Clouds", Sec. Symp. on Heavy Gases and Risk Asessment, May 1982, Battelle Inst., e.v. Frankfurt

3 K.Emblem and T.K.Fanneløp:"Gravitational Spreading of Heavy Gas Clouds, Part III." Lecture Series on Heavy Gas Dispersion, von Karman Inst. for Fluid Dyn., March 1982

4 J.A.Fay:"Gravitational Spread and Dilution of Heavy Vapor Clouds." Sec. Int. Symp. on Strat. Flows, Norwegian Institute of Technology, June 1980

5 J.A.Fay:"Unusual fire hazards for LNG tanker spills." Comb. Sci. and Tech., 7:47-49, 1973

6 A.P. van Ulden:"Simple estimates for vertical diffusion from sources near the ground." Atm. Env. 12: 2125-2129, 1974

7 L.C.Haselman:"Effect of humidity on the energy budget of a LNG vapor cloud." Liquefied Gaseous Fuel Safety and Environmental Control Assessment Program, DOE/EV-0085, Vol 2, U.S. Dept of Energy, Washington D.C. 1980

8 R.P.Koopman et al.:"Data and calculations of dispersion of 5 m^3 LNG spill tests." UCRL-52876, Lawrence Livermore Lab., Livermore, USA, 1979

9 J.S. Turner:"Buoyancy Effects in Fluids", Cambridge University Press, Cambridge, 1973. ISBN 0 521 08623 x

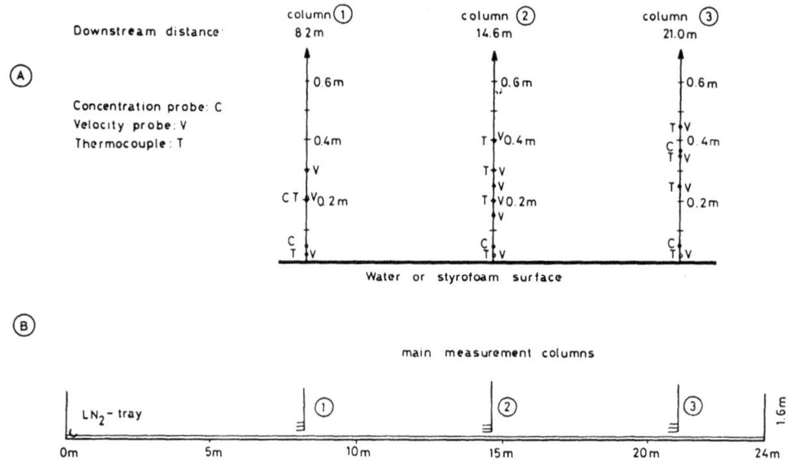

Fig.1 a) Positions of the different probes on the columns

b) Experimental duct with the main instrumentation columns for measurements of physical data

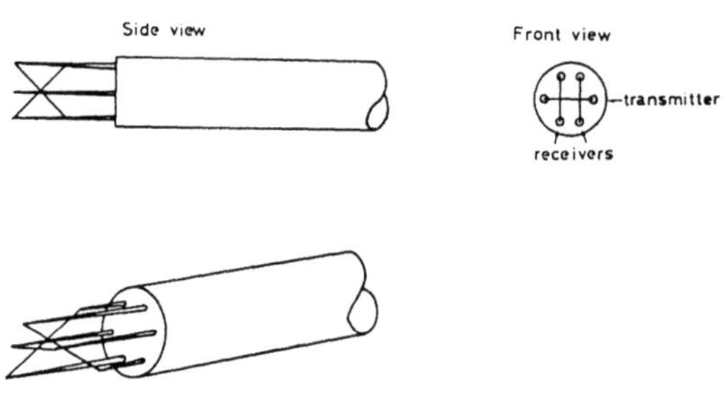

Fig.2 Hot-wire probe for velocity measurements

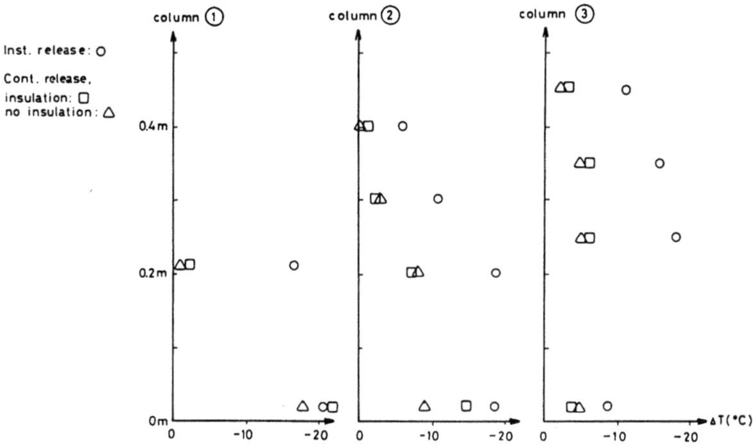

Fig. 3 Simultaneous data at different downstream positions for the difference between the temperature inside the cloud and the ambient when X_{LE} = 22 m.

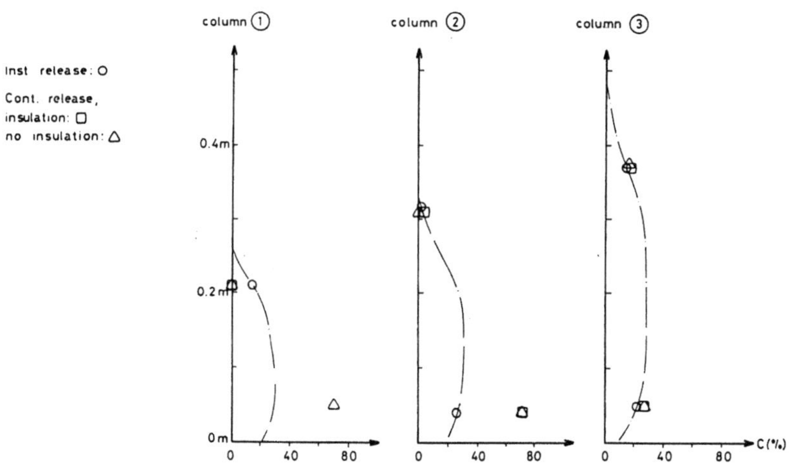

Fig. 4 Simultaneous concentration data at different downstream positions when X_{LE} = 22 m. Complete concentration profiles based on temperature recordings indicated for instantaneous release.

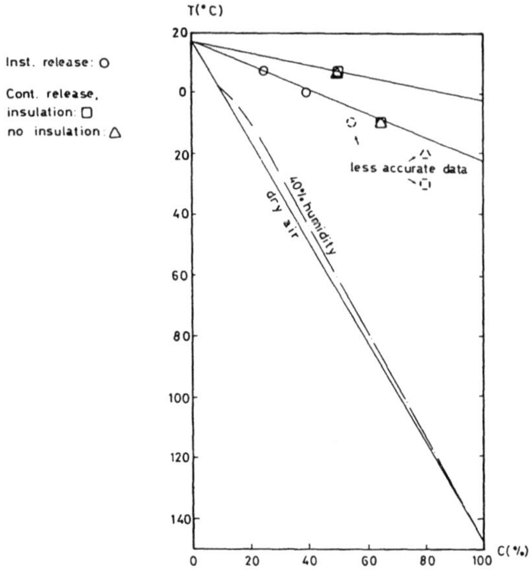

Fig.5 Data for mean concentration versus cloud temperature and curves for theoretical estimates of C(T) for nitrogen adiabatically mixed with dry air and with air of 40% rel. humidity.

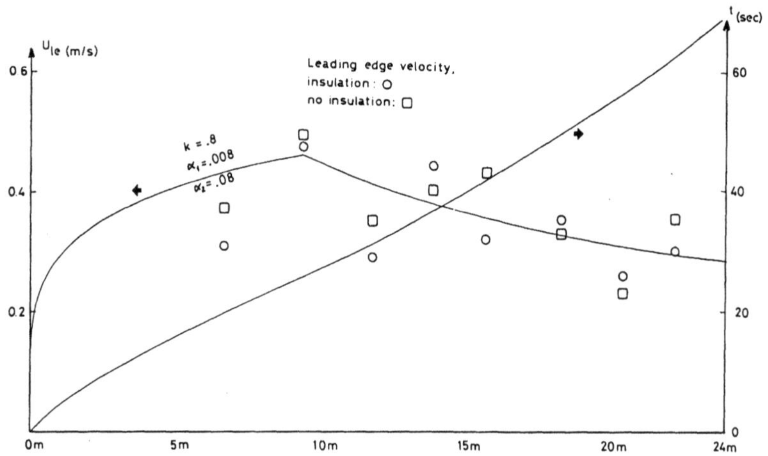

Fig. 6 a) Semicontinuous release. Data for leading edge velocity at different downstream positions of the leading edge and theoretical box-model calculations of this velocity. Also the relation $t(X_{LE})$ from the analysis is shown.

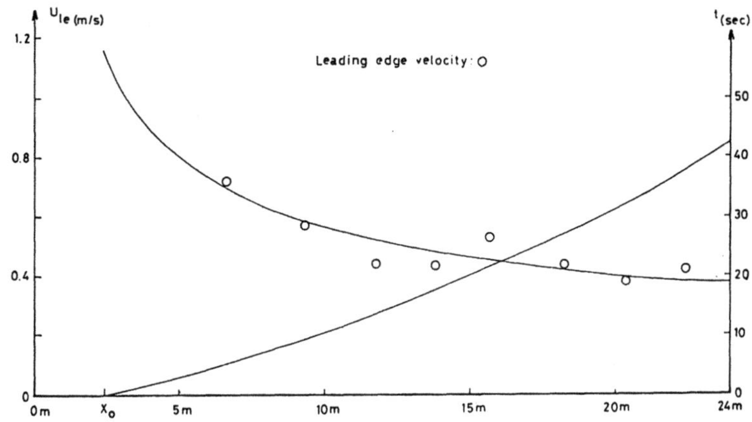

Fig. 6 b) Instantaneous release. For legend, see previous page.
$k=1.2$, $\alpha_1=0.023$, $\alpha_2=0.05$

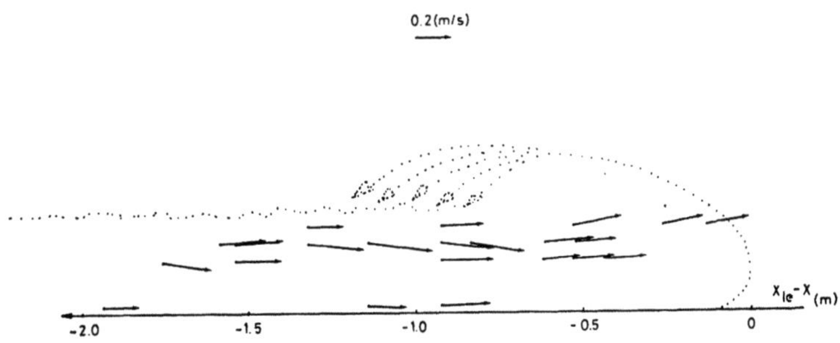

Fig. 7 Velocity vectors inside the cloud measured at column 2

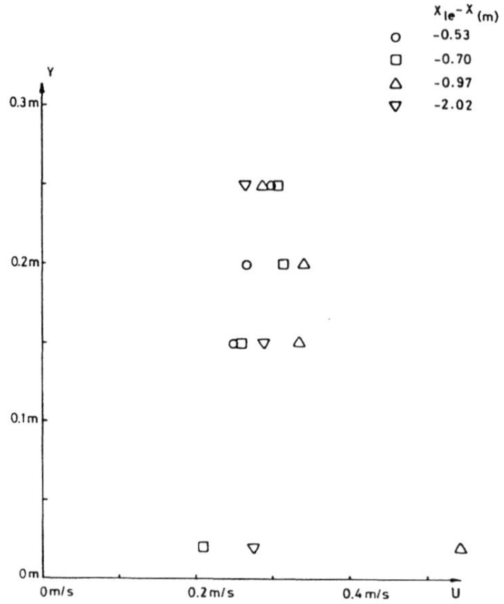

Fig. 8 Data for local velocity at different distances from the leading edge, measured at column 2

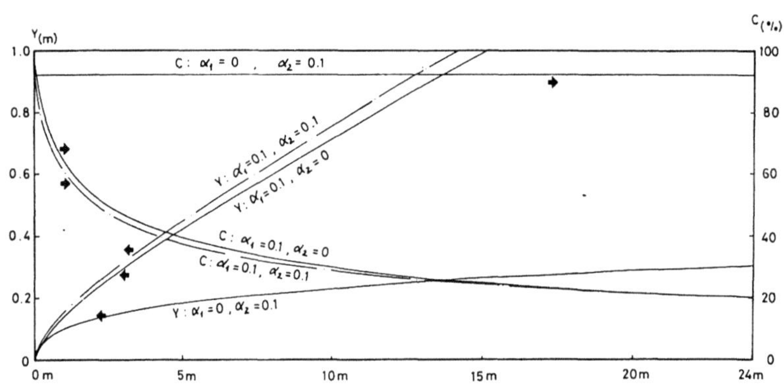

Fig. 9 Concentration and cloud height estimated from box model theory for different, extreme values of the entrainment parameters.

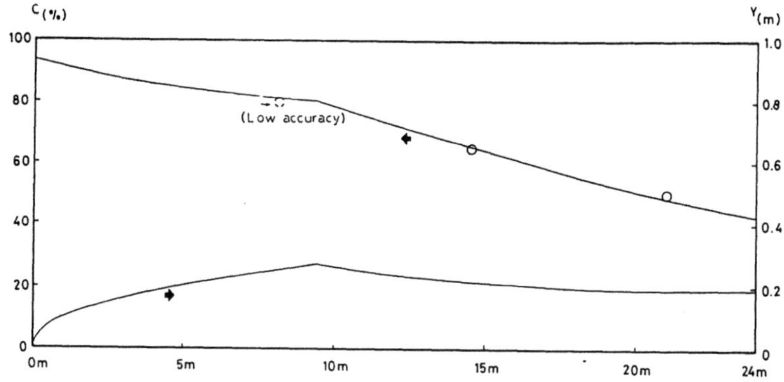

Fig. 10 a) Semicontinuous release. Experimental data and calculated curves for concentration versus distance.
$k=0.8$, $\alpha_1=0.008$, $\alpha_2=0.08$

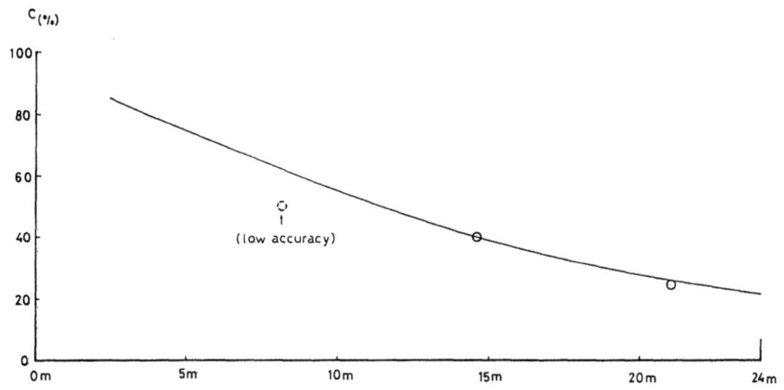

Fig. 10 b) Instantaneous release. Experimental data and calculations of concentration versus distance.
$k=1.2$, $\alpha_1=0.023$, $\alpha_2=0.05$

Water Channel Tests of Dense Plume Dispersion in a Turbulent Boundary Layer

T. B. MORROW, J. C. BUCKINGHAM, F. T. DODGE

Division of Engineering and Materials Sciences,
Southwest Research Institute, San Antonio, Texas, U.S.A.

Introduction

The behavior of dense gas or vapor plumes released into the atmospheric boundary layer is similar to the behavior of plumes of dense, water soluble chemicals discharged into a navigable river. In particular, the spatial concentration distribution for a dense chemical plume discharged into a slow flowing river shows the same gravity spreading and density stratification effects as are expected for a release of dense gas into the air at low wind speed.

The authors have performed a set of dense plume dispersion tests in a low speed, turbulent water channel to simulate the "near field" mixing and dilution of water soluble chemicals spilled in navigable rivers. The test data was used to validate improved models for the dispersion of soluble chemical spills for the U. S. Coast Guard Hazard Assessment Computer System [1].

This paper compares the predicted and measured concentration distributions obtained for the continuous release of dense fluid from a cylindrical pipe submerged in a turbulent boundary layer.

Experiments

The plume dispersion experiments were performed in a free surface water channel measuring 26m (length) x 1.52m (width) x 61 cm (depth) located in the Ocean Engineering Department at Texas A&M University. Water depth was maintained constant at 24 cm while the free surface water velocity was varied from 3.5 cm/s to 14.1 cm/s. Inlet baffles, weirs and screens were used to give a uniform free surface velocity profile across the width of the channel. Vertical velocity profiles, measured by hot-film anemometer, indicated a turbulent boundary layer in the channel with a ratio of shear velocity to free surface velocity of $u_*/U = 0.05$. Figure 1 shows a typical velocity profile plotted in semi-log coordinates for a free stream velocity of 14.1 cm/s.

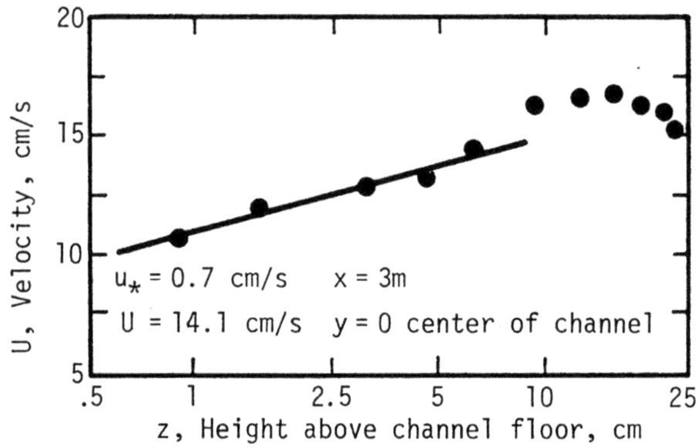

Fig. 1. Vertical Velocity Profile

Chemicals with specific gravities of $\bar{\rho} = 0.79$, 1.0, 1.05, and 1.40 were used in these experiments. A fluorescent tracer dye, Rhodamine WT, was mixed with the chemicals before release. The chemical stream was discharged along the downstream direction at a constant flowrate through a 1 cm diameter cylindrical tube located at channel mid-depth. Time average spatial concentration distributions were measured in the vertical and cross-stream directions at two locations, 1.2m and 3.0m downstream from the discharge plane. Samples of fluid were withdrawn from the plume for a period of 3 minutes through a rake of 0.3 cm sampling tubes. The dye concentration for each plume sample was measured by a spectro-fluorometer calibrated against standards of known dilution ratio. Tracer concentrations as low as 1 part per 2000 could be measured reliably by this technique.

Plume Dispersion Modeling

A numerical (computer) model for predicting the dispersion of buoyant, water-soluble chemical plumes was developed. This model is based upon Ooms' model [2] for predicting the trajectory and dispersion of buoyant plumes above the ground, and Colenbrander's model [3] for predicting the gravity spreading and dispersion of dense plumes over a level surface. Ooms' recommended values for α_1 and α_2, the entrainment coefficients due to shear and buoyancy, were used without modification. However, the model for entrainment due to ambient turbulence was modified. Ooms represents the rate of mass entrainment into the plume by turbulence as $\alpha_3 u'$. In this study, u' was equated to u_*, the shear velocity, and a value of α_3, the turbulence entrainment parameter, equal to approximately 0.5 was estimated from fitting the plume model predictions for concentration profile to the experimental data for a non-buoyant plume.

For plumes with moderate or strong buoyancy significant gravity spreading effects were observed when the plume reached the channel floor. Therefore, when Ooms' model predicted that the plume centerline reached the floor, a switch was made to Colenbrander's model. Colenbrander's recommendations for model parameters were followed with two exceptions. A value of 0.4 was used for the von Karman coefficient instead of 0.35. Also, the empirical function, $\phi(Ri_*)$ that represents the influence of plume density stratification on entrainment was modified from Colenbrander's recommended form

$$\phi(Ri_*) = 0.74 + 0.25\ Ri_*^{0.7} + 1.2 \times 10^{-7}\ Ri_*^{3} \tag{1}$$

to a similar form

$$\phi(Ri_*) = 0.62 + 1.39\ Ri_*^{0.7} \tag{2}$$

that appeared to improve the agreement between model predictions and concentration profile data.

Listings of the FORTRAN computer programs and details of the model matching conditions are given in [1].

Experimental Results

The plume dispersion tests reported in [1] were designed to simulate at a scale of 1:50 the discharge momentum and buoyancy conditions associated with a chemical cargo spill into a river. Thus, the test conditions were characterized by the values of the densimetric Froude number, Fr, and jet momentum ratio, J, which are defined* as

$$Fr = \rho_a\ U_a^2/(\rho_j - \rho_a)\ gD \tag{3}$$

*Subscripts j and a denote the jet discharge and ambient fluid flow conditions, respectively. D is the jet diameter.

$$J = \rho_j U_j^2 / \rho_a U_a^2 \quad (4)$$

For the range of conditions studied, the Froude number was the most important variable influencing plume behavior. The figures that follow compare the behavior of a non-buoyant plume (J = 16, Fr = ∞) with the behavior of plumes of moderate buoyancy (J = 17, Fr = 4.1) and strong buoyancy (J = 23, Fr = 0.51). The channel free stream velocity was 14.1 cm/s and the ambient turbulence level was approximately 5% for this set of experiments.

Figures 2 and 3 are flow visualization photographs of the non-buoyant plume (Fr = ∞). This plume spread symmetrically about its axis.

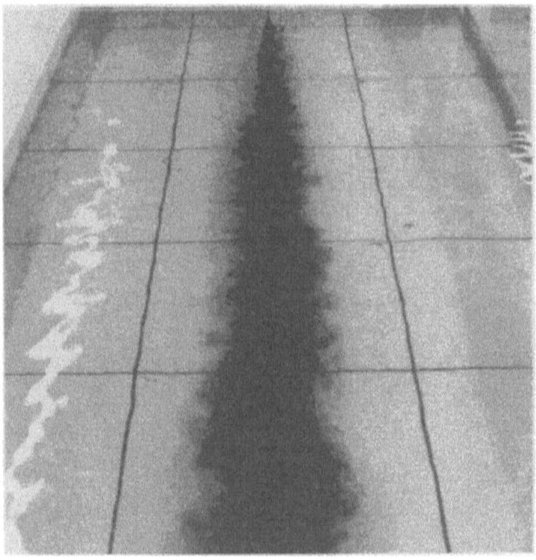

Fig. 2. Neutrally Buoyant Plume ($\bar{\rho}$ = 1.0, Fr = ∞)

Fig. 3. Neutrally Buoyant Plume, Side View ($\bar{\rho}$ = 1.0, Fr = ∞)

Figures 4 and 5 show that the plume concentration distribution was predicted relatively well by Ooms' model with a value of α_3 = 0.5.

Figures 6 and 7 are photographs of the moderately buoyant plume (Fr = 4.1). The plume dropped to the channel floor at a distance of 46 cm from the discharge plane. It did not form a pool on contact with the floor, but spread laterally as it continued to travel downstream.

Figures 8 and 9 compare the predicted and measured cross-stream concentration profiles for this plume at distances of 1.2m and 3.0m downstream of the discharge plane. Note that the plume was slightly bifurcared by contact with the channel floor. Despite the bifurcation, the overall agreement is good at both downstream locations.

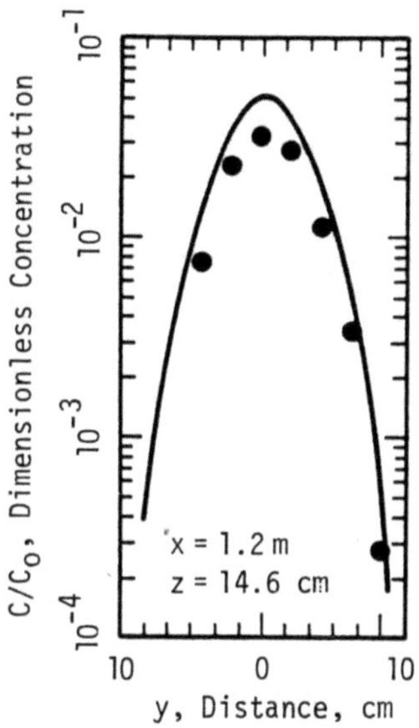

Fig. 4. Horizontal Concentration Profile for Neutrally Buoyant Plume (Fr = ∞)

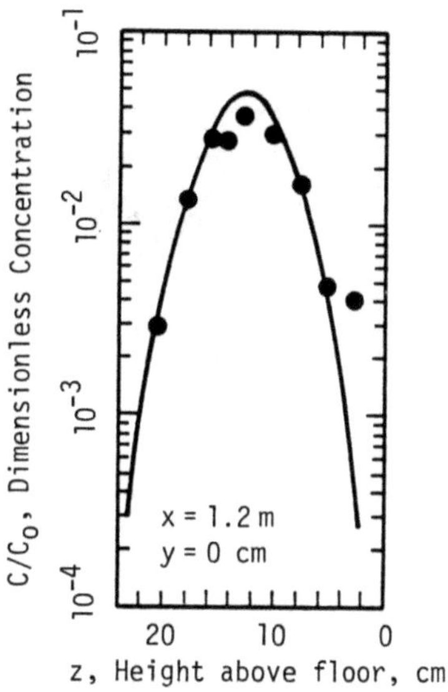

Fig. 5. Vertical Concentration Profile for Neutrally Buoyant Plume (Fr = ∞)

Fig. 6. Moderately Buoyant Plume, Top View ($\overline{\rho}$ = 1.05, Fr = 4.1)

Fig. 7. Moderately Buoyant Plume, Side View ($\overline{\rho}$ = 1.05, Fr = 4.1)

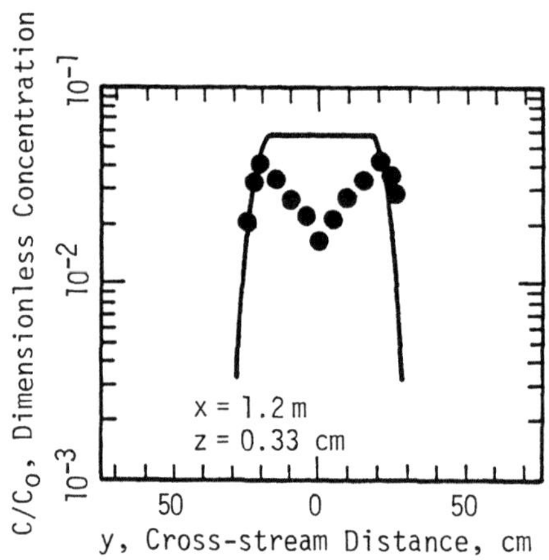

Fig. 8. Horizontal Concentration Profile for Plume with $\bar{\rho}$ = 1.05 (Fr = 4.1)

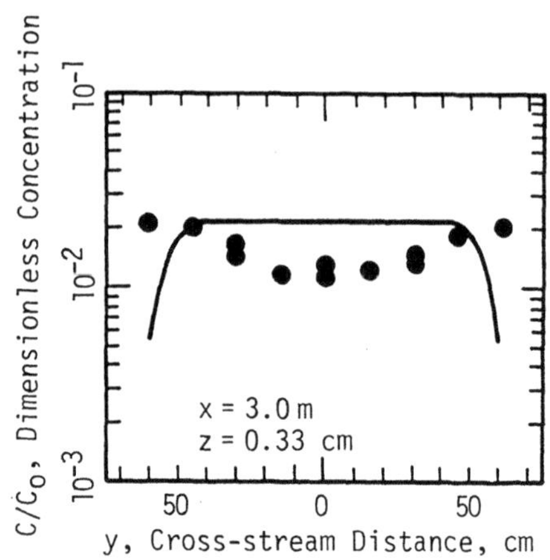

Fig. 9. Horizontal Concentration Profile for Plume with $\bar{\rho}$ = 1.05 (Fr = 4.1)

Figures 10 and 11 compare the predicted and measured values of plume centerline concentration and vertical dispersion coefficient. The measured values were determined from a fit of vertical concentration profile data to the assumed model equation

$$C(z) = C_A \exp\left(-(z/S_z)^{1+\alpha}\right) \quad (5)$$

A value of $\alpha = 0.14$ was used for the velocity power law coefficient. The behavior of S_z with increasing distance was sensitive to the plume entrainment function $\phi(Ri_*)$. Figure 11 shows that the agreement between model predictions and measured values of S_z was fairly good when Equation (2) was used for $\phi(Ri_*)$.

Figures 12 and 13 are photographs of the strongly buoyant plume (Fr = 0.51). The stream of chemical fell quickly to the channel floor within 15 cm of the discharge plane. The chemical stream formed a pool on the channel floor from which fluid was entrained to form the negatively buoyant plume shown in the figure. The plume spread laterally and reached the channel walls at a distance of only 40 cm downstream of the discharge plane. A reflection of the plume boundary from the channel wall propagated back across the plume, and reached the plume centerline at a distance of 2.4m from the discharge plane.

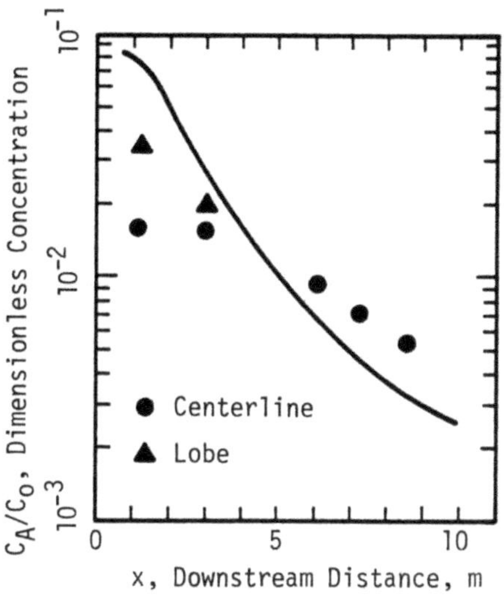

Fig. 10. Plume Centerline Concentration for Plume with $\overline{\rho}$ = 1.05 (Fr = 4.1)

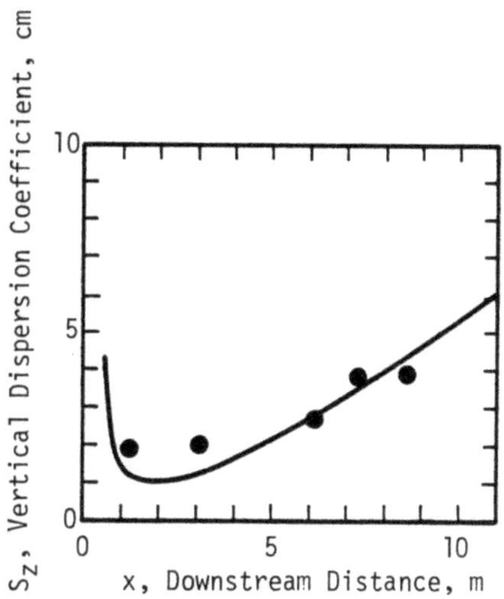

Fig. 11. Vertical Dispersion Coefficient for Plume with $\overline{\rho}$ = 1.05 (Fr = 4.1)

Fig. 12. Strongly Buoyant Plume Top View ($\overline{\rho}$ = 1.4, Fr = 0.41)

Fig. 13. Strongly Buoyant Plume Side View ($\overline{\rho}$ = 1.4, Fr = 0.41)

Figures 14 and 15 compare the measured and predicted cross-stream concentration profiles for this plume at distances of 1.2m and 3.0m downstream of the discharge plane. The effect of plume reflection from the channel walls is quite apparent in Figure 14.

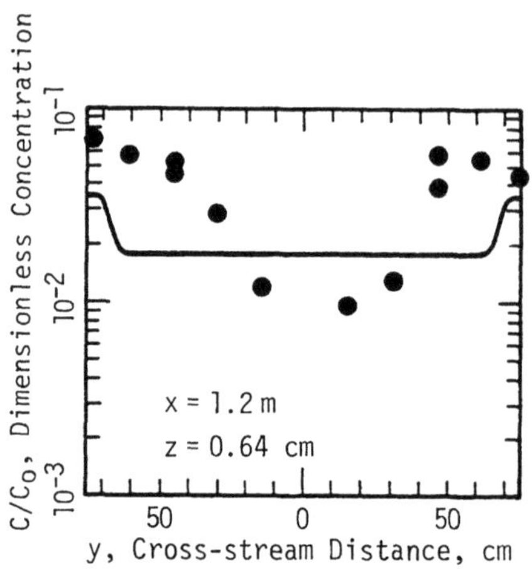

Fig. 14. Horizontal Concentration Profile for Plume with $\overline{\rho}$ = 1.4 (Fr = 0.51)

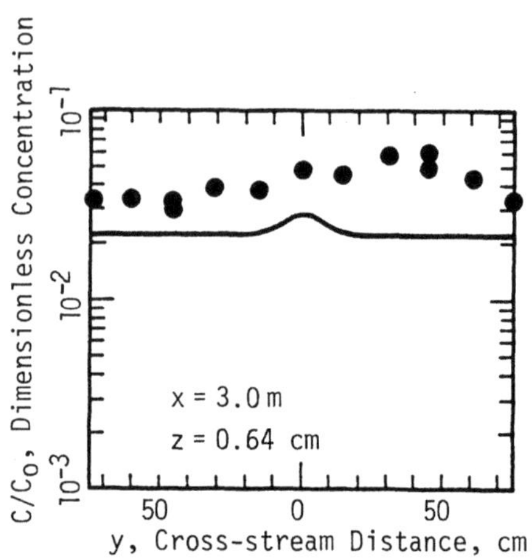

Fig. 15. Horizontal Concentration Profile for Plume with $\overline{\rho}$ = 1.4 (Fr = 0.51)

Figures 16 and 17 compare the predicted and measured values of the plume centerline concentration and the vertical dispersion coefficient as a function of downstream distance. Equation (2) was used to simulate the effect of density stratification on plume entrainment. In general, the predicted and measured values of concentration differed by less than a factor of 3. For the strongly buoyant plume, the model appeared to underpredict the plume centerline concentration and overpredict the amount of vertical dispersion. This could have resulted from overpredicting the plume entrainment in the free jet region between the discharge plane and the channel floor.

Conclusions

Analytical models based on Ooms' [2] and Colenbrander's [3] models for dense gas dispersion were developed and validated during a set of water channel tests of dense plume dispersion in a turbulent boundary layer. In general, the agreement between model predictions and experimental data was good. The main differences are the result of assumptions inherent in the plume models and experimental uncertainty in the measured data.

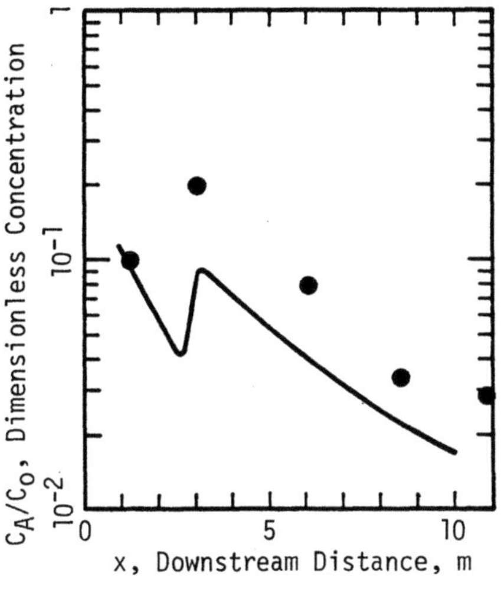

Fig. 16. Plume Centerline Concentration for plume with $\overline{\rho}$ = 1.4 (Fr = 0.51)

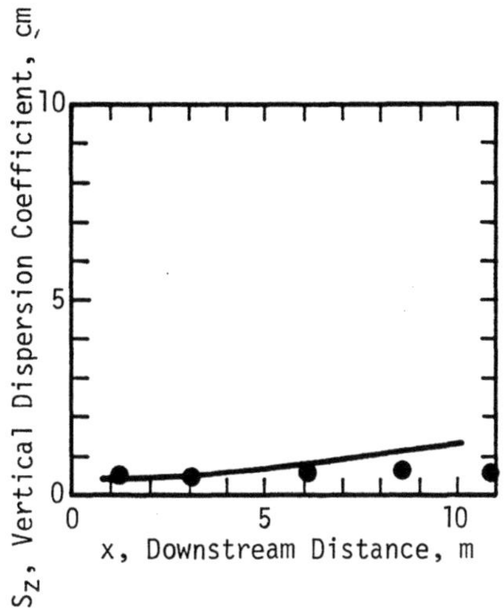

Fig. 17. Vertical Dispersion Coefficient for Plume with $\overline{\rho}$ = 1.4 (Fr = 0.51)

Acknowledgements

The authors are grateful to the U. S. Coast Guard for permission to publish this paper which is based upon work performed under Contract DOT-CG-920622-A. Thanks are due to Lts. M. F. Flessner and G. R. Colonna of the U.S.C.G. and Mr. H. Haufler of Southwest Research Institute.

References

1. Dodge, F. T., Buckingham, J. C., Morrow, T. B., "Analytical and Experimental Study to Improve Computer Models for Mixing and Dilution of Soluble Hazardous Chemicals," Final Report on Contract DOT-CG-920622-A, Southwest Research Institute, San Antonio, Texas, August 1982, NTIS AD A125649 and AD A126005.

2. Ooms, G., "A New Method for the Calculation of the Plume Path of Gases Emitted by a Stack," Atmospheric Environment, Vol. 6, pp. 899-909, 1972.

3. Colenbrander, G. W., "A Mathematical Model for the Transient Behavior of Dense Vapor Clouds," 3rd International Symposium on Loss Prevention and Safety Promotion in the Process Industries, Basle, 1980.

The Dispersion of Slightly Dense Contaminants

D.D. Stretch , R.E. Britter , J.C.R. Hunt

Cambridge University Engineering Department
Trumpington Street
Cambridge CB2 1PZ
ENGLAND

1. Introduction

Existing theoretical models of dense gas dispersion vary greatly in degree of complexity from simple layer averaged (integral equation) approaches to the use of complex turbulence models, the latter usually employing some form of eddy diffusivity closure approximation. However there remain several aspects of the problem which are poorly understood, and may therefore not be adequately modelled. For example the question of how "entrainment" (however it may be defined) or eddy diffusivities can be related to stability. Evidently a careful look at the dynamics of dense contaminant dispersion is called for. Our research, which we review here, is an attempt to study in some detail one aspect of the dispersion dynamics in particular, namely the effects of stable stratification, which may be set up in dispersing plumes or clouds, on the turbulence. We shall describe three approaches to the problem. Firstly, the use of Rapid Distortion Theory to investigate structural changes to homogeneous turbulence with varying degrees of stable stratification in the presence of a mean velocity gradient. Secondly, a Lagrangian dynamical model of fluid element motions (as previously employed in studies of mixing in homogeneous stratified turbulence) is introduced in the context of the present problem. Finally, an experimental program is described.

2. THEORETICAL ASPECTS

2.1 Aspects of the overall dispersion problem

We view the dispersion of dense plumes or clouds in terms of the following processes:

a. Mean flows driven by the negative buoyancy of the plumes or clouds

b. The dispersion of (density marked) fluid elements by virtue of continuous random turbulent motions.

c. Advection by the ambient mean flow.

d. There are stable density gradients within dispersing plumes/clouds, which suppress (in particular) the vertical motions of fluid elements and hence inhibit the dispersion process. This occurs on a time scale N^{-1}. On a longer time scale the reduction in the shear stress $-\overline{uw}$ also affects the mean velocity gradients in the clouds. These processes are indicated schematically in fig 2.1.1. There is some similarity with changes in the atmospheric boundary layer during an eclipse of the sun (Narasimha et al, 1981).

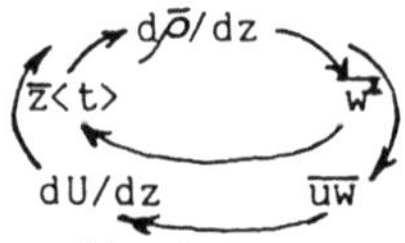

fig 2.1.1.

e. A further point, related to (d) above, is that with stable density gradients present, the generation of an internal wave field (frequencies $\leq$ N) is possible, which may in turn affect the dispersion process. The turbulent kinetic energy of fluid elements is transformed into more ordered, essentially non-diffusive wave motions.(Pearson,Puttock & Hunt,1983)

We have ignored accelerations and mixing near the source and we assume the validity of the Boussinesq approximation. That is, requiring $\Delta\rho/\rho_0 \ll 1$ always-the case of "slightly dense" contaminants.

2.2 Stages in plume and cloud development

A simple classification of different domains of plume or cloud development (in terms of dominant dynamical mechanisms) may be done using a bulk Richardson number parameter

$$R_i = g'h/u_*^2 \qquad g' = g\,\Delta\rho/\rho_0, \quad u_* = \text{friction velocity}$$

where

h is a vertical scale of the plume or cloud

u_* is taken as the relevant ambient turbulence velocity scale Thus when

a. Ri >> 1 negative buoyancy driven flows dominate the dispersion process over ambient turbulence effects- a gravity current phase. Usually there is a fairly sharp interface (with associated strong stable stratification) between the dense clouds and the ambient fluid. Furthermore, the dense cloud may not be fully turbulent and a major proportion of its turbulent kinetic energy may be derived from its negative buoyancy, converted from potential energy.

b. Ri $\sim$ 1 We expect a phase of stratification-inhibited turbulent dispersion. That is the turbulence in the cloud may be significantly affected by stable density gradients (although not suppressed altogether) which may in turn reduce the efficiency of the mixing process. Our research is directed at this stage.

c. Ri << 1 We expect a "passive" dispersion phase.

2.3 Theory of uniform shear, rapid distortion with stratification

In this section we discuss the use of rapid distortion theory (Townsend,1976 ; Hunt,1978) to investigate some, mainly qualitative, effects of stratification on the structure of homogeneous turbulent shear flows. The theory is expected to be a valid description of the distortion of the larger scales of high Reynolds number, low intensity turbulence, and for distortion times less than a typical eddy decay time. That is, we require

$$v/\mathcal{U} \ll 1 \qquad v/\alpha L \ll 1$$
$$\mathcal{U}L/\nu \gg 1 \qquad L^2\alpha/\nu \gg 1$$
$$\rho'/\rho_0 \ll 1 \qquad \rho'/L\partial\bar{\rho}/\partial z \ll 1$$
$$t_d < T_L \qquad (\text{lagrangian time scale} \sim \alpha^{-1})$$

where v is a turbulence velocity scale, ρ' is a density fluctuation scale, $\mathcal{U}$ a mean velocity scale (typically a difference between two values) and L is a turbulence length scale.

The relevant set of equations after linearization and making the Boussinesq approximation are (in usual notation)

$$\frac{\partial u_i'}{\partial t} + \mathcal{U}_\ell \frac{\partial u_i'}{\partial x_\ell} + u_\ell' \frac{\partial \mathcal{U}_i}{\partial x_\ell} = \frac{1}{\rho_0}\frac{\partial p}{\partial x_i} + g_i \frac{\rho'}{\rho_0} + \nu \frac{\partial^2 u_i'}{\partial x_i^2}$$
$$\frac{\partial u_i'}{\partial x_i} = 0$$
$$\frac{\partial \rho'}{\partial t} + \mathcal{U}_\ell \frac{\partial \rho'}{\partial x_\ell} + u_\ell' \frac{\partial \bar{\rho}}{\partial x_\ell} = 0$$

Following Townsend (1976), we employ a Fourier series decomposition of locally homogeneous fluctuating velocity and density fields. That is, define

$$u_i'(\underline{x},t) = \sum_{\underline{k}} a_i(\underline{k},t)\, e^{i\underline{k}(t).\underline{x}}$$
$$\frac{g}{\rho_0}\rho'(\underline{x},t) = \sum_{\underline{k}} b(\underline{k},t)\, e^{i\underline{k}(t).\underline{x}}$$

We consider a uniform plane shear and linear density gradient–

$$\mathcal{U}_1(x_3) = \alpha . x_3 \qquad -\frac{g}{\rho_0}\frac{\partial\bar{\rho}}{\partial x_3} = N^2 \qquad (N = \text{Brunt-Väisälä frequency})$$

The equations for the development of individual Fourier components, after eliminating the pressure terms and discarding the viscous terms are

$$\frac{da_1}{dt} = \alpha . a_3 (2k_1^2/k^2 - 1) + k_1k_3/k^2 . b$$
$$\frac{da_2}{dt} = \alpha . a_3 (2k_1k_2/k^2) + k_2k_3/k^2 . b$$
$$\frac{da_3}{dt} = \alpha . a_3 (2k_1k_3/k^2) - (k_1^2+k_2^2)/k^2 . b$$
$$\frac{db}{dt} = N^2 . a_3$$

and with the wave vectors varying as

$$\frac{\partial k_i}{\partial t} = -\frac{\partial \mathcal{U}_\ell}{\partial x_i} . k_\ell$$

The solutions of the rapid strain equations may be expressed in terms of the initial conditions as

$$\underline{a}(\underline{k},t) = \underline{A} . \underline{a}(\underline{k}_0,0)$$
$$b(\underline{k},t) = \frac{N^2}{\alpha} . a_3(\underline{k}_0,0) . B$$

where B and the matrix A are functions of the total strain $\beta = \alpha t$, and depend on the wave vector directions but not their magnitudes. The initial conditions were taken to be unstratified isotropic turbulence. That is with

$$\Phi_{ij}(k_0,0) = (\delta_{ij} - k_ik_j/k_0^2)\,\psi(k_0)$$

and with $\Psi(k_0)$ chosen as

$$\Psi(k_0) = (32\pi^3)^{-1/2} . \overline{u_{10}^2} . k_0^2 . L_0^5 . e^{-\frac{1}{2}k_0^2 L_0^2}$$

(L_0 is an initial integral scale) which corresponds to initial correlation functions of the exponential form.

Some preliminary results of the computations are shown in fig 2.2.1 The effects of stable stratification may be summarized as-

(a) Normal Stresses-The stratification initially damps the vertical fluctuations more than the other components ($\overline{w^2}$ increases again for Nt $\gtrsim$ 1.5).

(b) Reynolds Stresses-They are also damped by the stratification, due mainly to the strong effect on the vertical fluctuations.

(c) Density fluctuations and fluxes-Non-dimensionalised on initial variables these are decreased by the stratification. The ratio of the eddy diffusivities decreases with increasing stability.

(d) Spatial Correlation structure of velocity and density fields. The calculations show relatively small changes in this aspect of the turbulence structure, for the range of strains and Ri numbers used to date (i.e with $Nt_d < \alpha t_d$). This suggests that although the kinetic energy in individual eddies may be severely damped, particularly with repect to vertical motions, nevertheless they may retain the same basic structure. This tendency for the turbulence length scales to respond rather slowly to stratification is in contrast to their relative sensitivity to shear distortion.

2.4 A Lagrangian dynamical model of fluid element motions

An attempt has been made to derive a dynamical model of fluid element motions, following previous work on diffusion in homogeneous stratified turbulence (Csanady, 1964 ; Pearson, Puttock and Hunt, 1983). Kinematic random flight modelling (e.g Durbin, 1980) cannot be used in this problem because the length scales and variances are not known a priori. We consider the dynamics of control volumes (fluid elements), typical dimension ℓ of order the turbulence microscales, which move at the local fluid velocity. Only the vertical component will be considererd, for which the momentum equation, after making a Boussinesq approximation, is

$$\frac{dw}{dt}\langle t\rangle = -\frac{1}{\rho_0}\frac{\partial p'}{\partial z}\langle t\rangle - g\frac{\rho'}{\rho_0}\langle t\rangle + \nu.\nabla^2 w\langle t\rangle \qquad -(2.4.1)$$

where

p' is the pressure perturbation from it's hydrostatic value.
ρ' is the density perturbation from the local mean value.
$\langle\cdot\rangle$ denotes Lagrangian co-ordinates (following a fluid element)

The above equation is combined with the kinematical equations

$$dz\langle t\rangle = w\langle t\rangle . dt \quad \text{and} \quad dx\langle t\rangle = U\langle t\rangle . dt \quad \left(\frac{u'}{U} \ll 1\right)$$

in order to obtain the fluid element displacements. The viscous terms in equation 2.4.1 are approximated by

$$\nu\nabla^2 w\langle t\rangle \simeq -k_v . w\langle t\rangle$$

i.e. a drag linear in the velocity, and the pressure term by

$$-\frac{1}{\rho_0}\frac{\partial p'}{\partial z}\langle t\rangle \simeq -k_\omega . w\langle t\rangle + H\langle t\rangle$$

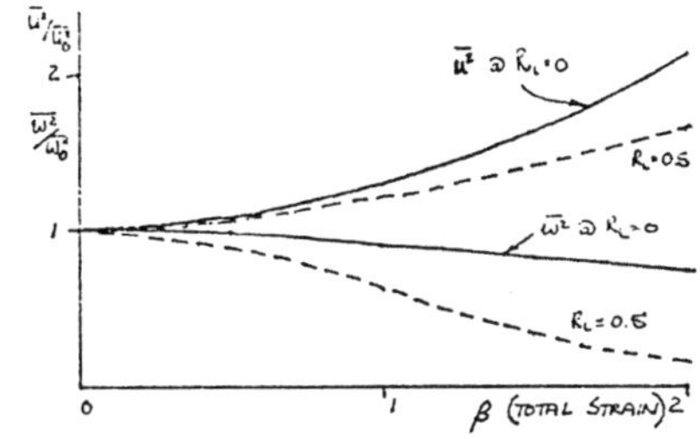

(a) Variances.

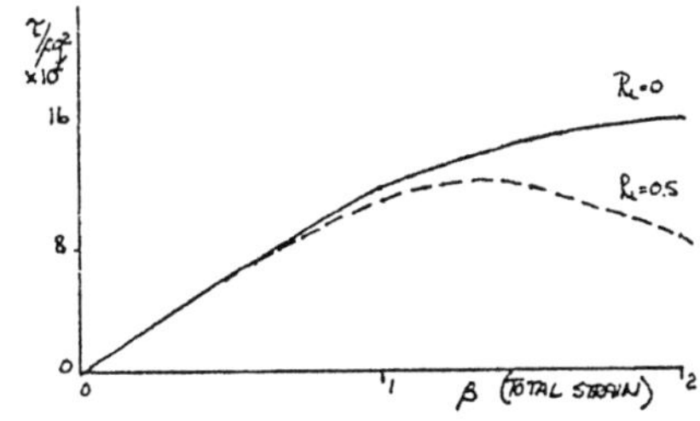

(b) Reynolds stresses.

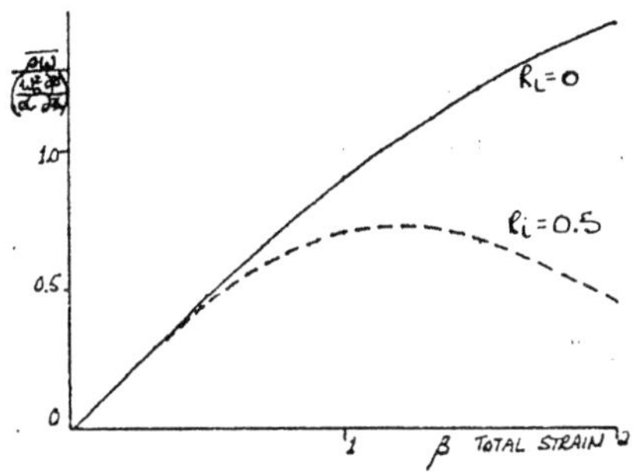

(c) Fluxes.

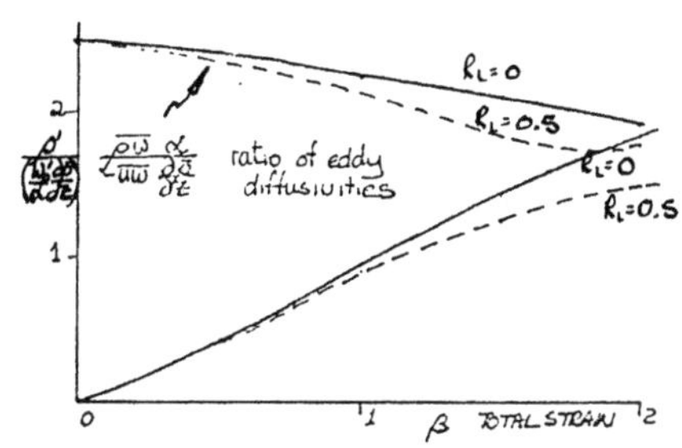

(d) Density fluctuations
(eddy diffusivity ratio)

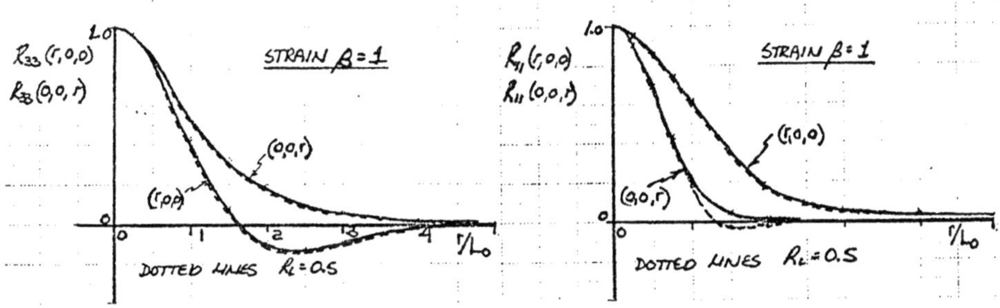

(e) Correlation functions.

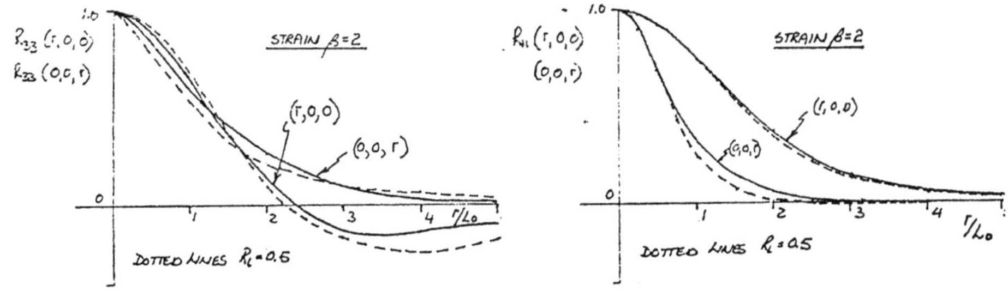

Fig 2.2.1 Results of Rapid Distortion Theory calculations

that is, the sum of a local damping due to internal waves (Pearson, Puttock and Hunt, 1983) which is linear in the velocity, and a random forcing H<t>. The parameters k_v and k_ω are estimated as

$$k_v \simeq 5\nu/\lambda^2 \quad (\lambda = \text{Taylor microscale}) \qquad \text{Krasnoff \& Peskin (1971)}$$

$$k_\omega \simeq 0.8N \quad (N = \text{Brunt-Väisälä freq}) \qquad \text{Pearson, Puttock \& Hunt (1983)}$$

Thus the approximate momentum equation for the fluid elements becomes

$$\frac{dw}{dt}\langle t\rangle = K\langle t\rangle\, w\langle t\rangle - g\frac{\rho'}{\rho_0}\langle t\rangle + H\langle t\rangle \qquad -(2.4.2)$$

with

$$K\langle t\rangle = -(k_v + k_\omega)$$

The stochastic differential equation may be "solved" by Monte Carlo simulation methods, as non-linearity renders it otherwise intractable. Probability distributions of fluid element displacements, and hence the mean concentration field, may be inferred from the simulations.

To estimate the properties of the random force H<t> we adopt the following reasoning. Batchelor and Townsend (1956) and Csanandy (1964) have previously argued that the pressure gradient fluctuations are the dominant contribution to fluid element accelerations. That is

$$\frac{dw}{dt}\langle t\rangle \sim H\langle t\rangle$$

Further, dimensional analysis suggests that the Lagrangian velocity spectrum in an inertial subrange should take the form (e.g. Tennekes & Lumley, 1972)

$$\Phi_w(s) = B.\varepsilon.s^{-2}$$

Hanna (1981) carried out a series of atmospheric measurements to verify this and obtained $B \simeq 0.40$. Hence using the above approximate relationship between w<t> and H<t>, it follows that in an inertial subrange

$$\Phi_H(s) = s^2.\Phi_w(s) = B.\varepsilon$$

i.e. a flat "white noise" shape of spectrum. Based on these arguments, the form of Φ_H chosen for initial experimentation was as indicated in fig 2.4.1.

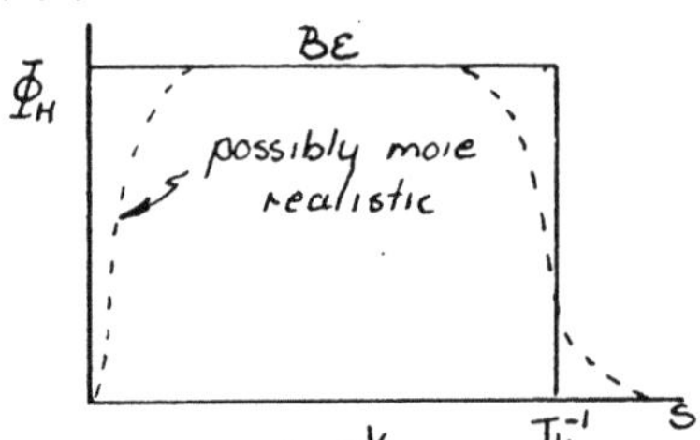

Fig 2.4.1 H-Spectrum

$$T_k = (\varepsilon/\nu)^{-1/2} \quad \text{Kolmogorov time scale}$$

The dissipation was estimated from the neutral case

$$\varepsilon = u_*^3/\kappa z$$

and equating this with the isotropic estimate

$$\varepsilon = 15\nu v^2/\lambda^2 \qquad v \sim u_*$$

provides an estimate of the length scale λ . The ambient mean velocity was taken as logarithmic

$$\mathcal{U}(z) = \frac{u_*}{\kappa} \ln(z/z_0) \qquad \kappa \simeq 0.4$$

No account has yet been made of the effects of stratification on Φ_H or U(z). Finally H<t> was assumed to be Gaussian distributed, with zero mean. Note however that the non-stationarity of H<t> arising from the inhomogeneity of the boundary layer gives rise to a non-zero mean Lagrangian vertical velocity (Monin & Yaglom, 1971 sec10.3)

Example simulations of the dispersion of passive and slightly dense clouds are shown in fig 2.4.2. A characteristic sheared puff shape is evident. The effect of increasing the cloud density is to concentrate material near the ground with reduced $d\overline{z}/dt$ and $d\sigma_z/dt$.

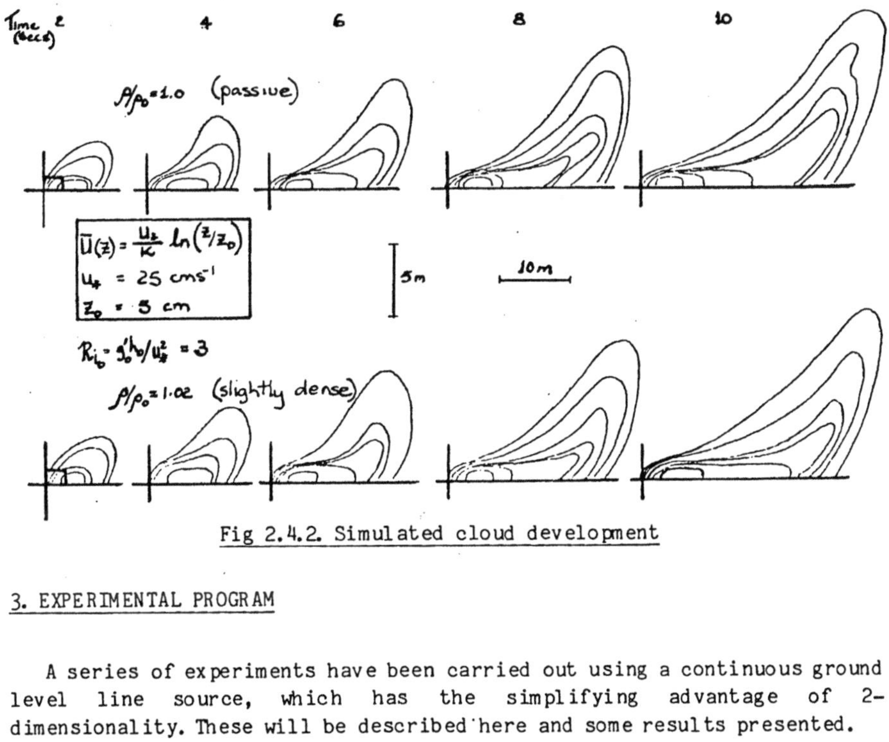

Fig 2.4.2. Simulated cloud development

3. EXPERIMENTAL PROGRAM

A series of experiments have been carried out using a continuous ground level line source, which has the simplifying advantage of 2-dimensionality. These will be described here and some results presented.

3.1 General description

The experiments were carried out in Cambridge using a low speed (0-2.5 m/s) smoke tunnel originally designed for visualization work on turbulent boundary layers. The facility is sketched in fig 3.1.1. The source consisted of a 100 mm wide slot in the floor from which CO_2 was emitted. An oil fog smoke was added to the source flow when required for visualization and a 5 W laser was used to provide sheet lighting.

Concentration measurements were made using a fast response F.I.D. system (Fackrell, 1979). Velocity measurements were made, with and without the dense plumes present, using a pitot tube and a pulsed wire anemometer system (Bradbury & Castro, 1971 ; Castro & Cheun, 1982). The two quantities varied in the experiments were the source flow rate and the free stream velocity and their physical effects may be characterised by the bulk Richardson number

$$Ri_p = q_o g'_o / U^3$$

q_o = source flow/unit width
$g' = g\,\Delta\rho/\rho_o$
$U \doteq 0.8{*}U_\infty$

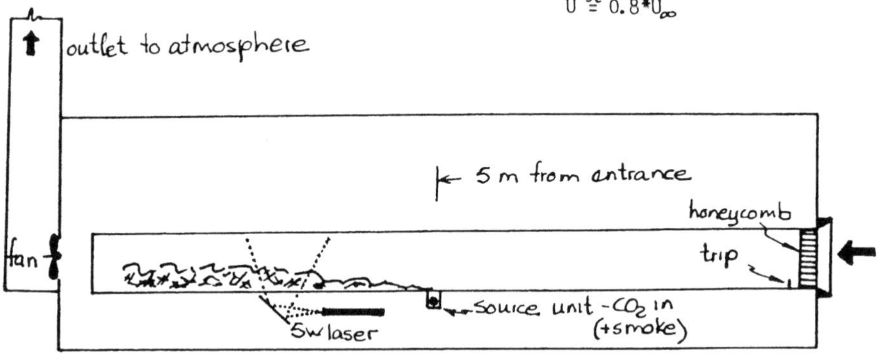

Fig 3.1.1 Schematic diagram of wind-tunnel facility.

3.2 Flow visualization results

A series of photographs are shown in fig 3.2.1 taken with longitudinal/ vertical sheet lighting. All were taken at the same downstream position from the source. The stability of the plumes was varied as indicated by the Ri_p values given in the figure.

The reduction in the vertical scale of the plumes with increasing stability is clearly evident. As Ri_p tends to a value near 0.01 , the plumes laminarize to form a stable layer at the wall. Internal waves may be discerned at the interface, driven by the free-stream turbulence impinging on the dense layer.

3.3 Measurements and analysis

(a) Mean concentrations

Typical mean concentration profiles for 3 downstream positions are shown in fig 3.3.1 , with the vertical position scaled by the plume half-height δ_z . Approximate similarity (in the form of an exponential profile) with respect to downstream development and for varying stability, is apparent from the plots. Notable deviations from similarity occur for the lowest Ri_p values at the position furthest downstream. At this position the plumes (except in the more stable cases) occupied the full extent of the boundary layer. The boundary layer height would then be a significant length scale with regards plume development. The exponential type profiles may be contrasted with the Gaussian profiles obtained e.g. by Fackrell & Robins (1982), for passive cases and measured in atmospheric boundary layer simulations. The difference may be explained by the reduced vertical velocity fluctuations in the near wall region of the smooth walled

$Ri_p = 0.003$

$Ri_p = 0.006$

$Ri_p = 0.009$

Fig 3.2.1 Smoke Visualisation results (1.5m downstream)

turbulent boundary layer of the present experiments, with a corresponding reduction of vertical mixing.

A simple analysis can reveal some of the effects of stability on the rate of plume growth defined by dh/dx, where h is a suitable vertical plume length scale (defined below). Ellison and Turner (1960) have previously argued on dimensional grounds, and tested experimentally, that far downstream from the source

$$\frac{dh}{dx} = f_1\left(\frac{g'h}{U^2}\right)$$

where

$g' = g\,\Delta\rho/\rho_0$

U is a mean velocity scale of the ambient flow.

The depth h and velocity U are defined such that

$$g_0' \cdot q_0 = g' h\, U$$

from whence

$$\frac{g'h}{U^2} = \frac{g'q}{U^3} = \frac{g_0' q_0}{U^3}$$

Integration yields

$$h = x\, f_1(g_0 q_0/U^3) + h_0$$

h_0 is virtual depth at $x=0$ which may depend on g_0', q_0, U and source width.

Writing

$$C_m \cdot U \cdot h = C_0 \cdot q_0$$

where Cm is the (max) ground level concentration, Co is the source concentration, it follows that

$$(C_m/C_0)^{-1} = \frac{Ux}{q_0} \cdot f_1 + f_2$$

so that data of (Cm/Co,Ux/q) may be used to infer dh/dx. Our measurements confirm this linear relationship. The data of Ri_p versus dh/dx are shown in fig 3.3.2. Also shown is Ellison & Turner's (1960) data, although their h was defined somewhat differently. The two sets of data are roughly consistent, and indicate a rapid decrease in dh/dx for $Ri_p >= 0.003$. For Ri_p values smaller than this, dh/dx evidently tends to a constant "passive" value consistent with Lagrangian similarity theory.

(b) Concentration fluctuations

Typical profiles of measured $\overline{c^2}$ are plotted in fig 3.3.3 with the height scaled on plume half-height and using the peak $\overline{c^2}$ as a scale for the abscissae. Similarity with respect to Ri_p number and downstream development is evident. As with the mean profiles, similarity fails far downstream and for the lower Ri numbers.

(c) Velocity measurements

Perhaps some of the most interesting measurements are those concerning the turbulence within the stable plumes. These were made at a fixed position downstream of the source and (to date) for two plume Richardson numbers. The results are shown in fig 3.3.4 . There is significant damping of streamwise and vertical velocity fluctuations and the Reynolds stresses. Approx profiles of gradient Richardson numbers are also shown, calculated from the mean velocity and concentration profiles. Note that

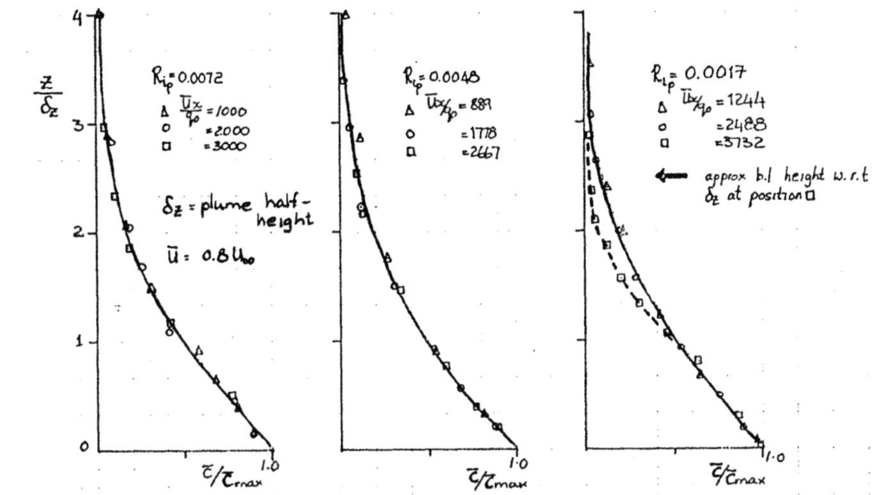

Fig 3.3.1 Measured mean concentration profiles at 3 downstream positions and for 3 bulk Richardson numbers.

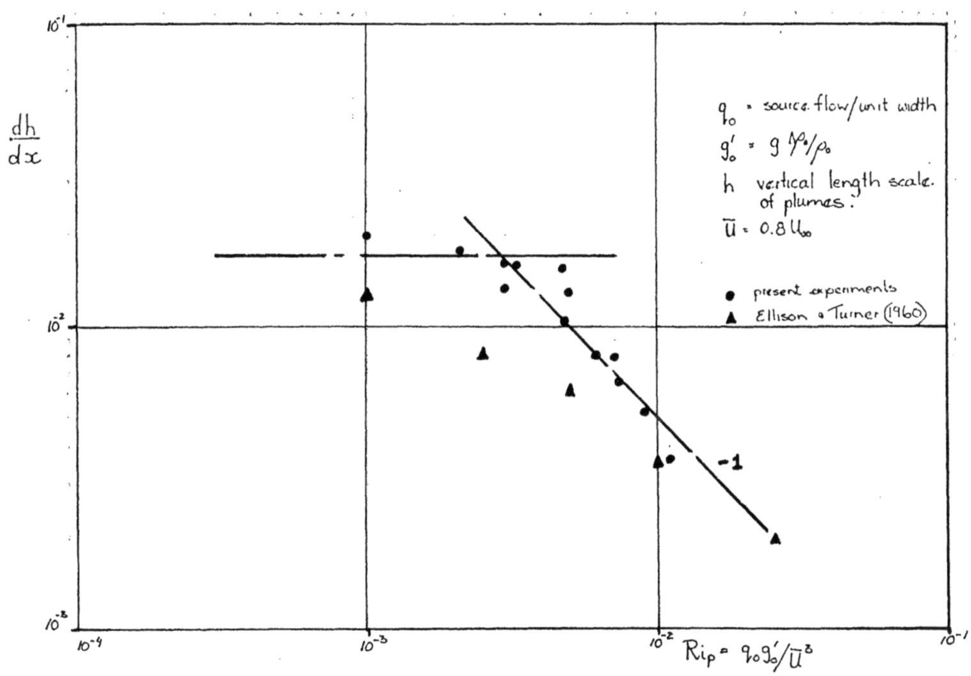

Fig 3.3.2 Plume vertical growth rate as function of bulk Ri number.

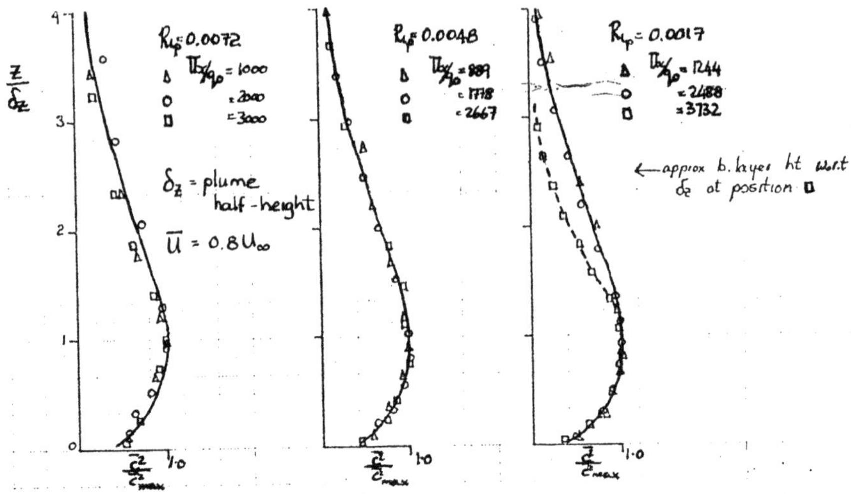

Fig 3.3.3 Measured concentration fluctuation profiles.

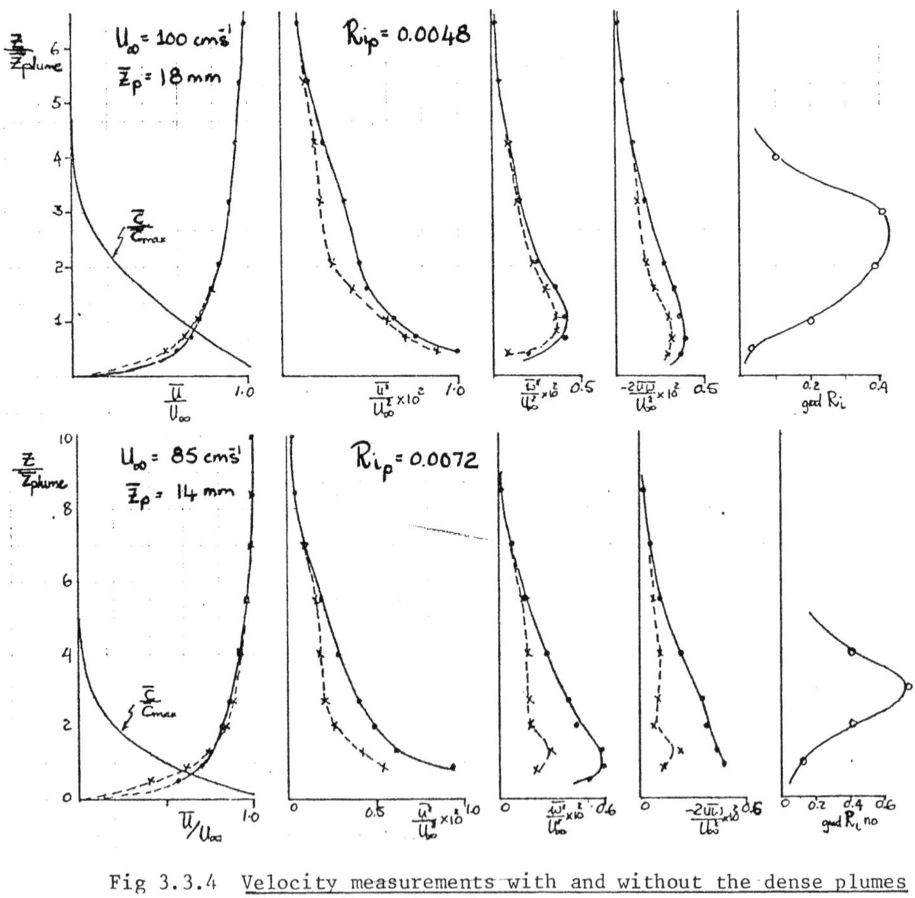

Fig 3.3.4 Velocity measurements with and without the dense plumes at 1m downstream of source.

the maximum gradient Ri numbers occur well above the mean height of the plumes in these cases - about 2 or 3 times $\overline{z}_{plume}$.

The mean velocity profiles (particularly noticeable in the more stable case) show signs of tending towards laminar profiles.

4. SUMMARY AND CONCLUSIONS

Three main aspects of our research have been reviewed. The application of Rapid Distortion Theory, Lagrangian "random flight" modelling, and finally an experimental program. Rapid Distortion Theory suggests that for a given strain and for $Nt < 1$ changes in turbulence structure are likely to be limited to the stress intensity ratios and Re stresses, with only small effects on the spatial correlation structure.

The Lagrangian model, although still rather speculative, seems to be potentially a useful tool in this context.

Finally, our experiments have provided further data on the relationship between the overall stability of 2-D plumes and their vertical growth rates. Furthermore the turbulence measurements within the stable plumes, although not yet analysed in detail, have indicated the effects to be anticipated.

Acknowledgement
We would like to acknowledge the support of the Central Electricity Generating Board for this research.

5. REFERENCES

Batchelor G.K and Townsend A.A (1956) Turbulent Diffusion. in Surveys in Mechanics, ed. Batchelor & Davies, C.U.P.

Bradbury L.J.S and Castro I.P (1971) A pulsed-wire technique for velocity measurements in highly turbulent flows. J.F.M, 49, 657-691.

Castro I.P and Cheun B.S (1982) The measurement of Re stresses with a pulsed-wire anemometer. J.F.M, 118, 41-58.

Csanady G.T. (1964) Turbulent diffusion in a stratified fluid. J. Atmos. Sci., 21, 439-447

Durbin P.A. (1980) A random flight model of inhomogeneous turbulent dispersion. Phys. Fluids, 23(11)

Ellison T.H. & Turner T.S. (1960) Mixing of dense fluid in a turbulent pipe flow. J.F.M., 8, 514-544.

Fackrell J.E. (1979) A flame ionization detector for measuring fluctuating concentration. J. Phys. E. Sci Instr, 13, 888-893.

Fackrell J.E. & Robins A.G.(1982) Concentration fluctuations and fluxes in plumes from point sources in turbulent boundary layers. J.F.M.,117,1-26.

Hanna S.R.(1981) Lagrangian and Eulerian time-scale relations in the daytime boundary layer. J. Appl. Met.,20,242-249.

Hunt J.C.R.(1978) A review of the theory of rapidly distorted turbulent flow and its applications. Fluid Dynamics Trans.,9,121-152.

Krasnoff & Peskin (1971) The Langevin model for Turbulent diffusion. Geophys. Fluid Dynamics,2,123-140

Monin A.S. & Yaglom A.M.(1971) Statistical fluid mechanics. M.I.T. Press,Camb.,Mass.

Narasimha R., Sethuraman S., Prabhu A., Rao K.N., & Prasad C.R.(1981) The response of the atmospheric boundary layer to a total solar eclipse. Fluid Mechanics Report 81 FM6, Indian Institute of Science, Bangalore

Pearson H.J., Puttock J.S. & Hunt J.C.R.(1983) A statistical model of fluid element motions and vertical diffusion in a homogeneous stratified turbulent flow. J.F.M.,129,219-249.

Tennekes H. & Lumley J.L.(1972) A first course in Turbulence. M.I.T. Press.

Townsend A.A.(1976) The Structure of Turbulent Shear Flow. C.U.P.,2nd edition.

The Initial Development of Gravity Currents from Fixed-Volume Releases of Heavy Fluids

JAMES W. ROTTMAN AND JOHN E. SIMPSON

Department of Applied Mathematics and Theoretical Physics
University of Cambridge, Silver Street, Cambridge CB3 9EW.

Summary

This paper describes some laboratory experiments of the initial development of gravity currents resulting from the instantaneous release of a fixed-volume of one fluid into a cross flow of another fluid of lesser density. Two limiting cases are considered in detail: the release of a cylindrical volume of neutrally-buoyant fluid into a uniform cross flow and the release of a cylindrical volume of heavy fluid into still surroundings. The results of the experiments are interpreted in terms of simple models.

1. Introduction

The work described in this paper is an attempt to determine the effects of the particular release conditions used in the Thorney Island field trials on the dispersion of the released heavy gas. As described more fully by McQuaid [1], each experiment at Thorney Island was begun by releasing 2000 m^3 of a heavy gas from an approximately cylindrical container 14m in diameter and 13m high. The sides of the container, made of plastic sheeting, were brought to the ground by elastic cords in less than two seconds, leaving an unconfined cylinder of gas at rest in an atmospheric cross flow. Similar release conditions were used in the earlier field trials at Porton Down, described by Picknett [2], and in the wind-tunnel simulations carried out by Hall [3]. More generally, the present work is an attempt to understand, by laboratory experiment and simple analysis, the physical mechanisms involved in the initial development of gravity currents by fixed-volume releases of heavy fluid into a cross flow.

There are two forces acting on the released volume of fluid during the early stages of motion after release: the drag associated with the cross flow around the contaminant vessel, and the buoyancy force due to the difference in density between the two fluids. An estimate of the relative magnitude of these two forces is given by a Richardson number, defined as

$$Ri_0 = g'_0 h_0 / U_0^2 \quad , \tag{1}$$

where $g'_0 = g(\rho_0 - \rho_a)/\rho_a$ is the reduced acceleration due to gravity, ρ_0 is the initial density of the released fluid, ρ_a is the density of the surrounding fluid, h_0 is a characteristic length scale for the initial height of the cloud, and U_0 is a characteristic mean velocity for the cross flow. For small values of this parameter, the cross flow determines the motion of the cloud and for large values, the buoyancy force is dominant. In a typical release both forces will act together, but to simplify the problem, and to isolate the different physical mechanisms so that their contribution at intermediate values of Ri_0 can be estimated, we performed experiments in the two limiting cases of small Ri_0 and large Ri_0. For small Ri_0 we performed laboratory experiments of the release of a cylindrical volume of neutrally-buoyant fluid into a uniform cross flow and for large Ri_0 we simulated in the laboratory the release of a cylinder of heavy fluid into still surroundings.

2. Release into a Uniform Cross Flow (small Ri_0)

The experiments of the release of a cylindrical volume of neutrally-buoyant fluid into a (nearly) uniform cross flow were performed in the continuous flow water channel described by Britter & Simpson [4]. The working section of this tank is 16.5cm wide and about 1m long. The available range for the mean speed U_0 of the flow in the channel is from 1 cm/s to 6 cm/s. We kept the depth of the flow in the channel at about 6cm. The first 40cm of the working section was illuminated from the side by a 1cm-wide slit of intense light about 1cm below the free surface.

A 4cm diameter cylinder was used, so the Reynolds number for these experiments ranged from 400 to 2400. The cylinder was placed in the channel about 25cm from the beginning of the working section, fluorescein was added to the fluid inside the cylinder, and then the cylinder was withdrawn vertically by hand. The resulting motion was photographed from directly overhead of the release point by a motor-driven camera at an exposure rate of 3 frames per second.

Sequential photographs of the motion of the initially cylindrical volume of marked fluid are shown in Figure 1. The Reynolds number for this flow is about 1600, with a mean flow speed of 4 cm/s, and the photographs shown were taken at about 1/3 s intervals. Immediately after release the marked fluid moves hardly at all in the streamwise direction but expands noticeably in the cross-stream direction, forming a roughly-shaped ellipse with major axis perpendicular to the mean flow direction. As the marked fluid begins to accelerate in the mean flow direction, the ellipse begins to bend into a 'horseshoe' shape with the open end of the horseshoe pointing in the downstream direction. The ends of the horseshoe tend to roll-up, generating turbulence, which in turn eventually leads to the break up of the entire structure. Similar observations have been reported by Picknett [1] of the cloud behaviour early after release in the Porton Down field trials.

The main features of the formation of the horseshoe structure are determined by potential flow theory. To show this, we used a two-dimensional vortex sheet method, similar to the method outlined by Baker, Meiron & Orszag [5], to compute numerically the motion of a cylinder of marked fluid released from rest into an irrotational flow field that has uniform velocity U_0 far from the cylinder. The vortex sheet method is particularly convenient for this calculation because the boundary of the released fluid is followed in time explicitly.

Interface contours for several times after release are shown in Figure 2. They form an ellipse immediately after the release (at t = 0) and at about $t = 0.5\,t_a$, where $t_a = x_0/U_0$ is a characteristic time and x_0 is the radius of the cylinder, the horseshoe shape begins to appear. In the calculation shown in Figure 2, the interface becomes unstable for $t \geqslant 0.8\,t_a$, just as the ends of the horseshoe begin to roll up. The instability appears first as a small sawtooth perturbation of the interface but this rapidly grows, resulting in a complete breakdown of the calculations.

Physically, the horseshoe shape is caused by the non-uniform pressure distribution about the cylinder. In the potential flow, the initial pressure distribution about the cylinder is symmetric, with high-pressure regions at the upstream and downstream stagnation points and low-pressure regions at the two intersections of the cylinder with a plane perpendicular to the cross flow, as indicated in Figure 2. This pressure distribution causes the initially cylindrical vortex sheet, which marks the interface between the released fluid and the external fluid, to deform into an ellipse with major axis perpendicular to the cross flow. This initial symmetric motion tends to concentrate the vorticity at the two ends of the major axis and in turn this causes an unsymmetrical 'roll-up' of the ends in a sense equal to the sign of the vorticity at each end (positive at one end and negative at the other). In the real flow the initial pressure distribution is unsymmetrical, with a 'dead zone' downstream of the cylinder, but this only enhances the basically inviscid process of the formation of the horseshoe shape.

The bulk motion of the unreleased fluid can be approximated by a vortex pair; the initial separation of the two vortices is such that their induced velocity exactly opposes U_0, but the initial pressure distribution forces the two vortices apart, thus reducing their induced velocity and consequently they are accelerated downstream by the external flow. Figure 3 is a plot of the position of the released fluid's centre of mass (in our numerical calculation) as a function of time

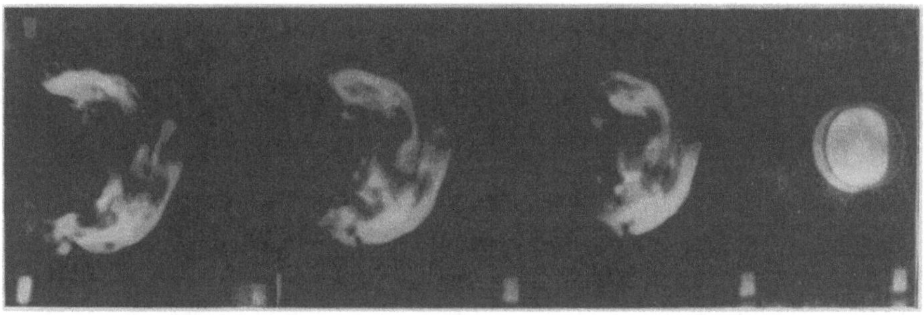

Fig.1 Sequential photographs of the motion resulting from the release of dyed fluid into a uniform cross flow. The mean flow speed is 4 cm/s (from right to left) and the cylinder diameter is 4cm. The second picture from the right was taken about 2.5 s after release and the subsequent pictures at 0.3 s intervals.

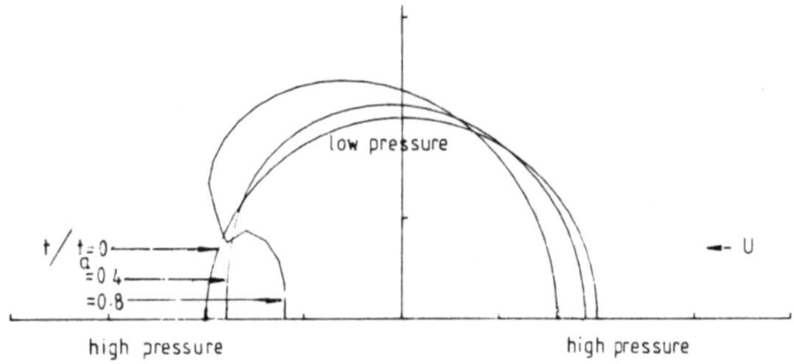

Fig.2 Computed interface contours at several times after release of a cylinder of fluid in a uniform cross flow (since the calculation is symmetric about an axis through the centre of the cylinder and parallel to U_0, only the upper half of the flow is shown).

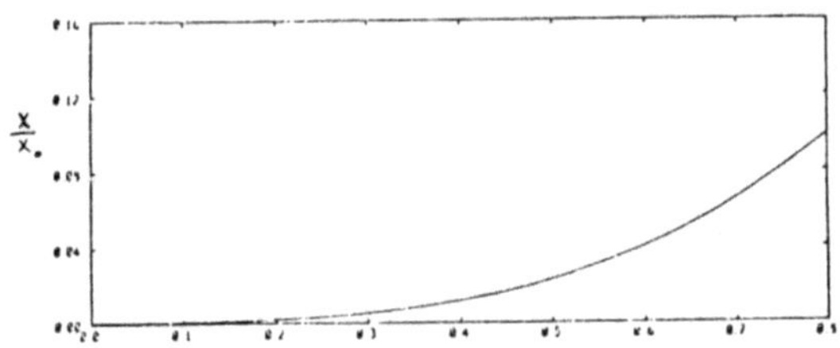

Fig.3 The computed downstream position of the centre of mass of the released fluid, corresponding to Fig.2, as a function of time.

after release, showing that its speed in the downstream direction increases slowly at first but in a time of order t_a is close to one-third of the cross flow speed (which in the plot would be a straight line with unit slope).

The instability in the numerical calculations is due to the growth of Kelvin-Helmholtz waves on the infinitesimally thin vortex sheet. This is a well-known difficulty of the vortex sheet method that is discussed in more detail by Moore [6]. In the real flow this instability is checked by a slight thickening of the interface due to mixing of the two fluids. But the experiments show that this mixing is small compared with the mixing due to the primary vortex roll-up of the released fluid that is in essence described by the vortex-sheet calculation.

3. Release into a Still Environment (large Ri_0)

The experiments of the release of a cylindrical volume of heavy fluid into a still environment were made in a sector-shaped Perspex tank, similar in design to that used by Britter [7] in his experiments on axisymmetric gravity currents with constant volume flux. The tank used in our experiments is 240cm long and 40cm high with an enclosed angle of about 10°. The sector-shaped tank allows the use of the shadowgraph technique to visualise the interior structure of a flow that simulates the flow produced in a circular tank of equal radius.

To perform an experiment, the tank was filled with tap water to a depth h_0 and then an aluminium gate was inserted in the tank at a radial distance x_0 from the vertex. A quantity of cooking salt was dissolved in the water between the gate and the vertex to achieve a desired density and finally more tap water was added carefully to both sides of the gate (if necessary) so that the total fluid depth everywhere in the tank was H. The parameter ranges for the density ratio, the total fluid depth and the ratio of the initial depth of the heavy fluid to the total fluid depth were: $0.95 \leqslant \rho_0/\rho_a < 1.00$, $30cm \leqslant H \leqslant 40cm$, $0.25 \leqslant h_0/H \leqslant 1.00$. The lock lengths used were x_0 = 60cm, 90cm and 120cm. The experiments were recorded on video tape, from which all measurements were taken.

Sequential shadowgraphs of the observed motion after the removal of the gate are shown in Figure 4 for the cases with h_0/H = 0.25 and 1.0. Shadowgraphs of similar releases in a rectangular channel are shown in Rottman & Simpson [8]. In all these flows, the front forms very soon after release (in a few tenths of a second) and the most intense mixing of the two fluids occurs near the front. The most striking difference between the axisymmetric and two-dimensional flows is the shape of the front and the intensity of rotational motion of the internal flow in the front. In the time the axisymmetric front travels a distance equal to about x_0, the majority of the fluid in the current becomes concentrated at the front (for small h_0/H) or in multiple fronts (for h_0/H near 1), leaving only a thin layer of heavy fluid behind the front (or fronts). The two-dimensional currents are more uniform in depth. In addition, the internal rotational flow and associated mixing in the axisymmetric fronts is more intense; indeed, the mixing appears to occur all the way down to the ground behind the front (or fronts) in these flows.

To obtain a better idea of the internal velocity field in the axisymmetric front, we mixed some aluminium powder with the heavy fluid and took photographs with a camera travelling at the front speed. The current was illuminated by a vertical slit of intense light aligned along a radial line through the centre of the channel. Sequential photographs are shown in Figure 5 for both cases h_0/H = 0.25 and 1.0. These photographs indicate that the rotational motion at the front is initiated by Kelvin-Helmholtz instabilities and maintained by vortex stretching as the ring expands outwards. It also appears that the multiple fronts, which occur when h_0/H is near 1, are caused by the large amplitude Kelvin-Helmholtz waves generated by the greater shear in these flows. When h_0/H is near 1, the light fluid that is displaced by the collapsing heavy fluid forms a gravity current (in the upper half of the

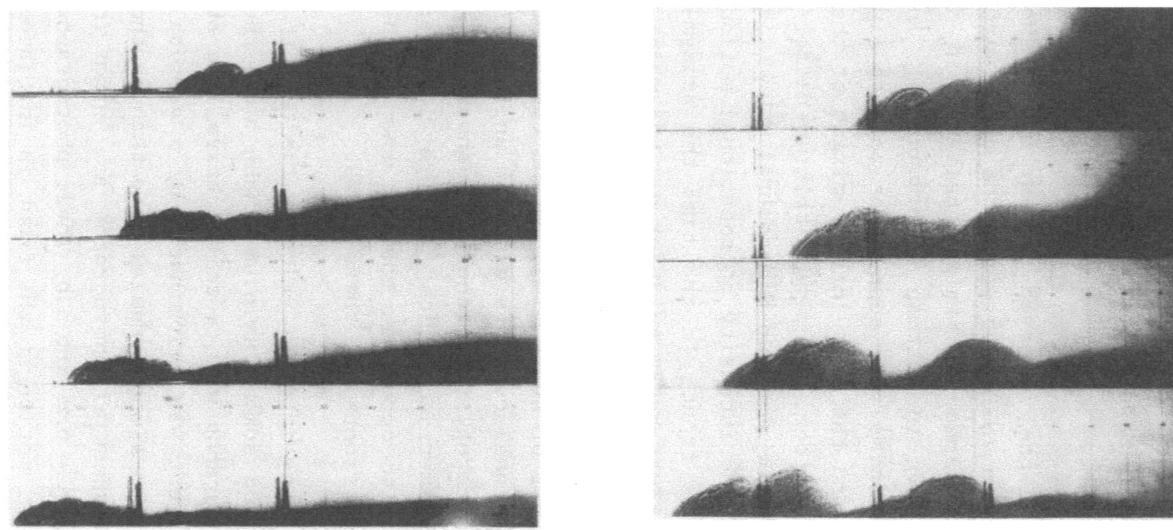

(a) (b)

Fig.4 Sequential shadowgraphs of a volume of salt water collapsing into fresh water in a sector-shaped tank, with x_o = 60 cm and g' = 47 cm/s^2. The numbers in the photographs are the radial distance in cm from the vertex of the tank:
(a) h_o/H = 0.25, and (b) h_o/H = 1.0.

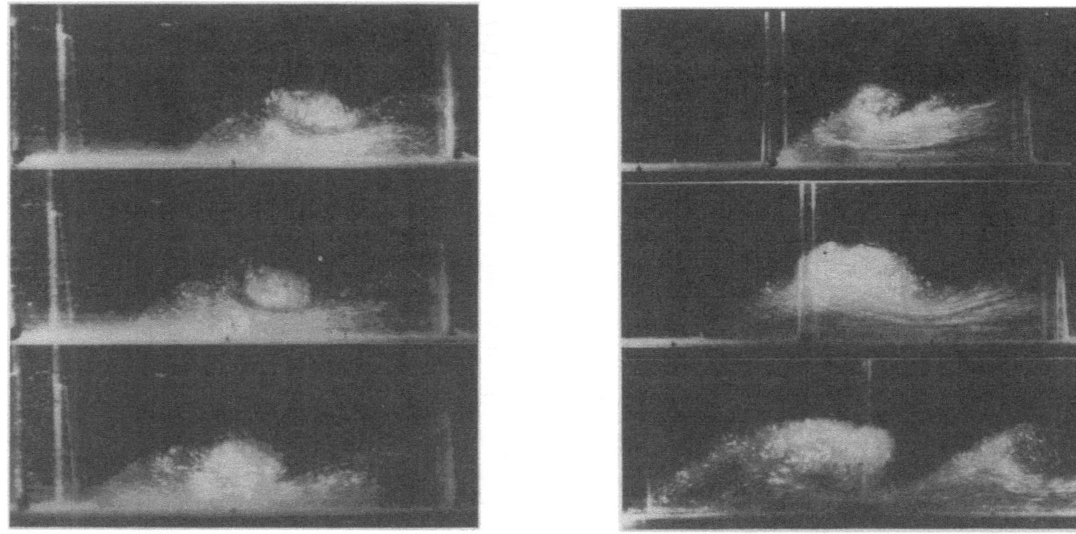

(a) (b)

Fig.5 Sequential streak photographs of the flows described in the caption for Fig.4.

tank) that travels towards the vertex, opposite to the direction of the heavy current. This opposing flow leads to greater shear at the interface than when h_0/H is small. Thus, a heavy gas cloud collapsing into a wind, as on the upwind side of a Thorney Island release, may develop multiple fronts.

To gain an understanding of why the two-dimensional and axisymmetric flows are so strikingly different, we solved the idealised 'dam-break' problem for both geometries. The 'dam-break' problem, described in more detail by (for example) Penney & Thornhill [9], is to determine the solution of the shallow-water equations for the heavy current depth h(x,t) and for the depth-averaged horizontal fluid speed u(x,t) given the discontinuous initial conditions

$$h(x,t=0) = \begin{cases} h_0 & 0 \leqslant x \leqslant x_0 \\ 0 & x > x_0 \end{cases} \quad ; \quad u(x,t=0) = 0. \qquad (2),(3)$$

In addition, we imposed the front condition

$$u_f^2 = \beta^2 g'_0 h_f \qquad (4)$$

where $u_f(t)$ is the front speed and $h_{f(t)}$ the front depth of the heavy current. This is needed because near the front, where vertical accelerations are large, the shallow-water equations are inaccurate. With $\beta^2 = 2$, (4) gives the theoretical front speed derived by Benjamin [10] for a long cavity. Based on the results of Rottman & Simpson [8], we chose $\beta^2 = 1$ for our gravity currents. This formulation is appropriate for the case with $h_0/H = 0$.

We solved this problem numerically using the method of characteristics. This method is both convenient, because the front boundary condition is easily incorporated, and informative, because the characteristic diagram immediately reveals the reason for the differences between the flows in the two geometries. The characteristic diagrams are shown in Figure 6 and the corresponding depth profiles, at several times after release, are shown in Figure 7. The two-dimensional flow consists of a steady-state front that has constant depth and

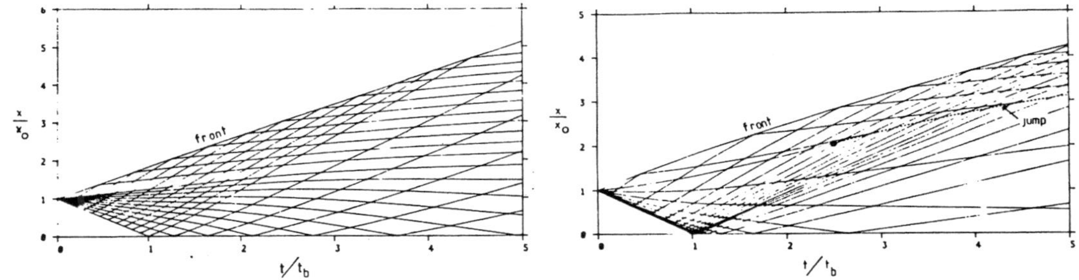

(a) (b)

Fig.6 Characteristic diagrams for the 'dam-break' problem with $h_0/H = 0$:
(a) two-dimensional geometry, and (b) axisymmetric geometry.

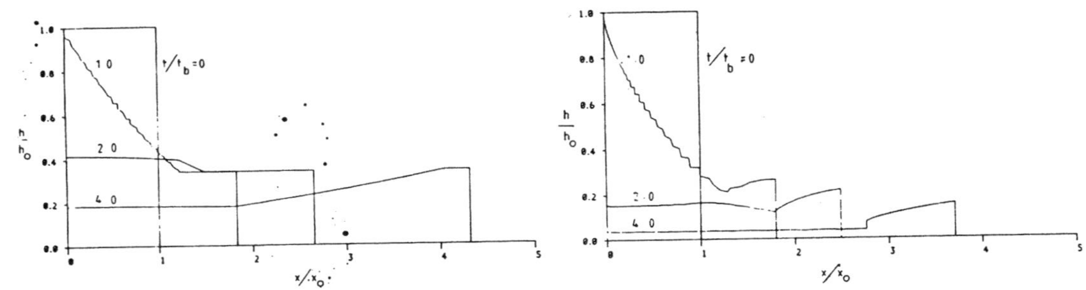

(a) (b)

Fig.7. The theoretical current depth profiles for the 'dam-break' problem with $h_0/H = 0$:
(a) two-dimensional geometry, and (b) axisymmetric geometry.

constant speed followed by an expansion wave that is centred about $x = x_0$. The axisymmetric flow for small time after release is identical, but within a time $t \approx t_b = x_0/(g'_0 h_0)$ the fluid at the leading edge begins to accelerate (indicated in the characteristic diagram by the curving of the negative characteristics in the direction of increasing x) due to the radial expansion of the current. However, the fluid just behind the front is inhibited from increasing its speed by the front boundary condition. Thus, disturbances (which travel along characteristic lines) that propagate back away from the front are overtaken by disturbances travelling on the leading edge of the expansion wave. This results in the formation of a hydraulic jump (indicated in the characteristic diagram by the coalescence of characteristic lines) at $t \approx 2t_b$ in the calculation. The formation of the jump is indicated in the diagram by a black dot and the path of the jump by a dashed line. Therefore, the concentration of heavy fluid at the rapidly spreading current front is a consequence of the radial geoemtry *in combination* with the front boundary condition.

4. Conclusions

In this paper we have concentrated on the initial phase of dispersion of a dense gas cloud produced by an instantaneous release into a cross flow. We have identified and developed simple models for two physical processes that determine the motion in this phase: (1) when Ri_0 is small the cloud rolls-up into a horseshoe shape on a time scale t_a, and (2) when Ri_0 is large the cloud forms a radially spreading vortex ring on a time scale t_b. For intermediate values of Ri_0, both mechanisms will be at work. We expect that for small Ri_0, the mixing due to the formation of the horseshoe shape will sufficiently dilute the cloud so that buoyancy will never be important. A discussion of this point (and of the other phases of dense gas dispersion) is given by Hunt *et al.*. [11].

Acknowledgement

This work was sponsored by the U.K. Health and Safety Executive under contract 1918/01.01.

References

1. McQuaid, J.: Large-scale experiments on the dispersion of heavy gas clouds. IUTAM Symposium on atmospheric dispersion of heavy gases and small particles, Delft, The Netherlands, August 1983.

2. Picknett, R.G.: Dispersion of dense gas puffs released in the atmosphere at ground level. Atmospheric Environment 15 (1981) 509-525.

3. Hall, D.J.: Validation of wind tunnel models of dense gas release. IUTAM Symposium on atmospheric dispersion of heavy gases and small particles, Delft, The Netherlands, August 1983.

4. Britter, R.E.; Simpson, J.E.: Experiments on the dynamics of a gravity current head. J. Fluid Mech. 88 (1978) 223-240

5. Baker, G.R.; Meiron, D.I.; Orszag, S.A.: Generalised vortex methods for free-surface flow problems. J. Fluid Mech. 123 (1982) 477-501.

6. Moore, D.W.: The spontaneous appearance of a singularity in the shape of an evolving vortex sheet. Proc. Roy. Soc. Lond. A364 (1979) 105-119.

7. Britter, R.E.: The spread of a negatively buoyant plume in a calm environment. Atmospheric Environment 13 (1979) 1241-1247.

8. Rottman, J.W.; Simpson, J.E.: Gravity currents produced by instantaneous releases of a heavy fluid in a rectangular channel. J. Fluid Mech. 135 (1983) 95-110.

9. Penney, W.G.; Thornhill, C.K.: The dispersion under gravity of a column of fluid supported on a rigid horizontal plane. Proc. Roy. Soc. Lond. A244 (1952) 285-311.

10. Benjamin, T.B.: Gravity currents and related phenomena. J. Fluid Mech. 31 (1968) 209-248.

11. Hunt, J.C.R.; Rottman, J.W.; Britter, R.E.: Some physical processes involved in the dispersion of dense gases. IUTAM Symposium on atmospheric dispersion of heavy gases and small particles, Delft, The Netherlands, August 1983.

Some Physical Processes Involved in the Dispersion of Dense Gases

J.C.R. HUNT, JAMES W. ROTTMAN

Department of Applied Mathematics and Theoretical Physics
University of Cambridge, Silver Street, Cambridge CB3 9EW.

R.E. BRITTER

Department of Engineering, University of Cambridge
Trumpington Street, Cambridge CB2 1PZ, U.K.

Summary

In this paper we attempt to provide a theoretical framework for the formulation of mathematical models of dense gas dispersion in the atmosphere. Our approach is to divide the evolution of a released dense gas cloud into four phases, during each of which different physical processes governing the behaviour of the cloud are most important. These processes are identified and recent attempts to understand them are discussed. We also give scaling arguments for the order and duration of the different phases as functions of the atmospheric and release conditions.

1. Introduction

In this paper we describe some physical processes that determine the dispersion in the atmosphere of dense (or heavier-than-air) gases released near the ground. We also discuss some recent theoretical, computational and experimental attempts to understand these processes. In our discussion, we mainly concentrate on the case where the gas is released rapidly (i.e., in a time small compared with the time for any ambient flow to pass around the released volume); however, the physical processes of dispersion are similar to those affecting gases released in other ways. Our main intention in this paper is to provide a theoretical framework that we hope will be helpful in formulating mathematical models of dense gas dispersion for a fairly wide range of atmospheric and release conditions.

Analysis and observation suggest that different physical processes are predominant at different times after release during the dispersion of a dense gas cloud, and so in our discussion we postulate the existence of four *phases* of dense gas dispersion, identified by the predominance of certain processes and the absence of other processes. Although a typical dense gas cloud passes through the different phases in the order in which we describe them, this is not necessarily true in general. Depending on the atmospheric and release conditions, one or even two of the phases may be bypassed as the cloud evolves and it is even possible for the cloud to go in reverse order through some of the phases. The main point is that the phases are defined by the predominant physical processes at work and not by a chronological sequence of development.

In the first phase, which we refer to as the initial phase, the inertia of the cloud and the mean atmospheric flow are the predominate forces that determine the motion. It is convenient to adopt a shorthand notation for the forces: let I denote inertia and XF denote the atmospheric (or external) mean flow. Similarly, the other forces are denoted by B for buoyancy and XT for external turbulence. In the second phase, which we refer to as the gravity-spreading phase, both buoyancy and the external mean flow are the dominant forces at work. In the third phase, which we refer to as the nearly-passive phase, the external turbulence in addition to the external mean flow and the cloud buoyancy, are the important forces. Finally, the fourth phase, commonly called the passive phase in the literature, is controlled entirely by the external turbulence and the external mean flow. For ease of reference, the four phases and the forces that predominate during each phase are listed in Fig.1, which provides an outline for the remaining sections of this paper.

To fully determine the evolution of a dense gas cloud, it is also necessary to know both the order in which a cloud passes through the different phases, and the time it spends in each phase. An answer to both these questions requires an understanding of the *transitions* between the different phases. In the remaining sections of this paper, we discuss the modelling of the dominant physical processes in each phase and describe how the ordering and timing of the transitions between the phases can be estimated. In our discussion we concentrate on the first three phases of dispersion, since the passive phase does not involve density effects and is a well-studied subject in its own right (although we do discuss the transitions to the passive phase). We particularly concentrate on the nearly passive phase as this is the most complicated and least understood of all the phases.

2. The initial phase (I/XF)

2.1 General description

The initial phase is usually excluded from most simplified models of dense gas dispersion. It is generally thought that this phase is too complex and that there are too many different modes of cloud formation for any simplified model to handle. However, we argue that the initial phase determines whether the cloud behaves as a dense gas or a passive tracer and therefore an understanding of this phase is essential for predicting the dispersion of a dense cloud.

The release of a dense cloud from some sort of container into an atmospheric flow usually involves a period of time during which the cloud is accelerated from rest (or near rest) by the external mean flow, a period of time when buoyancy is not important. The acceleration and deformation of the cloud during this period is mainly due to the non-uniform distribution of pressure remaining from the presence of the containment vessel in the mean atmospheric flow. Rottman & Simpson [1] have approximated the motion in this phase by computing the motion of a vortex sheet that separates the dense gas cloud from the external flow. The result for a circular cylinder of gas released in a uniform two-dimensional flow is sketched in Fig. 2. Since gravity can be ignored, this Fig. can be viewed in two ways: (i) as a plan view (Fig. 2a), it shows that the cylinder initially deforms into an ellipse, with major axis perpendicular to the mean flow direction,and eventually into a horseshoe shaped pair of counter-rotating vortices as the cloud is accelerated downstream; (ii) as a profile view (Fig. 2b), in which case the calculation represents the motion of an initially semi-circular ridge, it shows that the mean flow attempts to lift the cloud off the ground as it accelerates the cloud downstream. Both these aspects of the behaviour of a released cloud are observed in the wind tunnel experiments of Hall [2] and in the Thorney Island (McQuaid [3]) and Porton Down (Picknett [4]) field trials. In these cases, it is mostly the large-scale inviscid rollup of the vortex sheet that determines the mixing of the dense gas with the surrounding fluid. A similar behaviour occurs near the source of a continuous release into a crossflow, as sketched in Fig. 3.

As well as the mixing caused by the large-scale rollup of the released cloud, the heavy gas is also mixed with the external flow by the turbulence produced in the shear layers over the front and sides of the cloud. These shear layers are caused by the difference in velocity ΔU between the external flow and the heavy gas cloud. Since a planar shear layer typically has a thickness of about 1/10 of its length, and since the turbulence in a *curved* shear layer is significantly *less* than in a planar shear layer, the rate at which the heavy gas is mixed out of the cloud in these shear layers is no more than of order $\Delta U h/10$. Some of the 'detrained' heavy gas when it is swept to the lee is engulfed back into the cloud. The rest is advected into the turbulent wake of the cloud. So, although the bulk of the heavy gas moves at a speed less than U in a coherent cloud, some material is advected downwind with speed U.

However, there are several types of releases that do not involve any significant acceleration at the source. One example is that of a dense gas cloud produced by an explosion. As the gas explodes it mixes with the surrounding air giving the entire cloud the same mean velocity as the surrounding fluid. Another example, illustrated in Fig. 4, is the sudden release of a dense gas downstream of some type of barrier - a building, for example. This situation is initially similar to the release into a weak wind, so that the I/XF phase is nonexistent; the cloud passes immediately into a B phase, slumping sideways until it encounters the mean flow around the building and so into B/XF phase.

2.2 The transition from I/XF -> XF/XT

It is necessary to have an estimate for the duration of the period in which the inertia of the cloud significantly affects its behaviour. The simplest case to analyse is when the density of the released cloud is equal to the density of the surrounding fluid. For this case the question we seek to answer is: At what time after release does a neutrally buoyant cloud begin to behave like a passive tracer?

As described in Section 2.1, the cloud is initially accelerated and deformed by the mean flow. The mean flow accelerates the cloud to the mean flow speed in a time $t_a \approx h_o/U$, where $h_o = h(t=o)$ is the initial height of the cloud and U is the mean flow speed. The deformation of the cloud in this time generates internal turbulence which by analogy with wakes behind bluff bodies we may estimate as having

$$u_c \approx \frac{1}{3} U \tag{1}$$

(Castro and Robbins [5]), where u_c is a characteristic velocity of the turbulent fluctuations. For $t > t_a$ this turbulence is decaying but the cloud continues increasing in size due to turbulent diffusion (Mobbs [6]), such that

$$\frac{dh^2}{dt} \sim u_c \ell$$

where ℓ is a characteristic length scale of the turbulent fluctuations. By analogy with grid turbulence and wakes with weak velocity deficit, we know that the turbulence decays such that

$$u_c \sim U(tU/h_o)^{-1/2} \quad , \quad \ell \sim h_o(tU/h_o)^{1/2} \quad . \tag{2a,b}$$

So $u_c\ell$ is constant and therefore $h^2 \propto t$.

From (1) and (2a), we estimate

$$u_c \approx \frac{1}{3} U \, [t U/h_o]^{-1/2} \quad . \tag{3}$$

The cloud becomes passive when the external turbulence begins to determine the motion of the cloud; that is, when $u_a \approx u_c$, where u_a is a characteristic velocity scale of the external turbulent fluctuations. The time t_c when this occurs, is, from (3),

$$t_c \approx \frac{1}{9} \left(\frac{U}{u_a}\right)^2 \left(\frac{h_o}{U}\right) \quad . \tag{4}$$

For $t > t_c$, $h^2 \propto t^2$ as appropriate for passive dispersion (Csanady [7], Chapter 3). In the Thorney Island trials, $U \approx 2\,\text{m/s}$, $h_o \approx 10\,\text{m}$, and $U/u_a \approx 10$, typically, so $t_c \approx 50\,\text{s}$. In other words, the cloud is about 100m downwind from the release point before it behaves passively -- showing that the initial phase can have long-lasting effects.

The initial phase is also important in determining the ratio of the magnitude of concentration fluctuations c' to the mean concentration $\bar{c}$ in a passive cloud. Experiments and analysis show that $c'/\bar{c}$ decreases with the ratio of the scale $\ell(o)$ of the cloud, when it begins to disperse passively, to the scale of atmospheric turbulence L_t. See, for example, Fackrell & Robins [8].

2.3 <u>The transition from I/XF -> B/XF</u>

We now attempt to estimate the duration of the initial phase when density of the released cloud is greater than that of the surrounding fluid. Inertial and buoyancy forces tend to have somewhat opposite effects. As described in Section 2.1 and illustrated in Fig. 2, the inertial forces attempt to raise the cloud off the ground and to roll-up the vortex sheet that bounds the cloud in planes parallel to the ground. Buoyancy forces, which generate vorticity at the cloud boundary of *opposite* sign (on the downwind side) to that which exists due to the external flow, attempt to collapse the cloud, spreading it horizontally over ground. Which process dominates is a function of their relative strengths and also the rate of mixing between the two fluids.

Rottman & Simpson [1] present evidence that in the absence of buoyancy forces the cloud mixes with the external flow and accelerates to the mean flow speed in a time

$$t_a \sim h_o/U \quad , \qquad (5)$$

where h_o is a characteristic length scale for the cloud before release. This means that the mixing with the external flow is dominated by the vortex roll-up mechanism. In the absence of inertial forces (a release into calm surroundings), Rottman & Simpson [1] show that the buoyancy-driven flow develops an organized structure (a vortex ring or concentric vortex rings propagating radially outward) in a buoyancy timescale

$$t_b \sim (h_o/g\frac{\Delta\rho}{\rho})^{1/2} \quad , \qquad (6)$$

where we have assumed that the initial aspect ratio of the cloud is about unity.

Therefore, if $t_a \ll t_b$, the inertial forces have sufficient time to disrupt the developing buoyancy-generated flow and the cloud passes directly into a XF/XT phase. That is, buoyancy is not significant if

$$\frac{g\frac{\Delta\rho}{\rho}h_o}{U^2} \ll 1 \quad . \qquad (7)$$

When buoyancy is significant, the inertial effects are present only until the buoyancy-generated flow becomes organised, i.e. on a timescale t_b.

In this simple analysis and in the experiments of Rottman & Simpson [1], non-Boussinesq effects were not considered. Since many accidental releases and field trials involve gases that are two to four times the density of the surrounding air, non-Boussinesq effects, particularly in the initial stages of collapse, require further study.

3. The gravity-spreading phase (B/XF)

3.1 General description

In this phase the cloud spreads as a classical gravity current, as reviewed recently by Simpson [9]. Gravity currents are driven primarily by the excess hydrostatic pressure caused by their greater density than the surrounding fluid. The fluid motion within gravity currents is generally horizontal, except near the front where there is a recirculating flow with large vertical velocities. The front is usually deeper than the following heavy fluid and it is just behind the front where most of the mixing occurs, as a result of Kelvin-Helmholtz instability (as shown in the experiments of Britter & Simpson [10]). The mixed fluid is left behind as the front moves, and this lighter fluid, with a density intermediate between the ambient and the unmixed fluid, lies in an upper layer (U) above an inner layer (I) of unmixed dense fluid, as sketched in Fig.7a.

It seems fairly-well established that the rate of spreading of the cloud in this phase is described by self-similar solutions of the shallow-water equations. A table of these solutions, along with more references in which such solutions are derived, is given by Britter [11]. Rottman & Simpson [12] have established that the current front must travel about four initial radii after release before these self-similar solutions are approached. The results for the rate of spreading of a constant-volume axisymmetric current, which are of particular interest to us in this paper, gives

$$U_f \propto \left(h_o^3 g \frac{\Delta\rho}{\rho}\right)^{1/4} t^{-1/2} \quad , \tag{8}$$

where U_f is the radial rate of advance of the current front. Superimposed on this spreading motion is a horizontal translation of the cloud as a whole. The velocity translation is in the direction of the mean wind and its magnitude is some fraction of U.

The physical processes in this phase are fairly-well understood, at least for two-dimensional releases into calm surroundings, and it is this phase for which most of the commonly used 'box models' were developed. However, less is known about the effects of shear in ambient flow and the differences between two- and three-dimensional gravity currents.

3.2 The effects of shear in the external flow

Shear in the external flow has a significant effect on the shape of the gravity current produced by the release of a dense gas cloud. A typical cross section, as seen for example in the Thorney Island field trials, is sketched in Fig. 5. The most noticeable difference between this profile and that for a gravity current in a uniform external flow is the asymmetry between the upwind and downwind fronts of the current. The upwind front has the shape of a thin wedge with a lower angle to the horizontal than a normal gravity front, whereas the downwind front is deeper and steeper than the normal front.

A simple idealised analysis helps to explain the asymmetry in this flow. In a coordinate system fixed with the front of the current, as sketched in Fig. 6, we consider the solution of the equation

$$\nabla^2 \psi = -\omega , \qquad (9)$$

where ψ is the streamfunction and ω is the (assumed constant) vorticity for the flow outside the gravity current with the boundary conditions

$$\psi = 0 \quad \text{on} \quad \theta = \pi \quad \text{and on} \quad z = z_s , \qquad (10)$$

$$\text{and} \quad \tfrac{1}{2}(u_r^2 + u_\theta^2) + g\frac{\Delta\rho}{\rho} z = 0 \quad \text{on} \quad z = z_s \qquad (11)$$

$$\text{where} \quad u_r = \frac{1}{r}\frac{\partial\psi}{\partial\theta} , \quad u_\theta = -\frac{\partial\psi}{\partial r} . \qquad (12),(13)$$

Here, z is the vertical coordinate and z_s is the height of the interface between the two fluids of different densities and (r,θ) are the radial and angular coordinates. Boundary condition (11) requires the pressure to be continuous across z_s. We seek a solution valid near the stagnation point and therefore we are justified in ignoring the flow in the gravity current.

By considering small ω, z_s can be expressed as

$$z_s = z_0 + \Delta z_1 \quad (14)$$

where z_0 is the solution for the case with $\omega = 0$ (i.e. the well-known potential flow result with the front angle $\theta_0 = \pi/3$). When $\omega > 0$ the *perturbation* velocity increases with r up the face of the current. Thus, from (11) it follows that $\Delta z_1 < 0$. Conversely, if $\omega < 0$, $\Delta z_1 > 0$. This result is also suggested by the laboratory experiments of Simpson & Britter [13]. They developed a technique to maintain a steady gravity current in shear flows and their results show the front angle θ_s is larger for relative motion into shear flows and smaller for relative motion away from shear flows than for motion into uniform flows.

3.3 Three-dimensional effects

An interesting question about three-dimensional releases in a cross flow is: how will the initially circular-shaped cloud from a cylindrical release distort, if at all, in a cross flow? Box models assume that there is no distortion; that is, the cloud remains circular in plan form as it is carried downstream by the mean flow.

Any cloud that is released over a time scale t_R is elongated in the direction of the wind if the distance the cloud front travels during the release is significant compared with the lateral distance travelled by the cloud, i.e. $t_R U \sim t\sqrt{(hg\,\Delta\rho/\rho)}$. But, even if the gas container is removed instantaneously at $t = 0$, because of its inertia the bulk of the gas does not move immediately while the gravity current head moves upwind and downwind. Consequently the rarefaction wave (cf. [12]) must reach the upwind front first and slow it down before it reaches the downwind front. This leads to the alongwind dimension of a cloud being of a distance order $U[g\frac{\Delta\rho}{\rho}h]^{1/2}$ greater than the crosswind dimension.

Wind shear affects the speed of the upwind and downwind fronts differently and this may affect the shape of the cloud in a different way, perhaps providing the explanation of the occasionally observed stretching in the direction perpendicular to the mean wind.

4. The nearly passive phase (B/XF/XT)

4.1 The transition from B/XF → B/XF/XT

If a dense gas cloud is sufficiently dense initially that its spreading is controlled by inertia, buoyancy and the external mean flow, then eventually the difference in velocity between the cloud and the environment becomes small enough that it is comparable with a typical velocity (say, u_*, the friction velocity), of turbulent fluctuations in the atmosphere. This is the transformation from the phase B/XF to the phase B/XF/XT. Of course in the outer region of a cloud this is true at all phases of the evolution of the cloud, but we now want to consider the phase when the average velocity difference within the cloud is comparable with u_* (Fig. 7).

There are other situations where the initial density of the cloud (or a plume, for a continuous source) is small enough that the turbulent velocities are comparable with buoyancy or inertial induced velocities from the start. Then the cloud *begins* in the phase B/XF/XT. Such a situation is depicted in Fig. (8b) and discussed in Section 4.4.3

In Fig. (7a), we see how at the beginning of this phase for a dense gas cloud in a cross flow the turbulence within a cloud is largely generated by buoyancy forces in the presence of a lower surface and the external mean flow. In this figure h is the average height and L is the horizontal length of the cloud. Typically the velocity components of this turbulence are of the order of $\surd(g\frac{\Delta\rho}{\rho}h)$, i.e. the speed U_f of the gravity current front relative to the mean wind speed. This turbulence is the most intense and of the largest scale in the billowing motions near the edge of the cloud and is produced by the shear in this region (Britter & Simpson [10]). Some of the first measurements of such turbulence have been made recently in the head of a gravity current formed by drainage winds from the Rocky Mountains at Boulder, Colorado (see Hootman & Blumen [14]). Over the rest of the surface of the cloud there is an internal shear layer in which the mean velocity changes direction (relative to a coordinate system moving with the cloud). Because of the stable stratification the turbulence generated in this layer does not diffuse significantly above the height of the gravity current head (see Section 4.3).

If a cold dense gas passes over a warmer surface, such as the sea, there can be vigorous convection *within* the dense cloud. This is likely to generate an elevated stable layer within the dense cloud, similar to the elevated inversion at the top of the atmospheric boundary layer in convective conditions, as described, for example, in Kaimal *et al.* [15], Lumley *et al.* [16] and Colenbrander & Puttock [17].

When the internally induced turbulence decays to a value comparable with u_*, the external turbulence first affects the upper mixed layer where the internal turbulence and buoyancy-induced motion are weakest. With further decay, the turbulence and mean flow at the head are affected. When $u_* \geqslant U_f \approx \surd(g\frac{\Delta\rho}{\rho}h)$, then the outward motion of the head relative to the centre of the cloud is increased sufficiently to be greater than the mean buoyancy- driven outward flow in the lower unmixed layer of the cloud (which is of order $1.2\surd(g\frac{\Delta\rho}{\rho}h)$ (Britter & Simpson [10]). The recirculating motion in the head occurs as a consequence of continuity and only occurs when this outward inner-layer flow is greater than the outward flux in the head . So by amplifying this outward flux, external turbulence can eventually destroy the characteristic flow in the head of the dense gas cloud; this is the stage at which the spreading of the dense gas cloud changes from that characteristic of a gravity current to that of a cloud of slightly dense contaminant in a turbulent flow (Fig. 7b). The transition to the phase B/XF/XT occurs when $U_f \sim u_*$. An estimate of the time t_c for this transition is then obtained from (8), viz. $t_c \sim \beta\,(\frac{g\delta\rho_o}{\rho_o} h_o^3)^{1/2}/u_o^2$, where XT affects the region (U) and $\beta \geqslant 1$ when XT significantly affects the front.

When $u_* > U_f$, an important implication of the buoyancy driven outward flow being less than the outward flux is that there is a downward flux produced by the mean downward velocity $\bar{w}$. The growth of the cloud depends on how this compares with the *upward* flux of the dense gas by the diffusive action of the external turbulence. By continuity, $\bar{w} \sim U_f\, h/L$, and so the downward flux is of the order of $\Delta\rho\, U_f\, h/L$, while the upward turbulent flow is of the order of $\Delta\rho\, u_*$. Since, by hypothesis, $u_* > U_f$, and since $h/L \ll 1$, $\Delta\rho\, u_* \gg \Delta\rho\,\bar{w}$. Thus, at this stage, when the gas cloud has changed its form, the upward diffusive flux must be much *greater* than the downward flux produced by the mean sinking motion of the cloud. This argument also shows that any external turbulence can produce an upwards flux of the matter in the gas against the weak downflow long before the turbulence is strong enough to destroy the structure of the head.

The same transformation occurs in dense gas plumes emitted from a steady source (Fig. 8a). In the B/XF phase, there is a gravity current head upwind of the source and, if it is a point source, around the sides of the plume (Britter [18]). When the external turbulence is weak it is entrained into the turbulent front around the head and sides of the plume and its eddies simply impinge on the top of the plume as if it was a heavy liquid. But when the turbulence is stronger, or the plume has travelled further downwind of the source, the external turbulence interacts with and eventually mixes with the plume, just as it does with the upper stratified layer of the cloud (Stretch *et al.* [19]).

4.2 'Entrainment'

Our discussion of external turbulence has shown that it affects the dispersion of a dense gas by a number of mechanisms. Yet in many models and discussions of the problem, these are described by the single term 'entrainment'. To help clarify this aspect of dense gas dispersion we now consider the three main uses of this term, which in general correspond to different mechanisms of mass transfer in regions of inhomogeneous turbulence. Since the concept of *entrainment velocity* is particularly valuable when 'box models' are used (and therefore is in widespread use), it is important to define it clearly.

(i) In regions of shear flow and inhomogeneous turbulence, the gradients of Reynolds stresses, as well as acting on any mean flow to reduce the mean velocity gradients (like molecular viscosity) and thereby induce a mean flow in towards the turbulent region (such as in turbulent or laminar jets), can induce additional mean recirculating flows in planes perpendicular to the primary shear flow, such as in flow over surfaces with different roughness. These secondary flows are primarily caused by differences in the normal stresses and are unlike any effect caused by viscosity (Townsend [20], Chapters 6 and 7) The inwards velocity in either of these two flows is called an entrainment velocity $E^{(f)}$. The shear in a jet or at the head of a gravity current induce flows in towards the region of high turbulence with characteristic velocity $E^{(f)}$ proportional to the difference between the mean velocity inside and outside the turbulent region (Fig. 9a). This entrainment velocity of the fluid transfers the external uncontaminanted fluid into the gravity current; this definition of entrainment velocity is the same as that first introduced by Morton, Taylor & Turner [21].

(ii) The volumes occupied by regions of locally intense turbulence tend to increase as the eddies induce each other to spread outwards (Fig. 9b). This tends to disperse outward any contaminant together with the region of high turbulence. This outward velocity of the *boundary* of a region of high turbulence or of a region of significant concentration of a contaminant is also called an entrainment velocity $E^{(b)}$ or is defined by means of an entrainment parameter (Turner [22], Chapter 6). It is usually defined in some fixed coordinate system (then $E^{(b)} = -2E^{(f)}$ for a round jet), but perhaps more logically ought to be defined relative to the local mean flow (in which case $E^{(b)} = -3E^{(f)}$ for a round jet). Note that $E^{(b)}$ is usually in the opposite direction to $E^{(f)}$. This definition is then further generalised to include the rate of growth of a boundary of a region of marked fluid (e.g. by a contaminant) when the region of marked fluid does not contain turbulence with any greater intensity than the surroundings (e.g. Turner [22], Section 6.2 and Webber [23]). In such cases $E^{(b)}$ may be non zero, while $E^{(f)}$ is zero, as for example in a neutrally buoyant gas diffusing upwards in the atmospheric boundary layer where $E^{(b)} \propto u_*$ and $E^{(f)} = 0$ because there is no local turbulence associated with the cloud.

(iii) The third way for marked fluid or a contaminant to be transferred out of a particular region is by turbulence *outside* the region, as shown in Fig. 9c, when the region is stably stratified (Turner [22], Chapter 9). If this external turbulence is relatively vigorous, any contaminant diffusing out of the region is quickly dispersed so that the concentration outside the region is small. Then the effective boundary of the turbulent region remains at the interface between the contaminated region and the external region; the external turbulence does not penetrate the interface because of the stable stratification. In such cases the fluxes $F^{(c)}$, $F^{(\rho)}$ of concentration or density outwards across a fixed surface can be characterised by a flux entrainment velocity $E^{(F)} = F^{(\rho)}/\Delta\rho$. $E^{(F)}$ is independent of the difference in concentration across the interface. Note that $E^{(F)}$ may be in the opposite direction to $E^{(b)}$; when $E^{(b)} < 0$, this process is often referred to as 'erosion'. It is possible for $E^{(F)}$ to be a nonzero, while $E^{(b)}$ and $E^{(f)}$ are both zero.

4.3 Details of entrainment processes in dense gas dispersion

4.3.1 Turbulence outside a gravity current head

Both dense clouds or steady sources in cross flows have gravity current heads at the beginning of the B/XF/XT phase of development. In coordinates fixed relative to the head, the main external flow travels up and over the head, but some of the flow is 'entrained' *into* the head with a characteristic velocity $E^{(f)} \approx \frac{1}{7} \sqrt{(g \frac{\Delta\rho}{\rho} h)}$. When the characteristic velocity of the external turbulence u_* is less than $|E^{(f)}|$ the turbulent eddies cannot disperse any of the contaminant outwards because they are simply entrained into the gravity current with a velocity $E^{(f)}$. In the cloud they simply make a small addition to the already high level of turbulence there $(\sim \sqrt{(g \frac{\Delta\rho}{\rho} h)})$.

When the external turbulence is relatively strong enough for u_* to be of the same order as $|E^{(f)}|$, it can diffuse matter from the cloud outwards against the inward entrainment velocity. But there continues to be a sharp interface at the head of the gravity current, between the external flow and the dense fluid within the cloud. At this stage there can be a flux of contaminant *out* of the cloud, which can be quantified by a *flux* entrainment velocity $E^{(F)}$.

When $u_* / \sqrt{(g \frac{\Delta\rho}{\rho} h)}$ is larger still, the front of the cloud ceases to have the form of a gravity current and the *outwards* flux is better thought of as a velocity $E^{(b)}$ of the whole boundary of the cloud, where $E^{(b)} \sim u_*$. The first two of these stages are similar to the sequence of events recently explored in some detail by Thomas & Simpson [24]. As well as flow visualisation studies, they also measured the volume flow rate of dense fluid into the head of a steady gravity current in the presence of grid-generated turbulence in the external flow.

4.3.2 External turbulence above the stratified upper layer of a dense gas cloud

The external flow as it travels over the cloud interacts with the stratified flow in the upper layer of the cloud. This kind of interaction between a region of turbulent motion and an adjoining region of stably stratified fluid occurs in many natural flows, for example in the upper part of the atmospheric boundary layer in convective conditions and in the oceanic mixed layer. However, not only are there no qualitative theories for this interaction, there is not even a consensus as to what are the main mechanisms in any given situation.

Four main mechanisms, illustrated in Fig.10, have been proposed for the entrainment process under these conditions. We first consider the mechanisms in the *absence* of mean velocity gradients across the interface.

(i) Turbulent eddies impinge on the interface and generate sufficiently large fluctuating velocity gradients that the local Richardson number $\frac{g}{\rho}\frac{-\partial\rho/\partial z}{(\partial u/\partial z)^2}$ is small enough for Kelvin-Helmholtz billows to grow and break, inducing molecular mixing. One expects this process to be relevant when the stratification is strong enough to damp the vertical component of turbulence at the interface; several authors (e.g. Long [25]) have commented that in this limit fluctuations near the interface are expected to be similar to those in a turbulent flow near a rigid surface moving at the same velocity as the mean flow - such as a moving belt in a wind tunnel or a free surface of a turbulent liquid flow (Hunt & Graham [26]).

(ii) With strong stratification at the interface, impinging energetic turbulent eddies distort it sufficiently for fine filaments of the stratified fluid layer to be drawn into the turbulent region where again molecular diffusion completes the mixing process (Linden [27]). Linden hypothesised that such eddies are similar to vortex rings and then, by experiment and approximate analysis, was able to estimate entrainment rates across a density discontinuity.

(iii) Turbulent eddies distort the interface and set up internal waves in the stratified layer whose energy and form depend on the stratification. For a uniformly stratified layer, as shown in Fig. 9c, the waves propagate energy away

from the interface, but their amplitude at the interface may be large enough to cause mixing. If the stratified layer is strong near the interface and weak or nonexistent far above the interface, then trapped and resonant waves of large amplitude can be induced by relatively weak turbulence.
(iv) With weak stratification the turbulent layer can also grow and entrain by the same processes as occur at the edge of a turbulent boundary layer or wake in neutral stratification; the large eddies in the turbulent layer induce large random motions in the upper layer [$\mathcal{U}$] leading to the engulfment of external fluid (Townsend [20], Chapter 6).

In the presence of strong shear across the interface, billowing motions are generated - similar to Kelvin-Helmholtz billows - and largely control the mixing; i.e. (i) is the dominant mechanism under these circumstances. For the first three of these mechanisms, (i),(ii),(iii) there is a net flux of contaminant outwards across the interface, which is quantified by the flux entrainment velocity $E^{(F)}$. When any of these mechanisms control the upward flux, the interface is sharp and the upward velocity of the bounding surface $E^{(b)}$ is small. Most experiments and theoretical models show that in these situations $E^{(F)} = u_* f_1(Ri_*)$, where Ri_* is a local Richardson number that is a measure of the ratio of local buoyancy forces to inertial forces in the turbulence. If there is a sharp jump $\Delta\rho$ between the density in the upper layer of the cloud and that in the ambient flow, Ri_* is defined as $(L_t g \Delta\rho/\rho)/u_*^2$, where L_t is the scale of the external vertical turbulence. But if there only a gradual increase in density in $(\mathcal{U})$, with density gradient $(\partial\rho/\partial z)_{\mathcal{U}}$, Ri_* is defined as $L_t N_{\mathcal{U}}/u_*^2$, where $N_{\mathcal{U}}^2 = g(-\partial\bar{\rho}/\partial z)_{\mathcal{U}}/\bar{\rho}$. Various models have been advanced for the function $f_1(Ri_*)$; usually they have the form $f_1 \propto Ri_*^{-n}$ where $5/6 \leqslant n \leqslant 3/2$, but these models have not been tested in sufficient detail.

Any model for the flux across the interface should also include, we think, a prediction of the fluctuating velocity field near the interface. By comparing calculations with local velocity measurements we should be able to isolate the relevant mechanisms of entrainment in different situations. Carruthers & Hunt [28] have recently made an analytical calculation of the statistics of the turbulent velocity field across such an interface, when there is no shear at the interface, by matching the turbulence with the internal wave field set up in the stratified layer. They have compared their predictions of spectra with the measurements at the top of the convective boundary layer by Caughey & Palmer [29].

They find that the density fluctuations ρ' at the interface are controlled by how much of the energy measured at a point in the external turbulence is at frequencies close to the buoyancy frequency of the stratified layer $N_{\mho}$. Using the form of the Eulerian time spectrum given by Tennekes [30], they show that if

$$\frac{u_*}{L_t N_{\mho}} \lesssim 1/4 \quad , \quad \sqrt{(\overline{\rho'^2})} \sim \frac{u_*}{N_{\mho}} \left(\frac{u_*}{L_t N_{\mho}} \right)^{1/3} \left(-\frac{\partial \bar{\rho}}{\partial z} \right)_{\mho} \quad .(15),(16)$$

For the upper layer of a dense cloud $N \sim (g \frac{\Delta\rho}{\rho} h)^{1/2}$ and $L_t \sim h$. This theory has been extended to situations where there is a sharp change in ρ at the top of [$\mho$]. Comparing the prediction from this theoretical approach with the observed and computed forms of the vertical turbulent velocity fluctuation $\overline{w'^2}$ suggests that it is a promising way of analysing these interfaces between turbulent and stratified flows. The theory tends to indicate that mechanism (iii) is usually dominant in the absence of strong shear. With strong shear there is some uncertainty as to whether or when (i) or (iii) dominates.

4.3.3 External turbulence dispersing a slightly dense gas

In this Section we consider how the ambient or external turbulence disperses a cloud or plume of dense gas when the difference in density between the gas and its surroundings is small enough that buoyancy forces are comparable with or weaker than the inertial forces in the turbulent flow. This situation can arise in the later stages of the B/XF/XT phase when the cloud is no longer like a gravity current, or when a cloud or plume of gas is released into a turbulent air flow with a small initial density difference $\Delta\rho_0$. The latter situation can be defined rather precisely, whereas the former situation can only be defined after the previous stages have been analysed. Therefore to understand how a turbulent shear flow affects the dispersion of a slightly dense gas, we assume that the gas is released at a very low flow rate from a porous fence of height h placed in a neutrally stable turbulent boundary layer, as sketched in Fig.11. (An adaptation of Dr. Jensen's [31] experiment!)

The general questions about such a flow which need answering are: (i) Since we expect that the cloud must affect the vertical turbulence $\overline{w^2}$, the shear stress $\overline{-uw}$, and the mean velocity gradient dU/dz, on what time scales do these changes occur? How does this relate to the timescales for the growth of the buoyancy flux $g\overline{w\rho'}/\rho(0)$, and for the growth of the plume? Presumably if the latter timescale is small enough, the turbulence is unaffected. (ii) Is it appropriate to use a vertical diffusivity K_z to calculate the vertical flux of the gas in this problem and if so, how does K_z in the cloud differ from its upwind value? (iii) How do the height and the density concentration profiles in the cloud evolve as a function of the initial density difference and the *initial density profile*?

(a) Density flux and dynamical effects of the cloud flux

As the flow passes through the fence, the fluid elements are assumed to become mixed with the dense gas from the vertical source so as to increase their density by $\Delta\rho(z)$ without changing their initial random vertical velocity w_0. Therefore initially, at x = 0, there is no correlation between the vertical velocity and the density fluctuations, ρ', so $\overline{w\rho'} = 0$ and there is no density flux. As the turbulence carries the fluid elements up and down, $\overline{w\rho'}$ increases; initially $\overline{w\rho'} = -t\,\overline{w_0^2}\,\partial\Delta\rho_0/\partial z$. If the initial density profile is uniform and weak enough, and if the turbulence is homogeneous, then $\overline{w\rho'}$ reaches its equilibrium value of $-T_L\,\overline{w_0^2}\,\partial\Delta\rho_0/\partial z$

after about two Lagrangian time scales T_L. So until $t \approx 2T_L$, the eddy diffusivity $K_z = (-\overline{w\rho'}/\partial\rho/\partial z)$ is increasing with *time*.

The growth of the density flux means that heavy fluid elements rise and light elements fall; this buoyancy flux $g\overline{w\rho'}/\rho(0)$ or gain in potential energy of the flow is balanced by a loss in the turbulent kinetic energy. The rate of change for initially isotropic homogeneous turbulence (without shear) for the vertical ($\overline{w^2}$) and horizontal ($\overline{u^2}$) components are given for $t \ll T_L$ by

$$d\overline{w^2}/dt = -(8/5)t\, N^2 \overline{w_0^2}\ , \quad d\overline{u^2}/dt = -1/5\, t\, N^2 \overline{w_0^2}\ . \qquad (17)$$

(Stretch *et al.*[19]). Because of the isotropic nature of the pressure gradients, 20% of the kinetic energy loss comes from the horizontal components. A local analysis such as this omits the fact that about one third of the turbulent energy of the horizontal turbulence near the ground in a turbulent boundary layer is induced by the vorticity of large-scale eddies well above the ground, i.e. "inactive motions" (Townsend [20], p.123). The effect of even weak stratification is to cut off some of the energy supply (by the mechanism described in Section 4.3.2); this point and the effects of shear on (17) need further investigation.

After a time of order T_L, equation (17) is no longer valid because w_0 changes. But as the plume is growing in a turbulent boundary layer the value of T_L over the depth of the plume is also growing with time, so T_L remains of the same order as t (Chatwin [32] and Hunt & Weber [33]), and therefore (17) can still be used to *estimate* the change in $\overline{w^2}$. Hence the direct effects of stratification takes a time of order (N^{-1}) to affect $\overline{w_0^2}$, which is a factor of $Ri_0^{-1/2}$ longer than the Lagrangian timescale T_L. (Note that $Ri = N^2/(dU/dz)^2 = (N\, 0.4\frac{z}{u_*})^2 \approx (N\, T_L)^2$.)

The pressure gradients and other processes must also induce a decrease in $\overline{u^2}$ and $\overline{v^2}$. If the initial stratification is large enough that $N \gg T_L^{-1}$, then the timescale for $\overline{w^2}$ to be affected is also of order N^{-1}. According to the computer simulations of Riley *et al.* [34], the decay of $\overline{w^2}$ is oscillatory and is associated with internal gravity waves.

As $\overline{w^2}$ decreases, the Reynolds stress $\tau = -\rho\overline{uw}$ also decreases, but proportionately less than $\overline{w^2}$. From the momentum equation the change, $\frac{\partial \Delta U}{\partial z} = \Delta U'$, in the mean gradient is given approximately by

$$\frac{d\Delta U'}{dt} \sim \frac{\partial^2 \tau}{\partial z^2} \quad \text{, whence if } \overline{Z} \sim h(\text{ or } t \leqslant 10\,T_L)$$

$$\frac{\Delta U'}{U'(h)} \sim \frac{Ri_0}{5}(t/(h/u_*))^3 \quad . \quad \Sigma \quad , \tag{18}$$

where $\overline{Z}$ is the mean fluid particle displacement, Σ is the shape profile, $\Sigma \approx \frac{\partial^2}{\partial z^2}\left[R(t/T_L(z))\,(-\frac{\partial \Delta\rho}{\partial z})\right]/(\Delta\rho(0)/h^3)$, $\Delta\rho$ is the density profile and $R(t/T_L)$ is the Lagrangian autocorrelation function. A plot of a typical shape profile and density gradient profile at some downwind station in a dense gas plume with a sharp density interface is shown in Fig.12. Thus Σ changes sign from negative to positive near $z \sim h$, so the velocity gradients initially *increase* for $z/h < 1$ by an amount that depends quite sensitively on the initial density and turbulence profile. Further downwind this change of sign in $d\Delta U'/dt$ should occur where $z \sim \overline{Z}$, as is observed by Stretch *et al.* [19].

To answer our first question, it appears that the order of magnitude of the timescales for the growth of the flux, the decay of $\overline{w^2}$ and the change in velocity gradients are $T_L(h)$, $Ri_0^{-1/2}\,T_L(h)$ and $Ri_0^{-1/2}\,T_L(h)$. These estimates depend on coefficients which may have large numerical values.

Further downwind the changes in $U(z)$, $\overline{w^2}$, and $-\overline{uw}$ are all dependent on each other, so that the mean momentum equation and approximate equations governing the moments of turbulent velocity fluctuations have to be solved simultaneously. The forms of these turbulence equations are a matter of controversy (see, for example, Hunt [35]); for example the effect of stable stratification on the dissipation rate ϵ appears to be a sensitive function of the velocity gradient, but most models ignore this sensitivity (e.g. Britter *et al.* [29]). Also most models ignore effects of internal-wave motion which may be important if a sharp interface develops (cf. Section 4.3.2).

(b) Vertical diffusion of the cloud

The reason for discussing these dynamical effects of the density gradient in the cloud on the turbulence is that estimates of the vertical diffusivity K_z from the vertical turbulent $\overline{w^2}$ and the Lagrangian timescale T_L are needed to calculate the development of the concentration profile.

We first recall that in a *neutrally* stratified turbulent boundary layer, when a neutrally buoyant cloud or plume is released over a height h above the ground, K_z initially increases with time until $t \sim 2T_L(h)$ where $T_L(h) \sim h u_*$. During this time the mean height $\overline{Z}$ does not increase significantly, subsequently $\overline{Z}$ increases with travel time downwind, and with the gradient of K_z, approximately according to

$$\overline{Z} \approx h + \frac{dK_z}{dz}(t - 2T_L(h)) \tag{19}$$

Note that the solution to the diffusion equation based on the *local* K_z does not predict the delay in the rise of $\overline{Z}$, and so over predicts $\overline{Z}$ by 20% at $t \sim 10T_L(h)$ or about 40 heights downwind (Hunt [36]).

If the plume is slightly dense ($Ri_0 \ll 1$), then its upward diffusion and dilution of the gas is approximately the same as that of a neutrally buoyant plume and $\overline{Z}$ is given by (19). Since the maximum value of the density difference $\Delta\rho$ is proportional to $1/\overline{Z}$, the density gradient is proportional to $\frac{g}{\rho_0}\frac{\Delta\rho}{\overline{Z}} \approx N^2 \propto 1/\overline{Z}^2$.

So as the plume diffuses upwards, if $Ri_0 \ll 1$, (17) shows that

$$\frac{1}{\overline{w_0^2}} d\overline{w^2}/dt \sim -\frac{Ri_0 t}{(t + 2h/u_*)^2} \tag{20}$$

Thus the net effect of stratification on the rate of change of $\overline{w^2}$ decreases downwind, but integration of (20) shows that even if $Ri_0 \ll 1$, eventually $\overline{w^2}$ and $\overline{u^2}$ decrease by a significant proportion of $\overline{w_0^2}$. This mathematical singularity means that the turbulence in a boundary layer cannot disperse a line source of even very slightly dense gas without being itself affected. Of course when this change is significant (which takes a progressively longer time as Ri_0 decreases), the estimates for $\overline{Z}$ used in deriving (19) are no longer valid.

Consequently for line sources or extended area sources, even if $Ri_0 \ll 1$, the value of K_z must eventually decrease below its value in the upwind boundary layer. If the source is a point source or area source, $\Delta\rho$ and $\frac{d\overline{w^2}}{dt}$ decrease fast enough that the integration of (20) leads to a finite reduction $\overline{w^2}/\overline{w_0^2}$ of the order of Ri_0. So the width of a plume of slightly dense gas has a very considerable effect on its dispersion - an effect which needs quantifying.

Recent laboratory studies and theoretical analyses have shown that, in stably stratified turbulent flows, random vertical displacements Z and vertical diffusion are much more reduced than the variance of the vertical turbulence. The explanation advanced by Britter *et al.* [29] and Pearson *et al.* [30] is that the kinetic energy of the turbulence is not sufficient to create the potential energy needed for fluid elements to have large vertical displacements, unless the fluid elements mix with their surroundings and change their density. Since this mixing takes place in a much longer time (~ 5 times) than the time for the vertical turbulent fluctuations to displace fluid elements, the calculations based on no mixing provide estimates for the reduction in $\overline{(Z-\overline{Z})^2}$.

For a weak density gradient, imposed on a turbulent flow, the reduction in the vertical displacements is shown by the result

$$\Delta\,\overline{[(Z-\overline{Z})^2]} \sim t^3 T_L N^2 \overline{w_0^2} \quad , \tag{21}$$

to increase rapidly with time. Comparing (21) with (17) shows that the reduction in $\overline{Z^2}$ is significantly greater than the reduction in $\overline{w^2}$, as Britter *et al.* [37] found. So the reduction ΔK_z in the vertical diffusivity is such that

$$\Delta K_z/K_0 \sim -t^2 N^2 \quad , \tag{22}$$

so this perturbation analysis is only appropriate when

$$\left(\frac{t}{T_L}\right) \leqslant (T_L N)^{-1} \approx Ri_0^{-1/2} \ . \tag{23}$$

This is the same timescale on which $\overline{w_0^2}$ is affected by stratification. Note that the reduction in K_z does not appear to be a *local* function of Richardson number in this limit of weak stratification. With larger Ri, the limit of $\overline{(Z-\overline{Z})^2}$ does appear, in homogeneous turbulence, to be a function of the local stratification and turbulence (Britter *et al.* [37]).

In a growing slightly dense plume where the mean value of $T_L \sim t/10$, (for $t \geqslant 10\,T_L(0)$), and $N^2 \propto N^2(0)\, h^2/\overline{Z}^2$, (21) shows that the average decrease in K_z is given by $\langle \Delta K_z \rangle / \langle K_0 \rangle \sim -Ri_0$. This is approximately constant downwind.

(c) Effects of buoyancy forces and initial conditions on the evolution of the density profile

All the processes we have discussed so far affect the distribution of concentration in the plume. Visual observations and measurements show that plumes either tend to diffuse upwards with approximately the same profile of mean concentration

$C(z/\overline{Z})$ or their profiles change significantly with the formation of high gradients of concentration near the ground and some diffusion upwards of the material above the level of the high gradients. Definitive experiments and theory need doing to estimate the transition between these two types of behaviour, perhaps based on the ideas of Puttock [39] and Posmentier [40].

For the initial growth of the plume one or two results about the structure of line clouds can be derived from the diffusion equation written in the form:

$$\frac{d\Delta\rho}{dt} = \frac{\partial}{\partial z}\left(K_z \frac{\partial\Delta\rho}{\partial z}\right) \tag{24}$$

where $\Delta\rho$ is the density difference,

$$K_z = K_0 + \Delta K_z \quad ,$$

$K_0 = \kappa z u_*$ and ΔK_z is given by (22) ($\kappa \approx 0.4$).

Thence, following Chatwin [32], the rate of rise of the mean height $\overline{Z}$ of the cloud is given by

$$d\overline{Z}/dt = \frac{1}{M}\int_{-\infty}^{\infty}\int_{0}^{\infty}\Delta\rho\,\frac{\partial K_z}{\partial z}\,dz.dx = \frac{-1}{M}\int_{-\infty}^{\infty}\int_{0}^{\infty}K_z\,\frac{\partial\Delta\rho}{\partial z}\,dz.dx \tag{25}$$

$$\text{where } M = \int_{-\infty}^{\infty}\int_{0}^{\infty}\Delta\rho\;dz.dx \quad . \tag{26}$$

(To a close approximation this also gives the rate of rise of a plume (Hunt & Weber [33]).) Thus if $\partial\Delta\rho/\partial z < 0$, the mean height of the plume must rise, and in the limit of $\Delta K_z \to 0$, $d\overline{Z}/dt = \kappa u_*$. At the top of a plume, $\partial\Delta K_z/\partial z < 0$, which shows how the larger the initial density gradient, the slower the rise in $\overline{Z}$.

Differentiating (24) leads to the following equations for $(-\partial\Delta\rho/\partial z)$, or N^2, namely

$$\frac{dN^2}{dt} = \frac{\partial^2}{\partial z^2}(N^2K_z) \quad . \tag{27}$$

Since ΔK_z is proportional to $-t^2N^2$, the rate of increase of N^2, say at the height z_m where N^2 is a maximum, depends on the second derivative of N^2. This depends on the distribution of concentration in the cloud, and can be characterised by $\sigma = -(N^2)_{zz}(z_m)\overline{Z}^2/N^2(z_m)$. Note that $\overline{Z}$ increases with t as the cloud grows.

An approximate argument indicates that if $\sigma \gg 1$ and if $Ri_0 > 0.2$ the density gradient must steepen, and if $Ri_0 \leqslant 0.2$, or if $\sigma \sim 1$, then N_m^2 decreases, corresponding respectively to profiles (ii) and (i) in Fig.8b. Thus in a slightly dense step-like profile the maximum density gradient can steepen, at the same time as $\overline{Z}$ can increase, while on the other hand in a typical Gaussian profile the gradient decreases and $\overline{Z}$ increases. Obviously any quantitative estimates depend crucially on the estimates of ΔK_z in an evolving stratified plume or cloud. This is an important problem for further research.

5. Concluding remarks

The main general conclusion of this study is that the dispersion of a dense gas is a complicated and sensitive function of the particular atmospheric and release conditions. Therefore, it seems unlikely that a single simplified model, such as are in common use at the present, could accurately model the many different types of release conditions or indeed the different phases of a particular release. Models should be developed with particular release conditions in mind and attempt only to model particular phases of the evolution of a dense gas cloud.

Acknowledgement

This work was supported by the U.K. Health and Safety Executive under contract 1918/01.01. We are grateful for stimulating conversations with J.McQuaid and A.Mercer of HSE, and at Cambridge University D.J.Carruthers, J.E.Simpson, and N.H.Thomas of DAMTP, and D.Stretch of the Engineering Department.

References

1. Rottman, J.W.; Simpson, J.E.: The initial development of gravity currents from fixed-volume releases of heavy fluids. IUTAM Symposium on atmospheric dispersion of heavy gases and small particles, Delft, The Netherlands, August 1983.

2. Hall, D.J.: Validation of wind tunnel models of dense gas releases. IUTAM Symposium on atmospheric dispersion of heavy gases and small particles, Delft, The Netherlands, August 1983.

3. McQuaid, J.: Large-scale experiments on the dispersion of heavy gas clouds. IUTAM Symposium on atmospheric dispersion of heavy gases and small particles. Delft, The Netherlands, August 1983.

4. Picknett, R.G.: Dispersion of dense gas puffs in the atmosphere at ground level. Atmospheric Environment 15 (1981) 509-525.

5. Castro, I.P.; Robins, A.G.: The flow around a surface mounted cube in uniform and turbulent streams. J. Fluid Mech. 79 (1977) 307-335.

6. Mobbs, F.R.: Spreading and contraction at the boundaries of free turbulent flows. J. Fluid Mech. 33 (1968) 227-240.

7. Csanady, G.T.: Turbulent diffusion in the environment. D. Reidel Publishing Company 1973.

8. Fackrell, J.E.; Robins, A.G.: Concentration fluctuations and fluxes in plumes from point sources in a turbulent boundary layer. J. Fluid Mech. 117 (1982) 1-26.

9. Simpson, J.E.: Gravity currents in the laboratory, atmosphere, and ocean. Ann. Rev. Fluid Mech. 14 (1982) 213-234.

10. Britter, R.E.; Simpson, J.E.: Experiments on the dynamics of a gravity current head. J. Fluid Mech. 88 (1978) 233-240.

11. Britter, R.E.: The spread of a negatively buoyant plume in a calm environment. Atmospheric Environment 13 (1979) 1241-1247.

12. Rottman, J.W.; Simpson, J.E.: Gravity currents produced by instantaneous releases of a heavy fluid in a rectangular channel. J. Fluid Mech. 135 (1983) 95-110.

13. Simpson, J.E.; Britter, R.E.: A laboratory model of an atmospheric mesofront. Quart. J. Roy. Met. Soc. 106 (1980) 485-500.

14. Hootman, B.W.; Blumen, W.: Analysis of nighttime drainage winds in Boulder Colorado during 1980. Monthly Weather Review 111 (1983) 1052-1061.

15. Kaimal, J.C.; Wyngaard, J.C.; Haugan, D.A; Cote, O.R.; Izumi, Y.; Caughey, S.J.; Readings, C.J.: Turbulence structure in the convective boundary layer. J. Atmos. Sci. 33 (1976) 2152-2169.

16. Lumley, J.L.; Zeman, O.; Seiss, J.: The influence of buoyancy on turbulent transport. J. Fluid Mech. 84 (1978) 581-597.

17. Colenbrander, G.W.; Puttock, J.S.: Maplin Sands experiments 1980: Interpretation and modelling of liquified gas spills on the sea. IUTAM Symposium on atmospheric dispersion of heavy gases and small particles, Delft, The Netherlands, August 1983.

18. Britter, R.E.: The ground level extent of a negatively buoyant plume in a turbulent boundary layer. Atmospheric Environment 14 (1980) 779-785.

19. Stretch, D.; Britter, R.E.; Hunt, J.C.R.: The dispersion of slightly dense contaminants in turbulent boundary layers. IUTAM Symposium on atmospheric dispersion of heavy gases and small particles, Delft, The Netherlands, August 1983.

20. Townsend, A.A.: The structure of turbulent shear flows. Cambridge University Press 1976.

21. Morton, B.R.: Taylor, G.I.; Turner, J.S.: Turbulent gravitational convection from maintained and instantaneous sources. Proc. Roy. Soc. London A234 (1956) 1-23.

22. Turner, J.S.: Buoyancy effects in fluids. Cambridge University Press 1973.

23. Webber, D.M.: Gravity spreading in dense gas dispersion models. IUTAM Symposium on atmospheric dispersion of heavy gases and small particles, Delft, The Netherlands, August 1983.

24. Thomas, N.H.; Simpson, J.E.: Gravity currents in turbulent surroundings: laboratory studies and modelling implications. Oxford University Press (to appear).

25. Long, R.R.: A theory of mixing in a stratified fluid. J. Fluid Mech. 84 (1978) 113-124.

26. Hunt, J.C.R.; Graham, J.M.R.: Free stream turbulence near plane boundaries. J. Fluid Mech. 84 (1978) 209-235.

27. Linden, P.F.: The interaction of a vortex ring with a sharp density interface: a model for turbulent entrainment. J. Fluid Mech. 60 (1973) 467-480.

28. Carruthers, D.J.; Hunt, J.C.R.: Velocity fluctuations near an interface between a turbulent region and a stably stratified layer. Unpublished manuscript 1983.

29. Caughey, S.J.; Palmer, S.G.: Some aspects of turbulence through the depth of the convective boundary layer. Quart. J. Roy. Met. Soc. 105 (1979) 811-827.

30. Tennekes, H.: Eulerian and Lagrangian time microscales in isotropic turbulence. J. Fluid Mech. 67 (1975) 561-567.

31. Jensen, N.O.: On the dilution of a dense gas plume: experimental investigation of the effect of surface mounted obstacles. IUTAM symposium on atmospheric dispersion of heavy gases and small particles, Delft, The Netherlands, August 1983.

32. Chatwin, P.C.: The dispersion of a puff of passive contaminant in the constant stress region. Quart. J. Roy. Met. Soc. 94 (1968) 350-360.

33. Hunt, J.C.R.; Weber, A.H.: A Lagrangian statistical analysis of diffusion from a ground-level source in a turbulent boundary layer. Quart. J. Roy. Met. Soc. 105 (1979) 423-443.

34. Riley, J.J.; Metcalfe, R.W.; Weissman, M.A.: Direct numerical simulations of homogeneous turbulence in density stratified fluids. In 'Nonlinear properties of internal waves'. La Jolla Inst., Conference Proceedings of American Institute of Physics, No.76, New York 1981.

35. Hunt, J.C.R. (ed): The proceedings of the IMA conference on models of turbulence and diffusion in stably stratified regions of the natural environment. Oxford University Press (to appear).

36. Hunt, J.C.R.: Diffusion in the stable boundary layer. In 'Atmospheric turbulence and air pollution modelling'. Ed. F.T.M. Nieuwstadt & H. van Dop. Reidel, 1982.

37. Britter, R.E.; Hunt, J.C.R.; Marsh, G.L.; Snyder, W.H.: The effects of stable stratification on turbulent diffusion and the decay of grid turbulence. J. Fluid Mech. 127 (1983) 27-44.

38. Pearson, H.J.; Puttock, J.S.; Hunt, J.C.R.: A statistical model of fluid-element motions and vertical diffusion in a homogeneous stratified turbulent flow. J. Fluid Mech. 129 (1983) 219-249.

39. Puttock, J.S.: Turbulent Diffusion. Ph.D. dissertation, Cambridge University, 1976.

40. Posmentier, E.S.: The generation of salinity fine structure by vertical diffusion. J. Phys. Oceanogr 7. (1977) 298-300.

(1) Initial Phase	I/XF
(2) Gravity Spreading Phase	B/XF
(3) Nearly Passive Phase	B/XF/XT
(4) Passive Phase	XF/XT

Fig.1. The four phases of dense gas dispersion due to a sudden release. The predominant forces acting in each phase are given by the abbreviated notation: I, inertia; XF, external mean flow; B, buoyancy; XT, external turbulence.

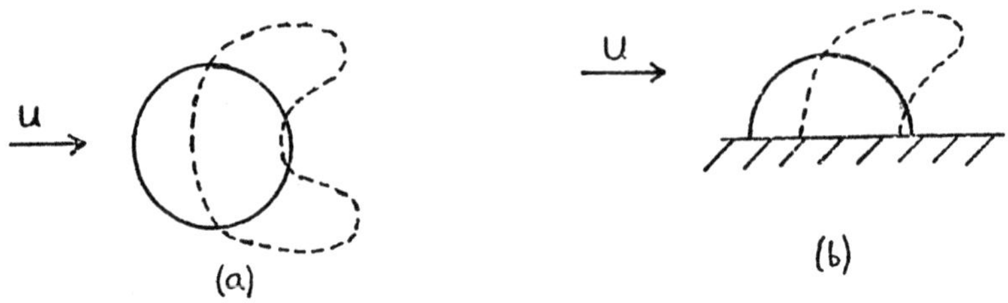

Fig.2 Sketches of boundary contour shapes for the release of a stagnant fluid into uniform cross flow: (a) release of a cylinder of fluid (plan view), (b) release of semi-cylindrical ridge of fluid (profile view).

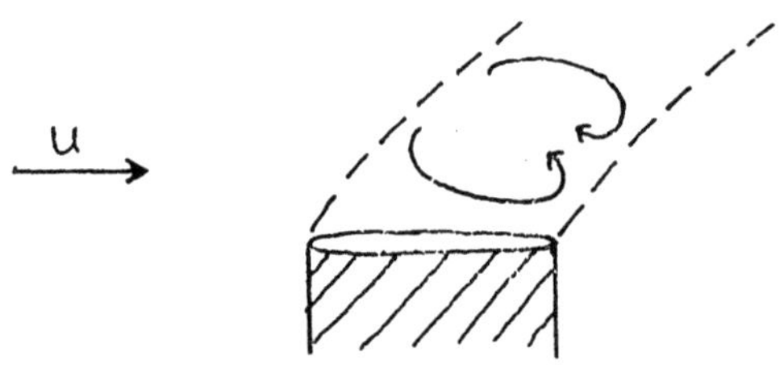

Fig.3 Sketch of a continuous release of a fluid into a cross flow, showing similar vortex roll-up as in Fig.2(a).

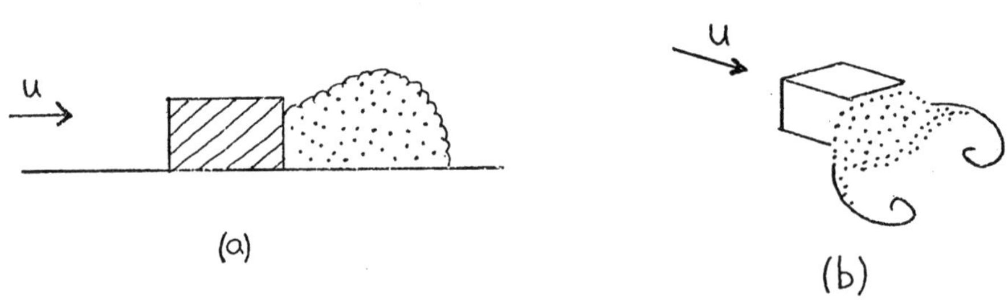

Fig.4 Sketches of the release of a heavy fluid downwind of an obstacle: (a) profile view, (b) three-quarters view.

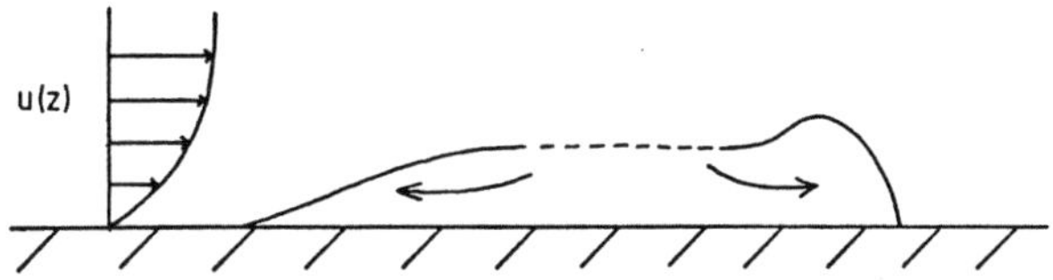

Fig.5. Profile-view sketch of a heavy fluid collapsing into a shear flow.

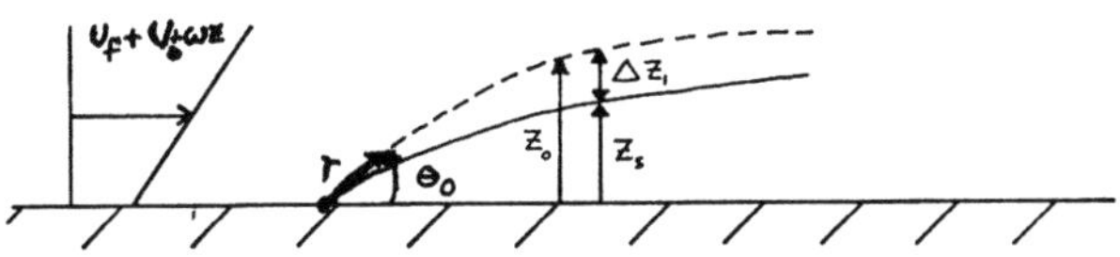

Fig.6 Diagram for an approximate calculation of shape of the head of a gravity current in a uniform shear flow.

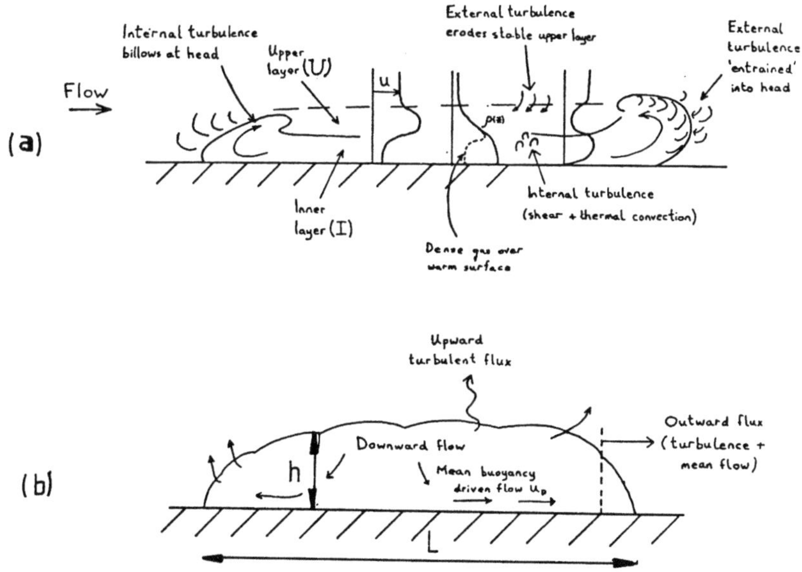

Fig.7. A heavy gas cloud in the nearly-passive phase (a) transition from gravity-spreading phase, (b) transition to passive phase. Arrows indicate motion in a frame of reference moving with the cloud.

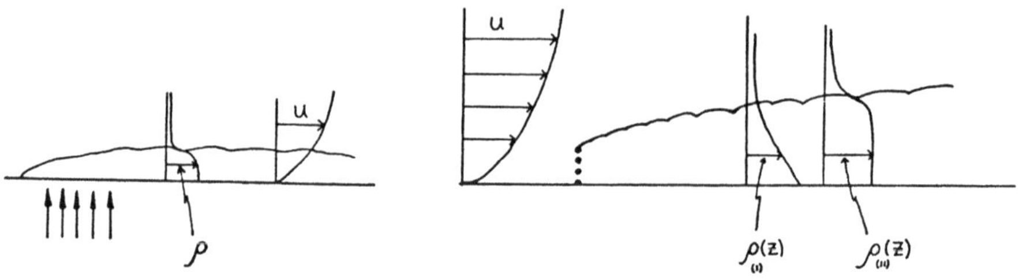

Fig.8 Two methods for generating dense gas plumes: (a) ground-level source, (b) 'fence' source.

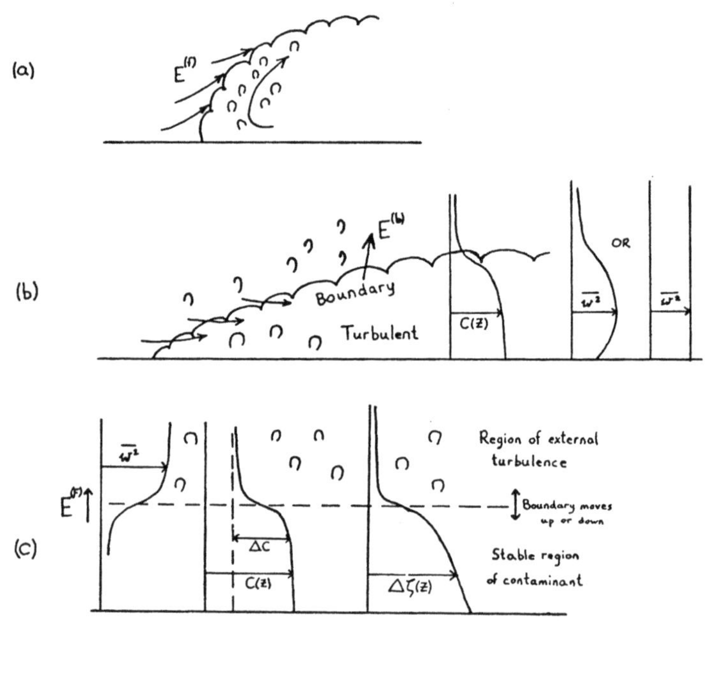

Fig.9. Three definitions of 'entrainment velocity': (a) an inviscid *flow* velocity $E^{(f)}$, (b) a *boundary* velocity $E^{(b)}$, (c) a *flux* entrainment velocity $E^{(f)}$.

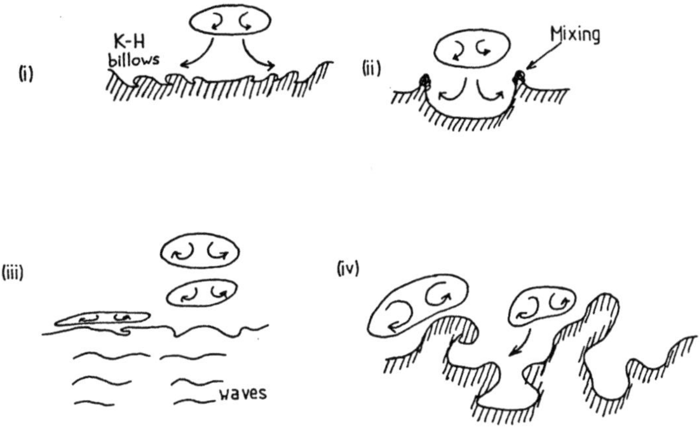

Fig.10. Four mechanisms for producing a flux of contaminant across a stable density interface.

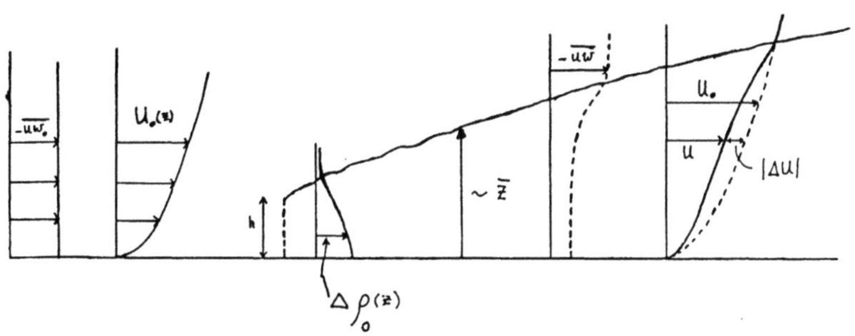

Fig.11. Dispersion of a slightly dense gas in a turbulent boundary layer.

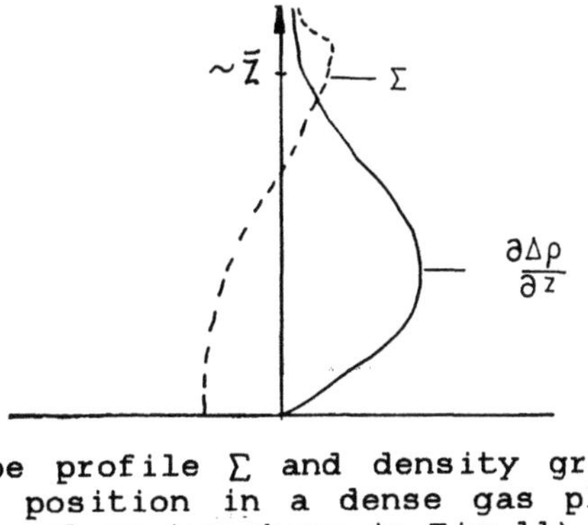

Fig.12. The shape profile Σ and density gradient at some downwind position in a dense gas plume in a turbulent boundary (as shown in Fig.11).

Gravity Spreading in Dense Gas Dispersion Models

D M WEBBER

Safety and Reliability Directorate
Wigshaw Lane, Culcheth, Warrington WA3 4NE, UK.

Summary

Different models of the spreading of a heavy gas cloud are compared within a unified analytic framework. Models agree on the radial spreading law but disagree on entrainment. It is shown how the box-model concept may be developed to give a smoother, more detailed description of the cloud.

Introduction

In modelling heavy gas cloud dispersion a major motivation is the desirability of having accurate safety analyses where an accidental release of a heavy cloud may cause a hazard due to fire, explosion, or toxicity. Currently, however, there is a wide variety of different models which give different predictions for gas dispersion. This is illustrated by the predictions commissioned by the Health and Safety Executive for the Thorney Island trials [1], an example of which is shown in Fig 1. With such model dependent results it is not clear which model the safety analyst is best advised to take. Indeed, as the results of the 3d codes are no more in mutual accord than are the box model predictions, it is not even clear which type of model is best adopted.

The discrepancies arise in turbulence or entrainment modelling. We shall examine some models here. We shall not consider 3d codes further, but shall take advantage of the inherent scrutability of box models in an attempt to expose the points of agreement and disagreement in the modelling. We shall consider only instantaneous, ambient temperature releases.

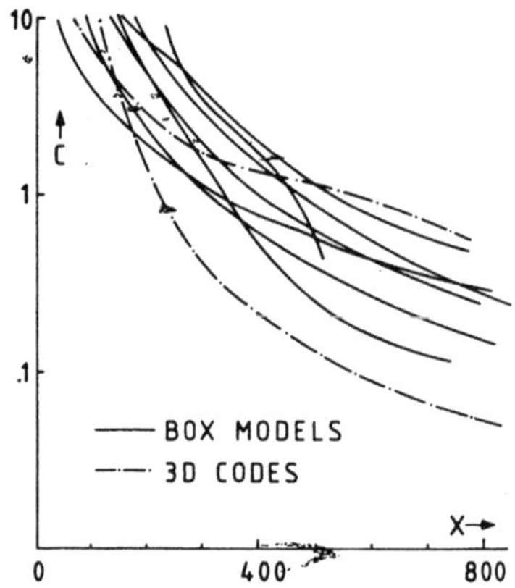

Fig 1.

Concentration vs downwind distance as predicted by various models for a specified Thorney Island Trial [1].

Structure of models

Box models of cloud dispersion, of which there are a number in the literature, picture the cloud as a cylinder of radius R(t) and height h(t), within which the density is ρ(t). The density relative to air can be expressed in terms of $\Delta \equiv (\rho-\rho_a)/\rho$ or $\Delta' \equiv (\rho-\rho_a)/\rho_a$. At ambient temperature the concentration of contaminant gas is proportional to Δ', and so the buoyancy variable $b \equiv gV\Delta'/\pi$ is conserved. Two other equations are required to describe the evolution of the cloud in terms of ρ(t), R(t), and h(t). These take the form

$$\frac{dR}{dt} = U_f \tag{1}$$

$$\frac{dh}{dt} = U_T - \frac{2h}{R}\left(U_f - U_E\right) \tag{2}$$

where U_f is the front velocity, U_T is a 'top entrainment' velocity and U_E is an 'edge entrainment' velocity. All of these must be modelled in terms of the independent variables.

The front velocity

The front velocity was introduced by Van Ulden [2] as $U_f = K\sqrt{gh\Delta}$. Subsequent authors have used either this or

$$U_f = K\sqrt{gh\Delta'} \tag{3}$$

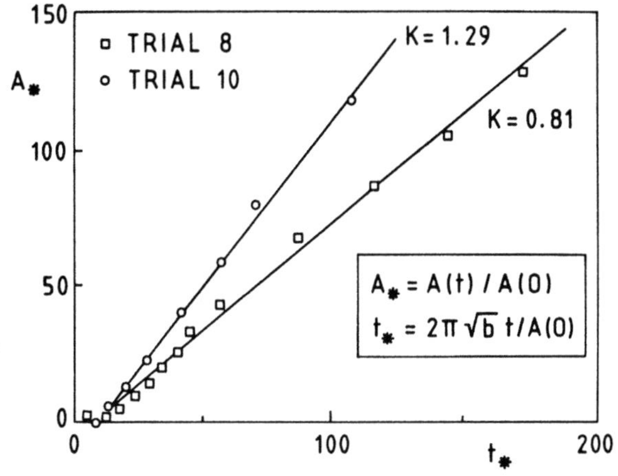

Fig 2 Top area vs time for two of the Porton Trials [3]. Values of K found from other trials lie mostly between these.

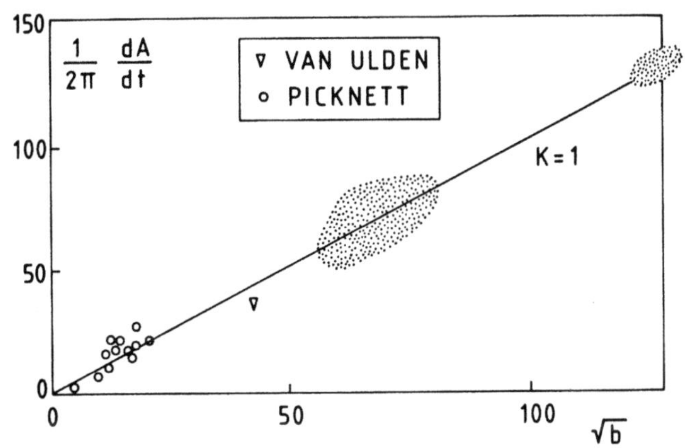

Fig 3 The area increase rate vs $\sqrt{b}$ (both in m^2/s). Existing data are consistent with K = 1. The shaded regions show where Thorney Island data are expected.

where K is a constant. For typical cloud densities the difference is not significant [2], particularly as some of the difference can be absorbed into a redefinition of K. We shall consider the form (3) throughout. This can be derived by balancing the mean hydrostatic pressure difference at the front $(\rho-\rho_a)gh/2$ with a resistance pressure $\rho_a U_f^2/2K^2$ due to the ambient air.

From (1) and (3) it follows that the top area $A = \pi R^2$ of the cloud behaves as

$$\frac{dA}{dt} = 2\pi K\sqrt{b} \tag{4}$$

This linear form for A(t) is in accord with Porton data [3], and that of ref [2] with $K \approx 1.0 \pm .2$ (see Figs 2,3). We conclude that in this aspect a wide variety of models [2-9] are in mutual agreement and indeed in reasonable agreement with data.

Entrainment

The modelling of the height of the cloud provides a different story. Models differ considerably. Knowing R(t) it is convenient to express h as a function of R. From (1) and (2) we have

$$\frac{dh}{dR} = \frac{U_T}{U_f} - \frac{2h}{R}\left(1 - \frac{U_E}{U_f}\right) \tag{5}$$

Using conservation of buoyancy the density can be eliminated from the velocity ratios leaving them as functions of R,h. In the models of refs [2-6] h and ρ only appear in the combinations $gh\Delta'$ leaving these ratios as functions of R only, and (5) is thus linear in h. In ref [7] the combination $gh\Delta$ is used. We can approximate this by $gh\Delta'$ without changing any essential physics and then equation (5) is linear here too. In the models of refs [8,9] equation (5) is intrinsically non-linear but the general solution can be found [10].

The edge entrainment term U_E/U_f is generally assumed to be zero [4,6] or constant [2,3,8,9] corresponding to entrainment caused by the gravity driven slumping of the cloud. In ref [7] U_E/U_f is taken as proportional to U_f.

For top entrainment U_T the models fall into two groups. The first [3,6,8,9] takes U_T to be due to the ambient turbulence, typified by the friction velocity u_*, but with a Richardson number suppression factor due to the stable density interface. The other group takes U_T to be dependent on the radial slumping velocity U_f with [7] or without [4] a Richardson number suppression factor. This is interpreted as being due to shear at the top of the cloud [4] or at the ground [7].

MODEL	U_E	U_T	TURBULENCE
VAN ULDEN 1974	αU_f	0	
GERMELES & DRAKE 1975	0	βU_f	TOP SHEAR
PICKNETT 1978	αU_f	$\frac{\beta u_*}{Ri_1}$	SUPPRESSED AMBIENT
{FRYER & KAISER / COX & CARPENTER} 1979	αU_f	$\frac{\beta u_*}{Ri_2}$	SUPPRESSED AMBIENT
EIDSVIK 1980	$\frac{\alpha U_f^2}{U_f(o)}$	$\frac{\beta v_E}{a + Ri_3}$ $v_E^2 \sim \frac{4}{9} U_f^2 + U_W^2$	SUPPRESSED GROUND SHEAR
FAY & RANCK 1981	0	$\frac{\beta u_*}{\sqrt{a^2 + Ri_1^2}}$	SUPPRESSED AMBIENT

$Ri_1 = \frac{gh\Delta'}{u_*^2}$	$Ri_2 = \frac{gl\Delta'}{u_*^2}$ $l = h.\left(\frac{h}{h_r}\right)^{\mu-1}$	$Ri_3 = \frac{gh\Delta}{v_E^2}$

Table 1. Entrainment models. α, β, μ, and a are constants. h_r is a constant height [14] and $\mu = 0.48$ in ref [8].

Of the first group three [5,8,9] take U_T proportional to u_*/Ri (but with different definitions of Ri) and have to cut the model off when Ri decreases to order one. An improvement is made in [6] where Ri is replaced by $\sqrt{(Ri)^2 + a^2}$ with constant a. This allows the model to be taken into the passive phase where $Ri \ll 1$.

Of the second group, [4] takes $U_T \sim U_f$, whereas [7] takes $U_T \sim u/(Ri + a)$ with $u^2 \sim \sqrt{U_w^2 + 4U_f^2/9}$, where U_w is the wind velocity. This latter model is also applied to the passive phase. The Richardson number is defined using the velocity scale u.

MODEL	h/h_0
VAN ULDEN 1974	$r^{-2(1-\alpha)}$
GERMELES & DRAKE 1975	$(1-\gamma)r^{-2} + \gamma r$
PICKNETT 1978	$(1-\gamma)r^{-2(1-\alpha)} + \gamma r^4$
{FRYER & KAISER, COX & CARPENTER} 1979	$\left[(1-\gamma)r^{-2\mu(1-\alpha)} + \gamma r^4\right]^{1/\mu}$
EIDSVIK 1980	$\frac{e^{2\alpha(1-\frac{1}{r})}}{r^2} + \frac{3\gamma\phi_1^2}{\phi_2^3}\frac{1}{r^2}\int_1^r dw\, e^{2\alpha(\frac{1}{w}-\frac{1}{r})}\frac{w^2(w^2+\phi_2^2)^{3/2}}{(w^2+\phi_1^2)}$
FAY & RANCK 1981	$r^{-2}\left[1 + \gamma\phi^2\left(r^2\sqrt{r^4+\phi^4} - \sqrt{1+\phi^4} - \phi^4 \ln\left\{\frac{r^2+\sqrt{r^4+\phi^4}}{1+\sqrt{1+\phi^4}}\right\}\right)\right]$

Table 2. Variation of height of cloud with radius. h_o and R_o are initial values and $r \equiv R/R_o$. α, γ, ϕ_i and μ are constants (see Tables 1,3).

The models and solutions are summarised in Tables 1,2 and 3. They behave quite differently. This is particularly evident in the limit of a calm atmosphere where u_* is zero.

The first group of models have U_T/U_f zero but in the second this ratio is a non-zero constant.

We see from (5) that at large enough R the sign of $^{dh}/dR$ is different in the two cases. In fact the latter case implies a non-conservation of energy [11]. It is clear from (5) that in a calm atmosphere we require $RU_T/U_f \to 0$ as $R \to \infty$.

MODEL	CONSTANTS
GERMELES & DRAKE 1975	$\gamma \sim R_0/h_0$
PICKNETT 1978	$\gamma \sim u_*^3 R_0 h_0^{-5/2} \Delta_0'^{-3/2}$
{FRYER & KAISER / COX & CARPENTER} 1979	$\gamma \sim u_*^3 R_0 h_0^{-3/2-\mu} \Delta_0'^{-3/2}$
EIDSVIK 1980	$\gamma \sim R_0/h_0$ $\phi_1 \sim \phi_2 \sim \sqrt{\Delta_0' h_0}/U_w$
FAY & RANCK 1981	$\gamma \sim u_*^3 R_0 h_0^{-3/2} \Delta_0'^{-3/2}$ $\phi \sim \sqrt{\Delta_0' h_0}/u_*$

Table 3. Scale dependence of the constants in Table 2 for the various models. Dependence on the friction velocity u_* or wind velocity U_w is also shown. The constants ϕ_i are unimportant in the gravity dominated phase, but determine where that phase will end (at $r \sim \phi$).

The scale dependence of the models is given by the behaviour of the constants γ and ϕ_i (in Tables 2,3) under changes in the initial values R_o, h_o, Δ_o' of R, h, Δ' and in u_* or U_w.

The models of references [6] and [7] offer an improvement over earlier ones by introducing a continuous transition to passive behaviour. The transition only occurs in the entrainment velocity U_T. Chatwin [12] has pointed out that the transition should also occur in the front velocity and multiplies (3) by a factor $(1 + \sqrt{a/Ri})$ where $Ri = gh\Delta'/u_*^2$ so that as Ri becomes small $U_f \to K\sqrt{a}\ u_*$.

Generalising the box model approach

One of the criticisms made of box models is that such models consider a grossly over simplified concentration distribution. On the other hand these models are at much the same level as simple passive dispersion models which assume a mean concentration distribution (gaussian rather than 'top hat') and model the time evolution of geometrical parameters. A more detailed concentration distribution can be considered* which allows one to unify the gravity dominated and passive dispersion models as follows.

Consider an axially symmetric cloud with density

$$\Delta'(t,\underline{x}) = \Delta'_c(t)f(r)g(z), \text{ where } f(0) = g(0) = 1 \text{ and } f(\infty) = g(\infty) = 0.$$

If f and g are Heaviside functions we have a 'top hat' model. If they are allowed to evolve in time into gaussians then a gaussian passive dispersion model can be recovered. Defining

$$H = \int_0^\infty g(z)dz \quad ; \quad W^2 = \int_0^\infty f(r)r.dr \tag{6}$$

the conserved variable is $b = g\Delta'_c W^2H$. The model thus requires equations for W, H and for the evolution of f and g from Heaviside functions into gaussians. To see how this works in the radial dimension take as an example $f(r) = 1$ for $r<R(t)$ and $f(r) = \exp\left(-(r-R)^2/2\sigma_E^2\right)$ for $r>R$. The effective radius is $W^2 = R^2 + \sqrt{2\pi}\, R\sigma_E + 2\sigma_E^2$. The equations

$$\frac{dR}{dt} \sim K\sqrt{g\Delta'_c H} \quad ; \quad \frac{d\sigma_E}{dt} \sim K\sqrt{a}\, u_* \tag{7}$$

give a very similar result to that of Chatwin [12] for the effective radius, but $Ri = gH\Delta'_c/u_*^2 \gg a$ now corresponds to a uniform concentration and $Ri \ll a$ to a gaussian. Similar ideas can be employed for the vertical concentration distribution, where the possibility opens up of modelling the behaviour of density interfaces as observed in mixing box experiments.

---

* The model described here is currently under development at SRD by the author in collaboration with C J Wheatley.

Such an evolution of concentration distribution would complete the smoothing of the transition to passive behaviour begun in [6,7] by having a smooth transition of U_T and continued in [12] by having a smooth transition of U_f.

Conclusions

We have examined assorted models of heavy gas cloud dispersion for instantaneous releases at ambient temperature. The models are essentially in agreement for the radial spreading law during the gravity dominated phase and are consistent with data. We have given solutions to the models (some of which are given by the original authors) for the evolution of the geometry of the cloud (see Table 2). The models are significantly different and scale differently (see Table 3).

More recent models [6,7] have included a smooth transition to passive behaviour in the entrainment velocity U_T. Such a transition is also expected in the front velocity U_f. This has been modelled in [12]. We have shown how the box model concept can be generalised to incorporate an evolving concentration distribution. The evolution of the entrainment and front velocities fall naturally into this scheme. Before developing an absolutely specific model of top entrainment within this framework it would be desirable to know which of the existing models best describes the behaviour of a cloud. Fortunately the data from the Thorney Island trials [1], involving detailed photographic and concentration records, will soon be available. Furthermore the calm atmosphere experiments of ref [13] should severely constrain models in the windless limit. There is in the near future, therefore, the prospect of being able to distinguish between existing entrainment models, and of having enough detailed experimental data to make the construction of more detailed models (such as the one described here) a viable proposition.

Acknowledgement

This paper relies heavily on work funded in part by the Commission of the European Communities (Indirect Action Programme on the Safety of Thermal Water Reactors). The C.E.C. takes no responsibility for the use to which this work may be put.

References

1. McQuaid, J. Contribution to this conference.

2. Van Ulden, A P. On the spreading of a heavy gas released near the ground. 1st Int.Symp. on Loss Prevention and Safety Promotion in the Process Industries, The Netherlands (1974).

3. Picknett, R G. Field experiments on the behaviour of dense clouds. Porton Down Report Ptn.1L/1154/78/1, (1978).

4. Germeles, A E and Drake, E M. Gravity spreading and atmospheric dispersion of LNG vapour clouds. US Coast Guard report CG-D-24-76 (1976).

5. Fay, J A. Gravitational spread and dilution of heavy vapour clouds. 2nd Int. Symp. on Stratified Flows, Trondheim (1980).

6. Fay, J A and Ranck, D. Scale effects in liquified fuel vapour dispersion. M.I.T. Report DOE/EP-0032 UC-11 (1981).

7. Eidsvik, K J. A model for heavy gas dispersion in the atmosphere. Atmospheric Environment 14 (1980) 769.

8. Fryer, L S and Kaiser, G D. DENZ, a computer program for the calculation of dense toxic or explosive gases in the atmosphere. UKAEA report SRD R152 (1979).

9. Cox, R A and Carpenter, R J. Further developments of a dense vapour cloud dispersion model for hazard analysis. Symp. on Heavy Gas Dispersion, Frankfurt (1979).

10. Webber, D M. The physics of heavy gas cloud dispersal. UKAEA report SRD R243 (1983).

11. Fay, J A. Some unresolved problems of LNG vapour dispersion. Gas Research Institute Workshop, MIT, (1981).

12. Chatwin, P C. The incorporation of wind shear effects into box models of heavy gas dispersion. Report on Contract 1189.1/01/01, RLSD, HSE Sheffield (1983).

13. Havens, J. Contribution to this conference.

14. Taylor, R J, Bacon, N E and Warner J. Scale length in atmospheric turbulence as measured from an aircraft. Quart. J R Met. Soc. 96 (1970) 750.

Experimental and Theoretical Studies in Heavy Gas Dispersion. Part II. Theory

Ø. Jacobsen and T.K. Fanneløp*

Division of Aero- and Gas Dynamics
Department of Mechanical Engineering
Norwegian Institute of Technology
Trondheim, Norway

*Now with the Swiss Federal Institute of Technology, ETHZ, CH-8092 Zürich, Switzerland.

Abstract

The motion and dilution of heavy gas clouds are studied on the basis of the thin-layer equations. The exact similarity solutions already developed by the authors for the case of instantaneous releases in the absence of wind with and without entrainment, are used herein to obtain solutions for heavy clouds and plumes moving with the wind from instantaneous and quasi-steady sources. Certain adjustments to correct for the nonsimilar initial conditions are proposed, and it is demonstrated that similarity prevails shortly after release also for "box-model" type starting conditions. The solution obtained provide all variables of interest (velocity, density, concentration) as function of space and time, and they are not limited to small density differences.

1. Introduction

For large-scale releases of hydrocarbons and other flammable heavy gases, it is essential to know, not only the average concentration as function of time, but also the distribution within the cloud. The secondary events which may follow the accidental release, i.e. fire and explosion, are controlled by the extent and location of regions with concentrations within the flammable range. For very large clouds, the portion having a concentration in the critical range, will be small in comparison with the total cloud volume. The present model represents an alternative both to the crude box model which provide only average concentrations, and the overly sophisticated numerical codes which show large discrepancies. (Havens [1,2]) The use of the thin-layer model leads to exact analytical solutions, as easy to interpret and to use as those of the box model, although the analysis leading to these results is considerably more complex. The analytic solutions are of interest both for actual prediction purposes and for testing the validity and accuracy of numerical codes. Since these similarity solutions do not take account of the initial release conditions, it is therefore of interest to study the importance of the initial conditions and the circumstances for which the similarity solution will have adequate accuracy.

2. Similarity Solutions Based on the Thin Layer Equations.

The thin-layer approach to the analysis of heavy gas clouds is similar to and builds on the known solutions for the spreading of oil slicks on water. The popular "box model" in its most primitive form is furthermore identical with the socalled "flat-slick approximation" for oil slicks.[3] Applications of this and similar 2-D models to predict the spread of wind-driven heavy gas plumes from steady sources, are preceded by published solution for the shape and thickness of oil plumes, from steady and timevarying sources, drifting with a steady current.[4] Common to all such models is the assumption that the flow occurs primarily in the horizontal direction and that the horizontal velocity and other relevant variables can be characterized by appropriate averaged values. The thin layer model is questionable in the front region where the vertical and horizontal velocities can be of the same order of magnitude. Empirical information indicates that the effect of ground friction on the spreading process is negligible. The remaining viscous stress opposing the outward motion can be expressed as $\tau_e = \rho_a v_e u$ where ρ_a is the density of air, v_e is the entrainment velocity and u is the local cloud velocity. Denoting the local thickness h and the density ratio of interest $r = (\rho - \rho_a)/\rho_a$, we obtain the relevant integral equations for the thin layer for radial (j=1) and channel (j=0) flows.

Continuity: $$\frac{\partial}{\partial t}[(r+1)h] + \frac{\partial}{\partial x}[(r+1)uh] + j(r+1)u\,\frac{h}{x} + v_e = 0$$

Momentum: $$\frac{\partial u}{\partial t} + u\,\frac{\partial u}{\partial x} + \frac{g}{2(r+1)h}\,\frac{\partial}{\partial x}(rh^2) - \frac{V_e}{r+1}\,\frac{u}{h} = 0$$

Concentration: $$\frac{\partial}{\partial t}[(r+1)ch] + \frac{\partial}{\partial x}[(r+1)cuh] + j(r+1)u\,\frac{h}{x} = 0$$

In addition we need to express the global conservation of heavy gas and the condition at the front as follows

$$\int_0^{x_L} (r+1)chx^j dx = \frac{M_g}{(2\pi)^j \rho_a} \quad , \quad \frac{dx_L}{dt} = kgr_L h_L^{\frac{1}{2}}$$

where M_g is the mass of heavy gas released and x_L is the coordinate for the leading edge of the cloud. The empirical coefficient k is presumed to include the effect of aerodynamic drag on the motion. In lieu of reliable information about k, one can use the "Bernoulli-result" k = 2, but experiments indicate a somewhat smaller value. We consider here the two cases: (1) no entrainment ($v_e = 0$), and (2) atmospheric turbulence ($v_e = -\gamma$ = constant). Thin-layer solutions for case (2) have also been published by Rosenzweig[6] in a parallel study, but only for flow situations where the Boussinesq approximation is valid. Rosenzweig obtains only the growing mode associated with the late-time motion of the cloud whereas our solutions include also the initial slumping motion. In accordance with Fanneløp and Waldman[3] the following similarity variables are introduced:

$$X = x/x_L \quad , \quad x_L = At^n$$

The transformed equations show that the velocity varies linearely within the cloud, and we obtain, using an isothermal mixing relation, the following expression for the height and density parameter:

$$h = H(X)t^{-n(1+j)} + \frac{\gamma}{n(1+j)+1} t \quad , \quad rhx_L^{1+j} = \Pi(X)$$

where the two terms in the expression for the height is associated with the slumping motion and the entrainment from atmospheric turbulence respectively. H(X), $\Pi(X)$, A and n are found from the equations using the auxiliary relations [5]. Inasmuch as the source conditions for an accidental spill is unlikely to be well represented by the similarity forms, the usefulness of the derived solutions will depend on the sensitivity to departures from similarity in the early stages of the spreading process. The effect has been studied through a numerical experiment (method of characteristics) where the exact solution for given "nonsimilar" initial conditions is compared with the similar solution at various instances in time after release. The initial condition considered is the homogeneous square cloud presumed by users of various box models, but it should be noted that this initial state is no more likely to occur in practice than the similarity forms. The calculations have been based on Case 1 conditions assuming entrainment and other viscous processes are relatively unimportant in the early stages of the spreading. The results are shown in Fig. 1. It is seen that the originally square cloud produces a cloud of similarity form in about the time it takes to spread ten times the initial length x_0. The error introduced in the later stages by the "nonsimilar" initial conditions therefore appears to be small.

During the later stages in the spreading process the density difference, between the cloud and the ambient, will be small and the Boussinesq-approximation should therefore be valid. Fig. 2, 3, 4 show the variation in frontal height with time as well as the relevant height and density profiles for different initial density ratios based upon the solutions for case 2.[3] (Section 3). The Boussinesq solution seems to be quite accurate in the early stages of the spreading process for $\Delta\rho/\rho_a \lesssim 1$. For larger values of $\Delta\rho/\rho_a$, the discrepancies become appreciable in the slumping period and beyond, and the Boussinesq solution should be used only for large times.

3. Instantaneous Source Moving with the Wind

The thin-layer similarity solutions discussed in the preceding section, can also be applied to heavy clouds from instantaneous sources drifting with the wind. The spreading process must then necessarily include an early acceleration phase during which the initially stationary gas mass attains, on the average, a near constant drift velocity dependent on both the local wind field and the cloud dimensions. For a cloud suddenly released from a container, the accelerating and slumping motion will occur simultaneously. In the presence of wind, that part of the solution which describes the slumping motion, is therefore of lesser interest than before due to our lack of understanding of the acceleration process.

We assume a constant drift speed. The coordinate $\xi = x \pm u_d t$ denotes the position within the moving cloud relative to its centroid with position $x_c = u_d t$. The coordinate transformation gives the same equations as derived for the case of no wind. We consider a constant entrainment rate and $j = 0$, as exact numerical results are available for this case in Rosenzweig's dissertation[6]. We then obtain

$$h = [D(1 + \frac{1}{2k} - X^2)^{-1.25} - \frac{8}{81}\frac{A^3}{g}(1 + \frac{1}{2k} - X^2)]t^{-\frac{2}{3}} - \frac{3}{5}v_e t$$

where

$$r = \frac{8}{9}\frac{A^2}{9}\frac{(1+\frac{1}{2k}-X^2)}{ht^{2/3}}$$

$$A^3 = \frac{27}{8}\frac{kg}{3+4k}V_0 r_0 \quad , \quad D = \frac{V_0}{2A}(1 + \frac{r_0}{9})\left\{\int_0^1 (1 + \frac{1}{2k} - X^2)^{-1.25}\right\}^{-}$$

and V_0, r_0 represent the initial volume per unit width and density parameters. For very large times for which the entrainment in h is dominant, the present result becomes equivalent to that of Rosenzweig.

It is possible to take account of the initial extent of the cloud assuming $\xi_L(t=0) = At_0^{2/3}$. The results will be the same as before using the transformed time $\tilde{t} = t + t_0$, except that there will be some change in the value of the constant D.

The exact numerical results given by Rosenzweig are presented through averaged values of r. We obtain the average $\bar{r}$ from

$$\bar{r}\int_{-\xi_L}^{\xi_L} hdx = \int_{-\xi_L}^{\xi_L} rhdx$$

which leads to the following expressions without and with "initial adjustment"

$$\frac{\bar{r}}{r_0} = \begin{cases} \left[1 + \frac{6}{5}\gamma \frac{A}{V_0} t^{5/3}\right]^{-1} & (1) \\ \left[1 + \frac{6}{5}\gamma \frac{A}{V_0}(\tilde{t}^{5/3} - t_0^{5/3})\right]^{-1} & (2) \end{cases}$$

Equations (1) and (2) are shown in Fig. 5 together with a simple Box-modell and compared with the results of Rosenzweig. The time is made nondimensional through $t_1 = tu/\ell$ where $\ell = \sqrt{V_0}$. The exact results are seen to be intermediate between those from (1,2) and the box model which gives a "shifted" asymptote for large times as expacted. Rosenzweig's analytic solution is accurate only for large times. But the similarity solution also predicts large variations in density within the cloud, of the order of 15:1 between the center and front for early times and about 5:1 for large times (Fig. 4). These differences have more important physical consequences than the somewhat smaller spatial variations in height (Fig. 3).

4. Steady Source in Wind

Exact similarity solutions to the thin-layer equations are obtainable only for special entrainment relations, and have therefore a limited applicability. One situation in which the assumption of uniform entrainment rate is reasonable, is that for which a steady (or quasi-steady) source of heavy gas combines with a constant wind to produce a plume along the ground. At some distance from the source the gravitational spreading velocity will be small in comparison with the wind and the plume can be considered the result of wind frift along the plume axis combined with a crosswise gravitational spreading velocity. The latter will be assumed to be independent of the drifting motion. But as the entrainment will be related primarily to wind shear, it is reasonable to assume constant entrainment velocity across the plume. The similarity solution of Section 2 with γ = const. and $j = 0$, can therefore be used for practical predictions of shape and concentration of steady plumes.

On denoting the longitudinal cloud dimension by y and the lateral by x, the concentration and cloud height for any cross-section y = const. are found from the similarity solution outlined in Section 2 by substituting the relevant time $t = Y/u_d$. The procedure can moreover be generalized to quasisteady sources as long as these can be approximated by a power-law variation in strength with time as shown by Waldman et al[4].

The details of the flow near the source are more difficult to analyze due to the interaction of the two flow fields. It is necessary to consider also the finite velocity from the source which is comparable in magnitude with the drift speed. For a constant source strength the radial spreading velocity u_r can be calculated simply, by means of the box model, or more rigorously using a complete similarity analysis analogous to that outlined in Section 2. On superimposing the wind drift speed u_d the velocity components at the slick contour can be expressed

$$u_y = u_{LE}\frac{y}{r} + u_d$$

$$u_x = u_{LE}\frac{x}{r}$$

The relevant similarity and box solutions for u_{LE} have the form

$$u_{LE} = nAt^{n-1} = nA^{\frac{1}{n}}\, r(1 - \frac{1}{n})$$

The differential equation for the slick contour thus becomes

$$\frac{dx}{dy} = \frac{nA^{1/n}\, x}{nA^{1/n}\, y + u_d r^{1/n}}$$

The stagnation point is located at

$$y_s = -\left(u_d/nA^{1/n}\right)^{\frac{n}{n-1}}$$

so that the complete contour is given by

$$\left(\frac{x}{-y_s}\right)^{\frac{1-n}{n}} = -\left(\frac{1}{n} - 1\right)\int_{\pi}^{\theta}\frac{d\theta}{(\sin\theta)^{2-\frac{1}{n}}}$$

in terms of the angular variable $\theta = \tan^{-1}(x/y)$.
This solution based on the pool spreading laws can be used to generate the flow field near the source; a region for which the "slab" approximation to the steady plume cannot be used.
In view of the crude approach, and the limited experimental information available, the box model appears adequate for the near field approximation. The box model for a constant mass source spreading radially yields

$$A = \left(\frac{k\Delta\rho g}{\rho_a}\,\frac{16}{9\pi}\,\frac{\dot{m}}{\rho_g}\right)^{1/4} \quad , \quad n = 3/4$$

The two solutions proposed, are matched at the point where the plume contours have equal slope.
A numerical example is presented in Fig. 6,7 where the coordinates are made nondimensional through the buoyancy length

$$L_B = g\ \Delta\rho/\rho_a\ \frac{\dot{m}}{\rho_g}\ \frac{1}{u_d^3}$$

Fig. 6 shows the development of the plume contour in the near-source and downwind regions. The corresponding variation in density across the plume at various downwind distances, is illustrated in Fig. 7.

References

1. Havens, J.A.: A description and assessment of the SIGMET LNG Vapor Dispersion Model. US Coast Guard Rep. CG-M-3-79, Feb. 1979.

2. Havens, J.A.: Heavy gas dispersion model evaluation. MIT-GRI LNG Safety and Research Workshop, 1982.

3. Fanneløp, T.K.; Waldman, G.D.: Dynamics of oil slicks. AIAA J. Col. 10, No. 4, pp 506-10, 1972. (Also AIAA Paper No. 71-14, 1971.)

4. Waldman, G.D.; Fanneløp, T.K.; Johnson, R.A.: Spreading and transport of oil slicks on the open ocean. Offshore Technology Conference, Houston, Texas, Paper No. OTC =548, 1972.

5. Fanneløp, T.K.; Jacobsen, Ø.: Gravitational spreading of heavy gas clouds instantaneously released. (Submitted for publication, ZAMP, 1983.)

6. Rosenzweig, J.J.: A theoretical model for the dispersion of negatively buoyant vapor clouds. Massachusetts Institute of Technology, Dissertation, 1980.

7. Fay, J.A.: Gravitational spread and dilution of heavy vapor clouds, in Second International Symposium on Stratified Flows. (ed. T. Carstens and T. McClimans), Norwegian Institute of Technology, June 1980.

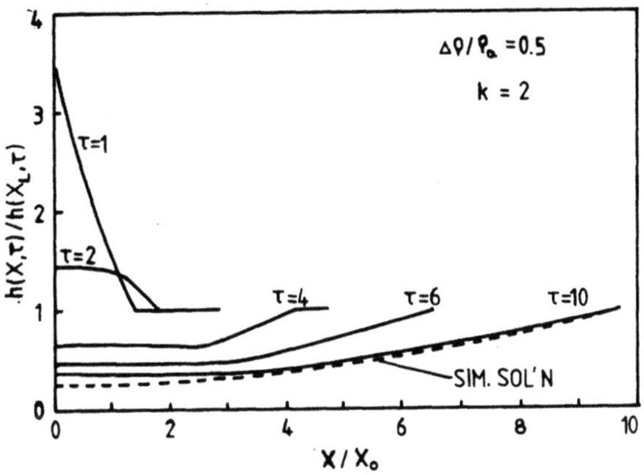

Fig.1. Numerical solutions of cloud cross-section at various early times compared with similarity solution; no entrainment. Nondimensional time $\tau = t\sqrt{g\Delta\rho/\rho_g h_0}/x_0$

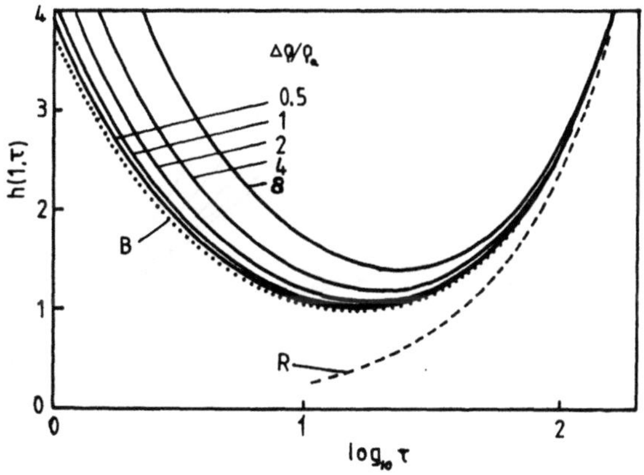

Fig.2. Variation in frontal height for various initial density ratios. $V_e = -0.05$, $k = 1.4$, B = Boussinesq approx., R = Rosenzweig analytical solution.

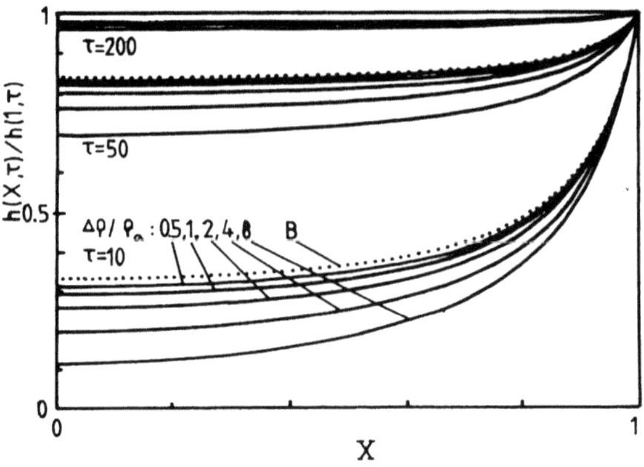

Fig.3. Cloud cross-section at various times for a range of initial density ratios. $V_e = -0.05$, $k = 1.4$, B = Boussinesq approx.

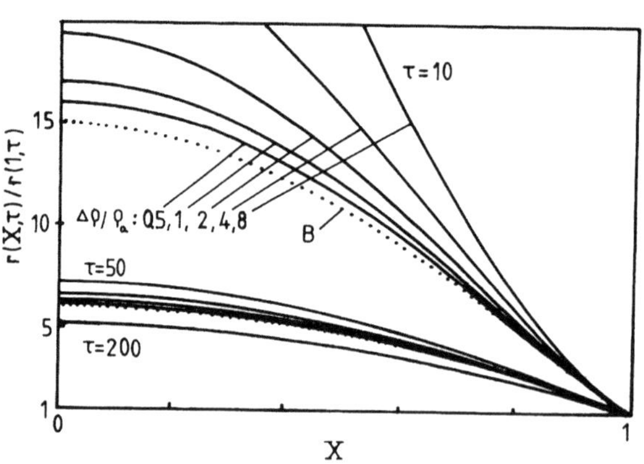

Fig.4. Density profile at various times for a range of initial density ratios. $V_e = -0.05$, $k = 1.4$, B = Boussinesq approx.

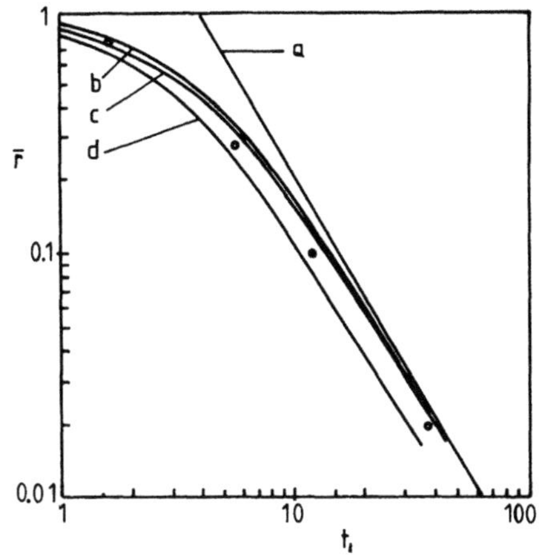

Fig.5. Variation of average cloud density with time.
a) Rosenzweig analytical solution, b) Eq'n (1), c) Eq'n (2)
d) Box-model, ∘ Rosenzweig exact numerical solution.
$r_0 = 1$, $k = 2$, $V_e = -0.1\ u_d$, $g\ell/u_d^2 = 1$.

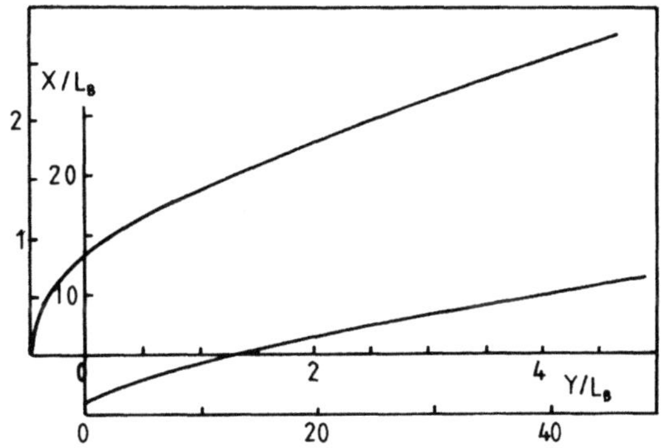

Fig.6. Plume contour downwind of constant source. k = 2.

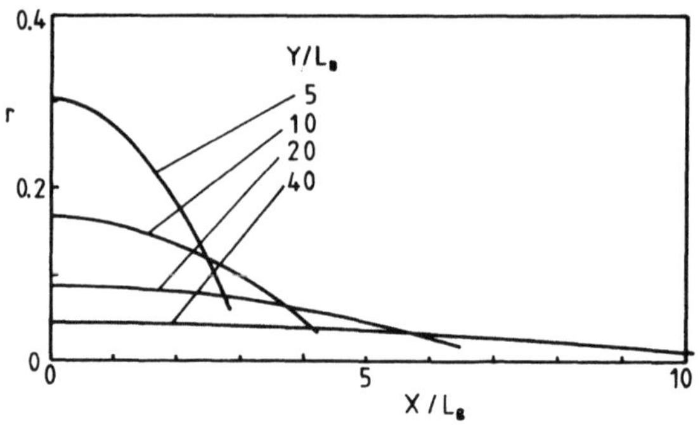

Fig.7. Concentration profiles across plume at various downwind stations. $k = 2$, $r_0 = 1$, $\dot{m}/\rho_g = 0.01\ m^3/s$, $u_d = 0.6\ m/s$, $V_e = -0.1\ u_d$

A New Bulk Model for Dense Gas Dispersion: Two-Dimensional Spread in Still Air

A.P. VAN ULDEN

Royal Netherlands Meteorological Institute,

De Bilt, The Netherlands

1. Introduction

Gravity currents exhibit the significant effect that buoyancy can have on fluid motion. A gravity current has a density that differs from that of the fluid in which it is embedded. Through the action of gravity the density difference leads to pressure forces that set the gravity current into a more or less horizontal motion relative to the ambient fluid. Examples of geophysical gravity currents are: atmospheric cold fronts, oceanic fronts, katabatic winds and snow avalanches. Man-made gravity currents may result from an accidental release of a dense gas in the atmosphere or of oil on water. Thus there is reason enough to study this interesting phenomenon. For a recent review of the general subject we refer to Simpson (1982).

In this paper we focus on horizontal bottom boundary currents that result from an instantaneous release of a dense fluid. In this class of currents falls an important part of man-made accidental releases that may form a hazard for the environment. For reasons of simplicity we will deal with the two-dimensional case only. The extension of this case to axi-symmetric currents is straightforward in principle (Hoult, 1972; Huppert and Simpson, 1980). We further assume that the ambient fluid is at rest and infinitely deep and that viscous forces are negligible. We thus exclude the "slumping regime" that is present in lock-exchange experiments (Barr, 1967; Huppert and Simpson, 1980; Simpson, 1982) and also the "viscous regime" that prevails when Reynolds numbers are low (Barr, 1967; Fay, 1969). Thus we deal with gravity currents that are in an "inertia-buoyancy regime", in which the dynamics of the current are governed by inertial and buoyant forces (Huppert and Simpson, 1980).

The built up of this paper is as follows. We start in section 2 with a review of studies made on "inertia-buoyancy currents" and discuss some general features of these currents. In section 3 we describe a new bulk model. This models consists of bulk equations for matter and momentum, that are derived from the basic equations of motion. In section 4 we present model results and an evaluation of these.

2. History and problem definition

2.1 The leading edge of gravity currents

Major information on the conditions of the leading edge can be found in Schmidt (1911), Benjamin (1968), Simpson (1972) and Simpson and Britter (1979). In figure 1 an arrested leading edge is shown.

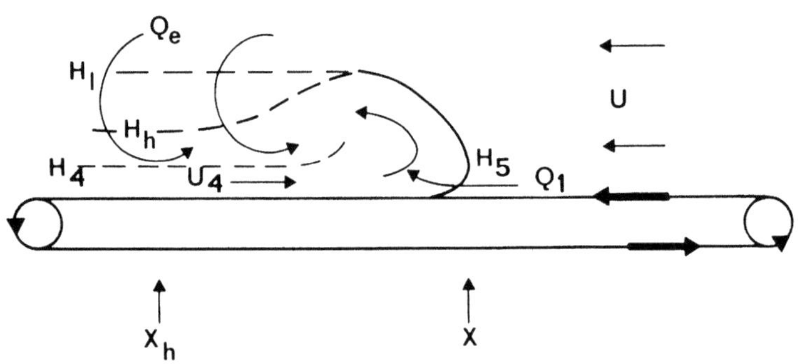

Fig. 1. The head of a steady gravity current (after Simpson and Britter, 1979).

Some characteristic features are:

1. At the leading edge a head is present with a depth H_1 that is about twice the depth H_h of the current behind the head.
2. An elevated forward stagnation point is present. Below this point an insignificant flux Q_1 of ambient fluid is entrained.
3. Behind the head a wake region is present in which significant mixing occurs. This leads to a vertical profile of the density surplus $\Delta\rho$ behind the head. Near the surface $\Delta\rho$ is fairly constant; in the wake region it can be described by a fourth order polynomial (S & B). A gaussian profile also is a fair approximation.

4. In the head a significant internal flow is present. Near the surface denser fluid moves towards the head with a velocity U_4. In the wake region a mixing layer moves away from the head.
   S & B found that

$$U_4 \simeq 0.2\ U, \tag{2.1}$$

   where U is the velocity of the current relative to the ambient fluid.
5. The relative velocity of the leading edge can be written as

$$U = C_h \sqrt{g' H_h}\ , \tag{2.2}$$

   where $g' = g\Delta\rho_h/\rho_a$ is the reduced gravity, $\Delta\rho_h$ the mean density difference behind the head, ρ_a the density of the ambient fluid and H_h the densimetric mean depth behind the head. A formal definition of the product of $\Delta\rho_h$ and H_h is (Fay, 1980):

$$\Delta\rho_h\ H_h = \int_o^{H_1} \Delta\rho(z)\ dz\ . \tag{2.3}$$

   With this definition the value of C_h may be computed from experimental data. From Schmidt (1911), Benjamin's (1968) review, S & B (1979), Fannelop et al. (1980), Huppert and Simpson (1981) it follows that

$$C_h = 1.15 \pm 0.05, \tag{2.4}$$

   both for steady and unsteady gravity currents, provided the Reynolds number $UH_h/\nu > O(10^3)$ and provided the current is deeply submerged.

2.2 Bulk properties of fixed volume releases

A major contribution to the understanding of the bulk properties of fixed volume releases has been given by Fannelop and Waldman (1971, 1972) and later by Hoult (1972). Their aim was to describe the spreading of oil on water. They assumed (2.2) as a leading edge boundary condition and used the shallow-water equations to describe the interior of the current. They showed that similarity solutions to this set of equations exist for the horizontal distributions of the layer averaged velocity $\overline{u}$ and the local depth h and for the dimensionless velocity

$$C_H \equiv U/\sqrt{g'H}\ , \qquad (2.5)$$

where $H = V/X$ is the mean depth of the current. Their results cannot be applied as such to the present problem, because entrainment and the internal flow were not included in their description. Also their solutions are only valid for great times, when the current has passed through its initial acceleration phase into the final deceleration phase. The problems of entrainment and initial acceleration have been dealt with - be it in a crude manner - by Van Ulden (1979). Van Ulden assumed a rectangular shape with a linear velocity distribution and derived a bulk equation for dU/dt. He included a static pressure force, a drag force and an effective stress due to entrainment in this equation. However objections may be raised against the way this equation was derived.

It is the purpose of this paper to improve on this. We will derive bulk continuity and momentum equations starting from the basic equations of motion. The now existing better-understanding of the leading edge conditions will be absorbed in the model. Also some features of the approach by Fannelop and Waldman (1971, 1972) will be included in the model. Furthermore a parameterization for top entrainment will be presented and some attention will be given to the problems that arize when the gravity current is not shallow.

3. A new bulk model for fixed volume releases

3.1 Introduction

In the remainder of this paper we deal with gravity currents that result from an instantaneous release of a volume V_0 (per unit width) of a fluid with density ρ_0 that is greater than the density ρ_a of the ambient fluid by an amount $\Delta\rho_0$. The two fluids are assumed incompressible and of equal temperature. The release is at the horizontal bottom at the beginning of an infinitely deep channel in which the ambient fluid is at rest. The initial volume has a length X_0 and a mean depth $H_0 \equiv V_0/X_0$. After the release a gravity current develops of the type that is shown in figure 2. This figure shows the characteristic length- and velocity variables that we will use in our model description. We distinguish between a head and a tail region. The vertical boundary between the two regions lies at a distance X_h from the origin. At this location the densimetric mean depth is H_h and the

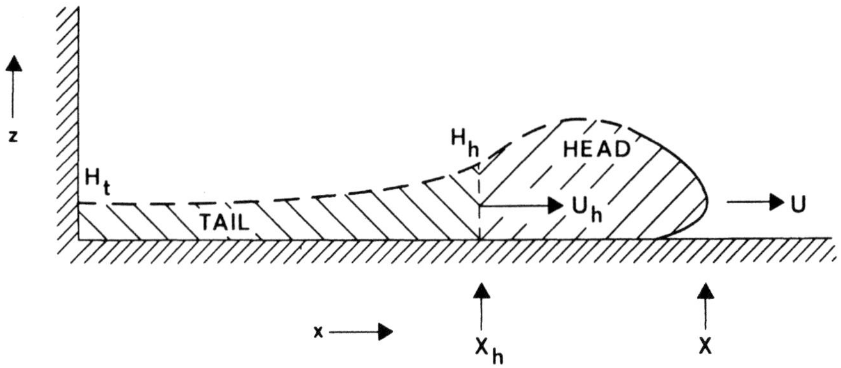

Fig. 2. The unsteady gravity current.

layer averaged fluid velocity U_h. The depth of the current in the origin is H_t. The distance from the origin of the forward stagnation point is X and the horizontal velocity of this stagnation point U. In terms of these variables we will derive bulk equations for the total current in section 3.2. In section 3.3 we will consider in some detail the depth and velocity distributions in the current. In section 3.4 we describe the head conditions. In section 3.5 we give a listing of the final model equations.

3.2 Bulk equations for the total current

3.2.1 Continuity equations

In this section we derive bulk equations for the volume and the density of the current. The volume is defined as:

$$V = \int_0^X h(x,t)\, dx\ , \tag{3.1}$$

where

$$h(x,t) \equiv 2 \int_0^\infty z\, \Delta\rho(x,z,t)\, dz \Big/ \int_0^\infty \Delta\rho(x,z,t)\, dz \tag{3.2}$$

is the local depth of the current. Thus the local depth is defined to be twice the local densimetric mean depth $\overline{z}(x,t)$. h(x,t) is allowed to vary with x and t and will be later expressed in terms of H_t and H_h. It is further useful to introduce the overall mean depth H that is defined as

$$H = \frac{1}{X} \int_0^X h(x,t)\, dx \tag{3.3}$$

Thus we may also write for the volume of the current

$$V = XH \tag{3.4}$$

The volume of the current increases with time due to entrainment:

$$dV/dt = Q_e \tag{3.5}$$

where Q_e is the volume rate at which ambient fluid is entrained.

We will now derive a scaling law for Q_e by considering the budget of potential energy. Some of the following arguments have been given by Van Ulden (1979). The potential energy of the current is

$$PE = \tfrac{1}{2}\, g\, \Delta\rho_c\, VH\ , \tag{3.6}$$

where

$$\Delta\rho_c = \frac{1}{V} \int_0^X \int_0^\infty \Delta\rho\, dx\, dz \tag{3.7}$$

is the mean density difference between the dense fluid and the ambient fluid. Because of the conservation of dense material $\Delta\rho_c\, V$ is a conserved quantity, independently from any mixing:

$$\Delta\rho_c\, V = \Delta\rho_o\, V_o\ , \tag{3.8}$$

where the subscript o denotes initial values. Because of (3.8) the rate equation for potential energy reads:

$$d\, PE/dt = \tfrac{1}{2}\, g\Delta\rho_c\, V\, (dH/dt) \tag{3.9}$$

In this equation dH/dt can be written as

$$dH/dt = W_H + W_e\ , \tag{3.10}$$

where

$$W_H \equiv - UH/X \tag{3.11}$$

is the mean downward motion of H due to slumping and where

$$W_e \equiv Q_e/X \tag{3.12}$$

is the mean upward motion of H due to turbulent entrainment of ambient fluid. It thus follows that slumping leads to a decrease of potential energy, while entrainment leads to an increase of it.

The loss of potential energy due to slumping goes into the production of mean kinetic energy. In its turn mean kinetic energy leads to shear stresses, that produce turbulent kinetic energy. This energy is partly dissipated, partly -due to entrainment- destroyed by transformation into potential energy. The latter proces is called buoyant destruction (Monin and Yaglom, 1971). Now we make the closure assumption that buoyant distruction is proportional to the production of turbulent kinetic energy. This assumption has led to successfull modeling of entrainment in the atmosphere and ocean (e.g. Tennekes and Driedonks, 1981). The production of turbulent energy in the present case can be derived from the analysis by Simpson and Britter (1979). They showed that shear production occurs mainly in the head region of the current, while destruction of eddies mainly occurs behind the head. Near the head eddies are created with an energy density of order $\frac{1}{2}\rho_a U^2$. This occurs at a volume rate of order $H_h U$. Thus the production rate should scale as $\frac{1}{2}\rho_a H_h U^3$. We find from (3.4)-(3.12) that the buoyant destruction equals $\frac{1}{2} g\Delta\rho_c H(dV/dt)$. Taking these factors proportional we find that

$$dV/dt = \varepsilon H_h U/Ri \tag{3.13}$$

where

$$Ri = g\Delta\rho_c H/\rho_a U^2 \tag{3.14}$$

is a bulk Richardson number and where ε an empirical coefficient. This result resembles the conventional scaling of side entrainment (Van Ulden, 1974; Fay, 1980) that reads for the two-dimensional case

$$dV/dt = C_x HU,$$

where C_x is a constant. The physical meaning of our result, is however completely different. In our model only the production of turbulent kinetic energy occurs at the leading edge, but the following entrainment occurs at the top of the current. Thus in our vue turbulent eddies are mainly created near the leading edge. While travelling away from the leading edge they lose their kinetic energy and increase the potential energy of the current.

The estimation of our entrainment coefficient ε is not easy. No data seem to be available for the 2-dimensional case. But data are available for the axisymmetric case on the equivalent value of C_x. Picknett (1978) finds $C_x \simeq 0.8$ from the Porton Down experiments. This result has been essentially confirmed by laboratory experiments Hall et al. (1982). With $H_h \simeq H$ and $1/Ri \simeq C_h^2 \simeq 1.3$ this leads to a first guess for ε of:

$$\varepsilon \simeq 0.6 \tag{3.15}$$

This value is much larger than the value found by Van Ulden (1974). This might be due to a misinterpretation of the visual data given in this paper. It should be noted that with $\varepsilon = 0.6$ the conservation law for potential and kinetic energy is not violated. Indeed from (3.9)-(3.15) we find that the potential energy decreases.

3.2.2 The momentum-integral equation

In its general form the horizontal component of the momentum-integral equation reads for an arbitrary volume V with boundary S (Batchelor, 1967, 3.2):

$$\frac{d}{dt} \int^{V} \rho u \, dv = \phi_x + \int^{S} \sigma_{xj} \, m_j \, ds \ . \tag{3.19}$$

Here ϕ_x is the net flux of momentum through the boundary S and σ_{xj} are the x-components of the stress tensor. In the present case V is a control volume that just includes the continuously changing volume of the current and S its outer boundary. The volume integral at the left side of (3.19) is the total horizontal momentum of the current:

$$M = \int_{o}^{X} \rho_c \, \overline{u}(x,t) \, h(x,t) dx \, , \tag{3.20}$$

where $\overline{u}(x,t)$ is the layer averaged flow velocity.
The evaluation of this integral we postpone till later. Further we deal in the present case with an ambient fluid that is at rest at some distance from the current. Therefore we assume that no significant entrainment of ambient momentum occurs and neglect the momentum flux ϕ_x. The momentum-integral equation then reduces to

$$dM/dt = \int^{S} \sigma_{xj} \, n_j \, ds \tag{3.21}$$

In this equation the surface integral of the stress tensor represents all horizontal forces that act on the current. Three types of forces can be distinguished (Batchelor, 1967). The first is the static pressure force, that -in the present case- is due to the negative buoyancy of the gravity current. By integrating the static pressure over the boundary of the current this force is readily found to be

$$F_p = \tfrac{1}{2} \, g \, \Delta\rho_c \, H_t^2 \, . \tag{3.22}$$

This result is exact and does not depend on the specific shape of the vertical density profile. F_p thus only depends on the density difference $\Delta\rho_c$ and on the depth H_t in the origin. The second force is the dynamical force due to the motion of the current relative to its environment. In the present paper we assume high Reynolds numbers and neglect the bottom shear stress. Further the upper boundary of the control volume is taken high enough that also there the shear stress can be neglected.
The dynamical force on the current then is the sum of the drag force on the head of the current and the lift force that may arize from asymmetry in the ambient flow around the head. The dynamical force may be written as

$$F_d = - \tfrac{1}{2} \, D \, \rho_a \, H_h \, U^2 \, , \tag{3.23}$$

where D is an effective drag coefficient that will be estimated later.

The third force is the most complicated one. It is due to horizontal and vertical accelerations of the current. These apply momentum changes to the ambient fluid and give rise to an acceleration reaction term (Batchelor, 1967; 6.4), that can be written as

$$G_i = -\frac{d}{dt}(\rho_a V \alpha_{ij} U_j) , \tag{3.24}$$

where α_{ij} is the tensor coefficient of virtual inertia. The present case resembles an accelerating elliptical cylinder with an aspect ratio H/X. For such a cylinder $\alpha_{11} = H/X$, $\alpha_{22} = X/H$, $\alpha_{12} = \alpha_{21} = 0$ (Batchelor, 1964, 2.6). Therefore in our case the horizontal component of the acceleration reaction should scale as

$$-\frac{d}{dt}(\alpha\rho_a H^2U) \tag{3.25}$$

where we have used the identity V = XH and where α is a coefficient of O(1) that accounts for the fact that our current is not an elliptical cylinder.

Also vertical accelerations may be important, because these lead to a non-hydrostatic pressure in the current. Because of the presence of a vertical wall in the origin this gives rise to a non-hydrostatic pressure force F_π that scales as HG_y/X. Since vertical velocities scale with $W_H = -HU/X$ the non-hydrostatic pressure force can be written as

$$F_\pi = -\frac{d}{dt}(\beta \rho_a H^2U) , \tag{3.26}$$

where β is another coefficient of O(1). From (3.25) and (3.26) we see that the net effect of horizontal and vertical accelerations is a force

$$F_a = - dM_v/dt , \tag{3.27}$$

where

$$M_v = (\alpha+\beta) \rho_a H^2U \tag{3.28}$$

is the virtual momentum of the current. Since $\alpha+\beta$ is of O(1) this corresponds with an added mass that is of the same order as the mass of the current itself when H/X and ρ_a/ρ_c are of O(1) and that vanishes when H/X vanishes. Thus F_a is only important when the current is not shallow. From (3.21)-(3.28) we find that the momentum-integral equation reads

$$d(M+M_v)/dt = \tfrac{1}{2}\, g\Delta\rho_c H_t^{\,2} - \tfrac{1}{2}\, D\rho_a H_h U^2 \;. \qquad (3.29)$$

This equation and the continuity equation (3.14) are the bulk rate equations for the total gravity current. In these equations the volume integral (3.1) and the momentum integral (3.20) still have to be specified. We will do so in the following sections.

3.3 The horizontal distributions of layer depth and layer averaged velocity in the tail of the current

It is the purpose of this section to evaluate for the tail of the current the volume-integral

$$V_t(t) \equiv \int_0^{X_h} h(x,t)\, dx \qquad (3.30)$$

and the momentum-integral

$$M_t(t) \equiv \int_0^{X_h} \rho_c\, \bar{u}(x,t)\, h(x,t)\, dx \qquad (3.31)$$

In particular we want to express these integrals in terms of the model variables H_t, H_h, X_h and U_h. In order to do so we need approximations to the functions $h(x,t)$ and $\bar{u}(x,t)$. During the early development of the current these are difficult to obtain, but quite soon the current becomes shallow enough that the shallow water equations are applicable to the flow in the tail of the current. In the present problem the following equations apply:

$$\frac{1}{h}\frac{Dh}{Dt} = -\frac{\partial \bar{u}}{\partial x} + \frac{w_e}{h} \qquad (3.32)$$

and

$$\frac{D\bar{u}}{Dt} = -g\,\frac{\Delta\rho_c}{\rho_c}\frac{\partial h}{\partial x} - \frac{\rho_a w_e \bar{u}}{\rho_c h} - \tfrac{1}{2}\,\delta\,\frac{\partial \bar{u}^2}{\partial x} \;. \qquad (3.33)$$

In these equations $D/Dt = \partial/\partial t + \bar{u}\,\partial/\partial x$, w_e is the local entrainment velocity and $\tfrac{1}{2}\,\delta\,\partial\bar{u}^2/\partial x$ a momentum flux gradient term that accounts for the fact that the vertical velocity profile is not uniform. δ is an empirical constant to be estimated later. It should be noted that (3.33) is fully consistent with our bulk equation (3.29) provided the current is shallow. Also in (3.29) an effective -entrainment related- stress gradient term similar to that in (3.33) is hidden in the dM/dt term. This can be checked by evaluating the time derivative of (3.20). The physical meaning of the decelerating stress gradient term is simply this. Entrainment does not affect the total momentum -since no momentum is entrained-, but it increase the total mass of the current. This necessarily causes a decrease in the mean velocity. Thus entrainment leads to a deceleration term in the equation for the mean velocity. We will now derive approximate solutions to (3.32) and (3.33) by making two similarity assumptions. The first is that the shape of the current is quasi-conserved in time i.e. that

$$\frac{1}{h}\frac{Dh}{Dt} \simeq \frac{1}{H}\frac{dH}{dt}, \tag{3.34}$$

virtually independent from x. The second -earlier made- assumption is that the layer averaged density difference remains horizontally uniform and equal to
$\Delta\rho_c$. This requires that

$$w_e/h \simeq W_e/H , \tag{3.35}$$

virtually independent from x (see also section 3.1). It then follows that

$$\partial\bar{u}/\partial x \simeq U_h/X_h \tag{3.36}$$

and that

$$\bar{u} \simeq x\, U_h/X_h \tag{3.37}$$

It further follows from (3.33)-(3.37) that $\partial h/\partial x$ is a linear function of x that vanishes in x = 0. Using the boundary conditions $h = H_t$ for $x = 0$ and $h = H_h$ for $x = X_h$, we now easily find that

$$h = H_t + (H_h - H_t)\,(x/X_h)^2 \tag{3.38}$$

The solutions for $\bar{u}$ and h happen to be of the same form as those obtained by Fannelop and Waldman (1971, 1972). However there are two differences. In our model H_t and H_h are independent variables that are determined by the dynamics of the gravity current. Further we use (3.37) and (3.38) only to estimate the volume and momentum integrals of the tail. From (3.30), (3.31) and (3.37), (3.38) we easily obtain

$$V_t = \frac{1}{3} (2 H_t + H_h) X_h \tag{3.39}$$

and

$$M_t = \frac{1}{4} \rho_c (H_t + H_h) X_h U_h \tag{3.40}$$

This completes our description of the tail.

3.4 The head of the current

The shallow water equations are not applicable to the head of the current. Instead we use the momentum-integral approach that we applied to the total current (3.19). The force balance for the head looks as follows. The static pressure force follows from the integration of the static pressure over the outer boundary of the head and equals:

$$F_p = \tfrac{1}{2} g\Delta\rho_c H_h^2 \tag{3.41}$$

The dynamic pressure force is

$$F_d = - \tfrac{1}{2} D \rho_a H_h U^2 . \tag{3.42}$$

Furthermore there is a momentum flux into the head due to the internal current in the head (figure 1). Near the surface the inward flow U_4 carries positive momentum into the head. The return flow U_3 carries negative momentum out the head. So the net effect of the internal flow is a positive momentum flux into the head. Assuming $U_3 \simeq U_4$ and $h_4 \simeq \frac{1}{2} H_h$ we find that this flux crudely is

$$Q_h = \rho_a H_h U_4^2 \tag{3.43}$$

Using $U_4 \simeq 0.2\ U$ (2.1) we may write this as

$$Q_h \simeq \tfrac{1}{2} \delta \rho_a H_h U^2 \tag{3.44}$$

where

$$\delta \simeq 0.08 . \tag{3.45}$$

Thus δ is an empirical coefficient that characterizes the non-uniformity of the vertical velocity profile. It has the same meaning as in (3.37).

We neglect the inertial terms in the momentum-integral equation for the head. It can be shown that these terms are normally small in comparison with the other terms. Thus we assume that the head is in a quasi-steady state. This assumption is supported by experiment. We have seen in section 2.1 that the dimensionless leading edge velocity $C_h \simeq 1.15 \pm 0.05$ both for steady and unsteady currents (2.4). Our momentum-integral equation now reads:

$$0 = \tfrac{1}{2}\, g\Delta\rho_c H_h^{\;2} - \tfrac{1}{2}\, D\rho_a H_h U^2 - \tfrac{1}{2}\, \delta\rho_a H_h U^2 \tag{3.46}$$

It follows from this equation that

$$C_h \equiv U \,/\, \sqrt{g'_c H_h} = 1 \,/ \sqrt{D-\delta} \tag{3.47}$$

where $g'_c = g\Delta\rho_c/\rho_a$. Since C_h and δ are known the value of D can be estimated from this equation. The result is

$$D = 0.84 \pm 0.07 \tag{3.48}$$

With (3.44)-(3.48) we have specified the important dynamical leading-edge boundary conditions.

We conclude this section with the specification of the volume and momentum of the head. We allow the volume of the head to vary in time, but assume that its shape remains unchanged. In section 2 we have seen that the depth H_1 of the head is about twice the depth H_h behind it. We also assume that its length scales with H_h. Thus we write

$$X - X_h = a\, H_h \tag{3.49}$$

Experimental data suggest that $a \simeq 2$. The volume of the head now is written as

$$V_h = b\, H_h^{\;2} , \tag{3.50}$$

where b ≃ 4. To estimate the momentum of the head we assume that -as in the tail- the layer averaged velocity increases linearly with x. It then follows that

$$U_h = UX_h/X \tag{3.51}$$

Using (3.49)-(3.51) we find for the momentum of the head.

$$M_h = b\rho_c \, UH_h^2 \, (X - \tfrac{1}{2} \, a \, H_h)/X \tag{3.52}$$

This completes our description of the head. Together the bulk equations derived in 3.1 and 3.2 and the equations for the tail and the head derived in 3.3 and 3.4 from a closed set. We will summarize the final set of equations in the next section.

3.5 Final model equations

In the final model equations we use the dimensionless density difference $\Delta = \Delta\rho_c/\rho_a$, the density ratio $R = \rho_c/\rho_a$ and the velocity integral $M^* = (M_t+M_h+M_v)/\rho_a$. The model has 8 variables i.e. X, V, M^*, U, H_h, H_t, Δ and R and 8 equations namely 3 rate equations and 5 diagnostic equations. These are the following. The first rate equation follows from the definition of U and reads:

$$dX/dt = U \tag{3.53}$$

The second rate equation follows from (3.11), (3.14) and (3.15):

$$dV/dt = \varepsilon \, H_h XU^3/g\Delta_o V_o \, , \tag{3.54}$$

where $\varepsilon \simeq 0.6$ is an entrainment coefficient, Δ_o and V_o are the initial values of Δ and V. The third rate equation follows from (3.29):

$$dM^*/dt = \tfrac{1}{2} \, g\Delta \, H_t^2 - \tfrac{1}{2} \, D \, H_h U^2 \, , \tag{3.55}$$

where $D \simeq 0.84 \pm 0.07$. These equations determine the development of X, V and M^*. The other variables follow from the diagnostic equations. From (3.4), (3.28), (3.40), (3.49), (3.51) and (3.52) we find that the leading edge velocity is:

$$U = M^*/\{\frac{R}{X}[\tfrac{1}{4}(H_t+H_h)\ (X-aH_h)^2 + bH_h{}^2(X-\tfrac{1}{2}\ aH_h)] + c\ \frac{V^2}{X^2}\}, \quad (3.56)$$

where a ≈ 2, b ≈ 4 and c ≈ 2. The depth of the leading edge follows from (3.47):

$$H_h = (D-\delta)U^2/g\Delta\ , \quad (3.57)$$

where $\delta \simeq 0.08$. The depth in the origin follows from (3.39), (3.49) and (3.50):

$$H_t = \frac{3}{2}\ (V - bH_h{}^2)/(X-aH_h) - \tfrac{1}{2}\ H_h\ . \quad (3.58)$$

From (3.4) and (3.11) we get the relative density difference

$$\Delta = \Delta_o V_o/V \quad (3.59)$$

and from (3.16) the density ratio

$$R = 1 + \Delta \quad (3.60)$$

This completes our model. It should be noted here that our model is non-hydrostatic. Furthermore we have nowhere applied the Boussinesq-approximation. So the model is suited to describe gravity currents with a high density ratio and currents that are not shallow.

4. Evaluation of the model

4.1 Some general model results

In this section we describe some model results. We will present our computations in dimensionless form. From the momentum equation it follows that the appropriate dimensionless time τ is

$$\tau = t/t_*\ , \quad (4.1)$$

where

$$t_* = X_o\ /\ \sqrt{g\Delta_o H_o} \quad (4.2)$$

is a time scale. The subscripts o denote initial values. In the time scale t_* two length scales are involved i.e. X_o and H_o. Thus our scaling differs from that by Fannelop and Waldman (1972) and by Hoult (1972), who used only one length scale $L_o = \sqrt{V_o}$ and neglected variations in the initial aspect ratio H_o/X_o. It is clear that this ratio should be taken into account, because it affects the amount of potential energy that is present in the initial state.

We will present some computations made with the values of the empirical constants given in the last section, for a release with $H_o/X_o = 2/3$. In our model this corresponds with initial values for H_t and H_h of 1 and 0 respectively. In figure 3 we show the dimensionless velocity

$$C_H = U/\sqrt{g\Delta H} \tag{4.3}$$

as a function of the dimensionless time τ for Δ_o =0.1, 2 and 4. It appears that for Δ_o = 0.1 the velocity approaches rapidly an asymptotic value C_∞ = 1.094. In terms of C_H the initial acceleration of the current is almost completed when $\tau \simeq 2$. Thus t_* is a characteristic time scale that crudely separates the acceleration phase from the deceleration phase (Because U decreases when C_H is constant). For high initial values of Δ, the acceleration phase lasts longer, but after it the dimensionless velocity reaches temporarily higher values. For very great times C_H again approaches the asymptotic value C_∞. The latter feature can be attributed to entrainment, that leads to vanishing Δ's at great times.

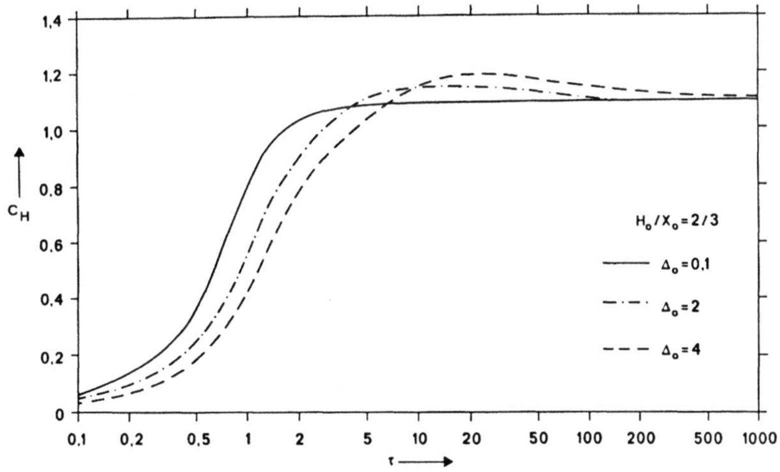

Fig. 3. The dimensionless velocity as a function of dimensionless time.

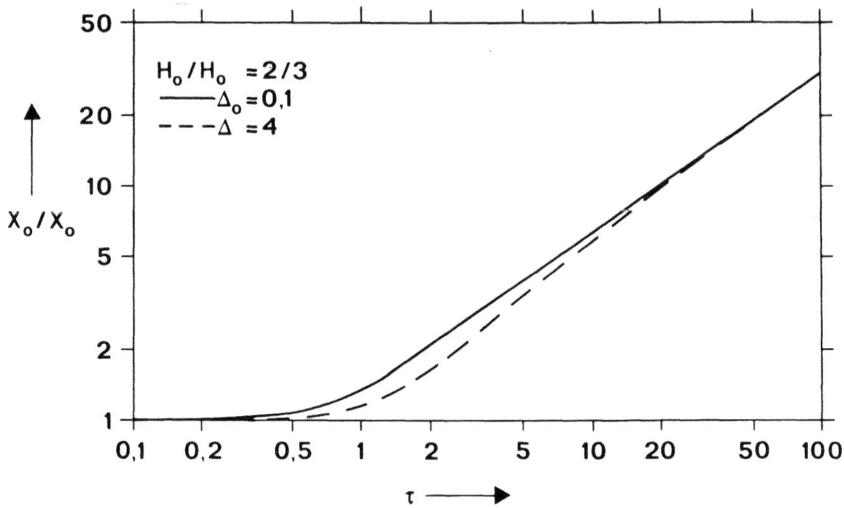

Fig. 4. The dimensionless current length as a function of dimensionless time.

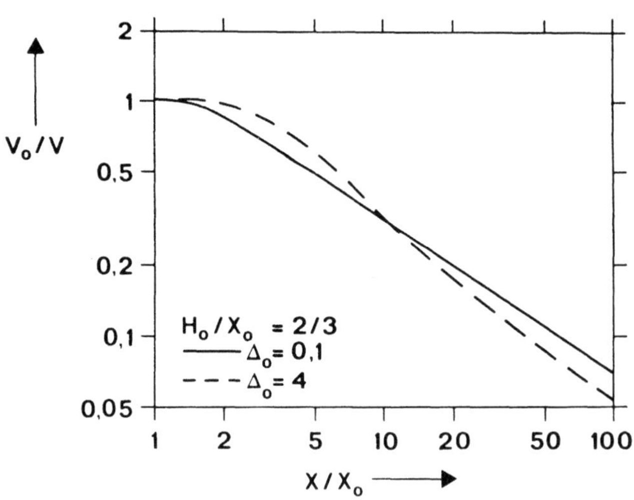

Fig. 5. The dimensionless volume concentration as a function of dimensionless current length.

In figure 4 we give the dimensionless current length X/X_o. The effect of Δ_o in X/X_o is moderate. In figure 5 we present the dimensionless volume concentration V_o/V as a function of X/X_o. Also here the effect of Δ_o is moderate. More interesting is that entrainment does not start immediately with the spreading of the dense fluid. This is due to the initial delay in the creation of turbulence.

We have also tested the sensitivity of the model to variations in the head size coefficients a and b and in the coefficient c for the virtual inertia. It appears that the model results are almost invariant for changes in a and b even as large as a factor 2. Changes in c have some effect. E.g. using $c = 1$ instead of $c = 2$ leads to higher velocities for $\tau < 1$ and lower velocities for $1 < \tau < 10$. There is no net effect on X/X_o for $\tau > 10$. On the other hand there is a notable effect on the concentration even for great times. For $\tau > 10$ the concentration is about 10% higher with $c = 1$ than with $c = 2$. This is due to a lower entrainment rate in the period $1 < \tau < 10$.

The model sensitivity to D, δ and ε will be discussed in the next section.

4.2 The momentum budget for great times

Sooner or later - this depends on the value of Δ_o - the dimensionless velocity will approach its asymptotic value C_∞. This occurs when $\Delta \ll 1$ and $H/X \ll 1$. When this is the case the velocity integral M^* reduces to

$$M^* = \tfrac{1}{4}\left(\frac{H_t}{H} + \frac{H_h}{H}\right) VU . \tag{4.4}$$

Furthermore when C_H is constant also H_t/H and H_h/H are constant. It then follows that

$$\frac{dM^*}{dt} = \tfrac{1}{4}\left(\frac{H_t}{H} + \frac{H_h}{H}\right) V \frac{dU}{dt} + \tfrac{1}{4}\left(\frac{H_t}{H} + \frac{H_h}{H}\right) U \frac{dV}{dt} . \tag{4.5}$$

The first term at the right side is the inertial force due to the deceleration of the current and the second term the vertical stress gradient due to entrainment. With this result the momentum equation (3.55) can be transformed into an equation with a single unknown i.e. C_∞. Dividing (3.55) by $\frac{1}{2}\, g\Delta H^2$, using (3.57) and (3.58) to eliminate H_h/H and H_t/H and using (4.3) we arrived at the following equation:

$$- \frac{1}{8} [3 + (D-\delta)C_\infty^2 + \frac{1}{4} \varepsilon(D-\delta) [3 + (D-\delta)C_\infty^2] C_\infty^6$$

$$= \frac{9}{4} - \frac{3}{2} (D-\delta)C_\infty^2 + \frac{1}{4} (D-\delta)^2 C_\infty^4 - D(D-\delta)C_\infty^4 \qquad (4.6)$$

The first term at the left represents the inertial force i, the second term the stress gradient s, the first three terms at the right the static pressure force p and the last term the dynamic force d. For given values of D, δ and ε (4.6) can be solved for C_∞^2. For $D = 0.84$, $\delta = 0.08$ and $\varepsilon = 0.6$ the solution is $C_\infty^2 \simeq 1.2$ or $C_\infty = 1.094$. The magnitude of the various terms of (4.6) is i = -0.58; s = 0.76; p = 1.09; d = -0.91. It further follows that $H_t \simeq H_h \simeq H$. This implies that the static pressure varies little over the current and that the static pressure force mainly acts at the leading edge, where it is approximately balanced by the dynamical force on the head. In the bulk of the current the inertial force is approximately balanced by the shear stress due to entrainment. Thus the shear stress has a significant effect on the dynamics of the gravity current.

We may also use the asymptotic form of the momentum equation (4.4) to evaluate the sensitivity of the model to variations in the empirical constants D, δ and ε. It appears that C_∞ is most sensitive to variations in de value of $D-\delta$. A 10% increase in $D-\delta$ corresponds with a 4% decrease in C_∞. Further we found that a 10% increase in ε corresponds with a 1% decrease in C_∞. The model is quite insensitive to variations in δ. A 100% increase in δ -with $D-\delta$ kept at the same value- leads to a 1% decrease in C_∞.

5. Summary and conclusions

We have developed a model for the spreading of a dense fluid in an infinitely deep channel. The model consists of rate equations for the length, the momentum-integral and the volume of the resulting gravity current. Diagnostic equations describe the shape of the current and the velocity distribution in it. The model is non-hydrostatic. The Boussinesq approximation has not been made. This makes the model suited to desribe currents with a density that is considerable higher than that of the ambient fluid. A new parameterization of entrainment is proposed, that does not violate the conservation law for potential and kinetic energy. The model contains 6 empirical coefficients that have been estimated as well as possible from literature data. A new dimensionless representation has been proposed. In this representation the dimensionless velocity, length and volume are within moderate limits universal functions of the dimensionless time. The model is mathematically rather simple and physically rather complete. A definite test of the model against well documented experimental data still has to be made. From a physical point of vue the model is a significant advance over the similarity approach by Fannelop and Waldman (1972) and over the crude dynamical approach by Van Ulden (1979).

Acknowledgements

The author wishes to thank Th.L. van Stijn and B.J. de Haan for taking care of the numerical computations in this paper and for helpful discussions.

References

Barr, D.J.H., 1967. Densimetric exchange flow in rectangular channels. Houille Blanche 6/67, 619-631.

Batchelor, G.K., 1967. Fluid Dynamics, Cambridge University Press.

Benjamin, T.B., 1968. Gravity currents and related phenomena. J. Fluid Mech., 31, 209-248.

Fanneløp, T.K. and Waldman, G.D., 1971. The dynamics of oil slicks - or "creeping crude". 9th Aerospace Sciences Meeting A.I.A.A., New York, January 1971.

Fanneløp, T.K. and Waldman, G.D., 1972. The dynamics of oil slicks. A.I.A.A. Journal, 10, 506-510.

Fanneløp, T.K., Krogstadt, P.A., Jacobsen Ø, 1980. The dynamics of heavy gas clouds. Norwegian Institute of Technology. Report J.F.A.G. B-124.

Fay, J.A., 1969. In "Oil on the Sea", ed. D.P. Hoult, 53-63, New York, Plenum.

Fay, J.A., 1980. Gravity spread and dilution of heavy vapor clouds. Proceedings of 2nd int. symp. on stratified flows, 471-494, Trondheim, June 1980.

Hall, D.J., Hollis, E.J., Ishaq, H., 1982. A wind tunnel model of the Porton dense gas spill. Warren Spring Laboratory, Report LR 394 (AP), May 1982.

Hoult, D.P., 1972. Oil spreading on the sea. Ann. Rev. Fluid Mech., 4, 341-368.

Huppert, H.E. and Simpson, J.E., 1980. The slumping of gravity currents. J. Fluid Mech., 99, 785-799.

Monin, A.S. and Yaglom, A.M., 1971. Statistical Fluid Mechnics Vol. I. M.I.T. Press, Cambridge, Mass.

Picknett, R.G., 1978. Fluid experiments on the behaviour of dense clouds. Part I, Main Report. Ptn 1 L 1154/78/1, Chemical Defense Establishment, Porton Down, U.K.

Picknett, R.G., 1981. Dispersion of dense gas puffs released in the atmosphere at ground level. Atm. Environment, 15, 509-525.

Schmidt, W., 1911. Zur Mechanik der Böen. Meteorologisches Zeitschrift, August 1911.

Simpson, J.E., 1972. Effects of the lower boundary on the head of a gravity current. J. Fluid Mech. 53, 759-768.

Simpson, J.E. and Britter, R.E., 1979. The dynamics of the head of a gravity current advancing over a horizontal surface. J. Fluid Mech., 94, 477-495.

Simpson, J.E., 1982. Gravity Currents in the Laboratory, Atmosphere and Ocean. Ann. Rev. Fluid Mech., 1982, 14, 213-234.

Tennekes, H. and Driedonks, A.G.M., 1981. Basic entrainment equations for the atmospheric boundary layer. Boundary Layer Meteorology, 20, 515-531.

Van Ulden, A.P., 1974. On the spreading of a heavy gas released near the ground. Int. Loss prevention Symp., The Netherlands. C.H. Buschman ed., Elsevier, Amsterdam.

Van Ulden, A.P., 1979. The unsteady gravity spread of a dense cloud in a calm environment. 10th Int. Techn. Meeting on Air Pollution Modeling and its Applications. NATO-CCMS, Rome, October 1979.

MIX
Papier aus verantwortungsvollen Quellen
Paper from responsible sources
FSC® C105338

If you have any concerns about our products,
you can contact us on
ProductSafety@springernature.com

In case Publisher is established outside the EU,
the EU authorized representative is:
Springer Nature Customer Service Center GmbH
Europaplatz 3, 69115 Heidelberg, Germany

Printed by Libri Plureos GmbH
in Hamburg, Germany